The Path Not Taken

Transformations: Studies in the History of Science and Technology

Jed Z. Buchwald, general editor

The Path Not Taken

French Industrialization in the Age of Revolution, 1750–1830

Jeff Horn

The MIT Press
Cambridge, Massachusetts
London, England

© 2006 Massachusetts Institute of Technology

All rights reserved. No part of this book may be reproduced in any form by any electronic or mechanical means (including photocopying, recording, or information storage and retrieval) without permission in writing from the publisher.

For information on quantity discounts, email special_sales@mit press.mit.edu.

Set in Sabon by SPI Publisher Services. Printed and bound in the United States of America.

Library of Congress Cataloging-in-Publication Data

Horn, Jeff.
The path not taken : French industrialization in the age of revolution, 1750–1830 / Jeff Horn.
    p. cm.
Includes bibliographical references and index.
ISBNs: 0-262-08352-3, 978-0-262-08352-2 (alk. paper)
1. Industrialization—France—History—18th century. 2. Industrialization—France—History—19th century. 3. Industrialization—England—History—18th century. 4. Industrialization—England—History—19th century. I. Title.

HC275.H67   2006
338.094409'033—dc22   2006044260

10 9 8 7 6 5 4 3 2 1

Contents

Acknowledgments

This book is the result of a lot of help and support. Initial financial assistance came from Margaret C. Jacob's National Science Foundation grant for "The Cultural Origins of the Industrial Revolution" (SBR-9310699), for which I was a researcher, and from Stetson University. Another NSF grant (SBR-9810232) provided funds for a vital year in France. Manhattan College funded further research and permitted me to take a year at the Dibner Institute for the History of Science and Technology at the Massachusetts Institute of Technology, a welcoming and stimulating place to write. The entire staff there was amazingly helpful, but Carla Chrisfield, Rita Dempsey, Bonnie Edwards, and Trudy Kontoff deserve my special thanks. The faith of these individuals and institutions in my work is greatly appreciated. Peg Jacob, Ed Hackett, and George Smith contributed enormously to getting this project started and enabling me to complete it.

Thinking through the problems associated with French industrialization was a wonderful way to spend 14 years. At the University of Pennsylvania, Lynn Hunt and Jack Reece completed the professional training begun by Bob Forster and Orest Ranum at Johns Hopkins. They imparted a holistic approach to French history and instilled a deep appreciation of archival research. They have never failed to inspire and encourage me.

In France, innumerable archivists and librarians in Amiens, Bordeaux, Paris, Rouen, Saint-Étienne, and Troyes made this project both possible and enjoyable. Pascal Dupuy generously provided me with access to the library at IRED at the Université de Rouen during my teaching stints there. He also referred me to a number of somewhat obscure but highly useful secondary sources. Daniel Fauvel allowed me to consult the archives of the Société d'Emulation of Rouen. Régine Beaudoin helped me get access to a number of documents in the departmental archives of the Seine-Maritime. Joëlle Luhmann at IRED provided access to useful materials. David Gaultier at CNAM was exceptionally helpful and competent. François Crouzet shared his wine and his critique of my conclusions. Visiting France to research this book has been a pleasure and a passion.

Various groups have encouraged my attempts to figure out what all my evidence meant. My colleagues taught me a great deal when I presented my work to the Johns Hopkins European History Colloquium, the New York Area French History Seminar, the Society for French Historical Studies, and the Western Society for French History. A number of friends have read parts of this manuscript to help me refine my arguments including Cyndy Bouton, Larry Epstein, and Jeff Ravel. Their comments have helped me more than they know. Steven Clay, Suzanne Desan, and Tip Ragan continually reassured me that my data supported my conclusions. From Pat Malone, I learned about water power and American industrialization. Steve Kaplan urged me to dig deeper into the relationship between labor and the French state. Joel Mokyr has been a consistent source of encouragement and insights into comparative history. Daryl Hafter has shared her wise counsel and appreciation of the ancien régime. Her challenges have led to useful reformulations of old questions. Roe Smith persuaded me to rethink the relationship between state and society during early industrialization. He has been a wonderful mentor and advocate. From our first meeting hunched over administrative reports in the Archives Nationales a decade ago, Len Rosenband has been an intellectual sounding board, a dogged reader, and a writing teacher: his ongoing help has been invaluable to the conceptualization of this project and to its current form. I also appreciate the practicality and professionalism of series editor Jed Buchwald. Sara Meirowitz has been an extremely helpful editor.

My colleagues in the History Departments of Stetson University, George Mason University, and Manhattan College have helped and encouraged me. The interlibrary loan departments of these schools tracked down many wonderfully obscure references for me.

My family and friends have looked on this seemingly never-ending project with a mixture of incomprehension, amazement, and forbearance. My son David helped me to conclude the book—so I could spend more time with him.

This book is dedicated to Julie, who puts up with me.

The Path Not Taken

1

Divergence, Convergence, and the French Path to Industrial Development after 1750

This book is about industrial prosperity: how to get it and when it emerges. Standard descriptions of the trajectory of comparative industrialization, such as David Landes' classic *The Unbound Prometheus*, depict French industry as unable to compete with laissez-faire Great Britain until after 1850 at the earliest.[1] Budding French attempts at industrial competition were unsuccessful, and later efforts were rapidly overshadowed by two emerging industrial powerhouses: an arrogant Imperial Germany and the upstart United States. In many historians' accounts, the French were permanently relegated to a kind of industrial purgatory. For Landes, the nature of French entrepreneurialism was to blame. The French emphasis on family firms, an "outdated" form of organization that championed austerity, frugality, and high profit per unit sold at the expense of profit maximization and relentless expansion of output, consigned the French to perpetual second-class status. Landes attributed these business choices to cultural preferences; material constraints, exogenous technical considerations, dissimilar labor relations, and political pressures were at best secondary factors in his evaluation of entrepreneurial decision making and the course of French industrialization.

For those who follow this interpretation, French industrial development was held back by regressive institutions like the *corporations* of the ancien régime. Moreover, the French state's emphasis on military conquest and its dirigiste approach to oversight of the economy hindered efforts to imitate the classically liberal economic policies that had brought extraordinary industrial success to their rivals across the English Channel. Yet, even as an undergraduate, I remember being struck by the fact that most historians report that France's per capita income was roughly comparable to Great Britain's by the outbreak of World War I.[2] How did France come to be competitive with the predominant industrial nation? When did France escape from industrial purgatory? What social or economic groups led the way? What was the nature, the style, and the pace of French mechanization? Was an active state role necessary for nations playing industrial "catch up"? Why was an economic strategy that proved successful over the long term so thoroughly denigrated in the historical community?

The more deeply I read the secondary literature relating to comparative industrial development, the more thoroughly I was struck by its fundamental Anglocentrism. England (later Great Britain) was seen as following the only legitimate market-based route to industrialization; divergences from that path were, by definition, considered to be detours from the path to "true" productivity, rather than alternative passages. Aggregate statistics, which are essential to a "pure" economic historian, were shunted aside by historians of "political economy." Unmistakable signs of sectoral, regional, or national economic competitiveness or industrial success in other countries during the age of British ascendancy were minimized by references to the greater incidence of poverty on the Continent. In contrast to the wealthier and more market-oriented British, Landes persistently evoked an image, adopted from Guy de Maupassant, of a frugal and impoverished French peasant going out of his way to collect a piece of string.[3] Such interpretations of the relative wealth of the two societies have also been based on an overemphasis on easily accessible sources, such as the travel journals of agronomist Arthur Young, who concentrated on the poor conditions of large-scale, overtaxed French farms to underscore the wealth created by the disappearing self-sufficient family farms of England.[4] The fundamental Anglocentrism of these visions of conditions on the Continent frustrated my initial attempts to understand the course of French industrial development.

The Success of the French Path to Industrial Society

During the revolutionary era (1750–1815), French economic growth did not equal Great Britain's. When the revolutionary and Napoleonic wars ended, in 1815, the British were approximately a generation ahead in industrial technology and in the elaboration of the mechanized factory. Britain reached its peak of economic power in the period 1830–1850, achieving the enviable position of "the workshop of the world" displayed so proudly at the Crystal Palace Exposition of 1851. This textbook version of economic history has been largely unchallenged for generations, but it glosses over important developments that took place on the Continent.

At the height of British industrial dominance relative to the rest of Europe—the period from 1815 to 1850—French industry grew rapidly, if not so rapidly as Britain's. Anglocentrists overlook the fact that France remained the largest industrial nation in the world until 1820, at least in terms of gross output. Even more impressively, this growth took shape in a depressed and diminished state. In 1815, much of the eastern part of the country was occupied by enemy troops, and France owed the victorious allies vast sums. Commodity deflation seriously limited sales of manufactured goods in a largely agricultural nation.

Despite these constraints, France posted a particularly high annual rate of growth—3.7 percent—from 1815 to 1820. Population growth from 29.3 million in 1815 to a little less than 32.4 million in 1830 spurred this recovery. Despite high tariffs, French foreign trade recouped, reaching 13 percent of the gross national product by 1830. France steadily decreased imports of manufactured goods and forcefully increased its export of manufactures, retaining its status as the world's number-two exporter of industrial goods. Overall, material output rose doggedly, surmounting the booms and busts of the international economy. Estimates of industrial expansion range from 2.5 percent to 3.4 percent. The agricultural sector complemented this growth with 1.2 percent annual increases during the period 1820–1870. French society and its structures changed dramatically as a result of industrial development. France had an industrial revolution, albeit more gradual and less abrupt than Great Britain's.[5]

France enjoyed impressive long-term growth, both overall and per capita, particularly in light of its slower population expansion, limitations in factor endowments, dilatory growth of the domestic market, restricted imperial advantages, circumscribed capital stock, and fragmented transport system. Between 1815 and the First World War,

France's average annual increase in per capita economic growth was 1.4 percent. In material output, France's per capita increases roughly matched Great Britain's.[6] In large part because of a spurt during the Belle Époque, by 1914, on a per capita basis, France's economic performance was indeed broadly comparable to Great Britain's.[7] Britain's per capita income remained higher by about 20 percent, but the gap between Britain and France remained relatively constant.[8]

When the material and technological bases of industrial success changed at the end of the nineteenth century, France outperformed Britain handily. This later expansion must not be permitted to overshadow the industrial strengths France developed up to 1850 and maintained successfully in the face of British preeminence. As a number of historians have observed, the economic structures that made Great Britain successful in the first half of the nineteenth century later held it back.[9] Martin Daunton writes: "Arguably [after 1850], a gap was opening up between the production *institutions* which were developed in Britain—small family firms, a reliance on subcontracting between and within firms, a highly formalized system of collective bargaining—and the needs of production *technology*."[10] Daunton highlighted the divide between organizational changes and technological development, but these factors are often conflated in explanations of the sources of economic growth. Over the long term, France survived the onslaught of British dominance in a position to take full advantage of the new technologies, industries, and sources of power that emerged later in the century as part of the "Second Industrial Revolution." This work focuses on the sources of France's later successes and on the wellsprings of the French perseverance that emerged during the first Industrial Revolution.

The outlines of this analysis of comparative economic performance were suggested by Patrick O'Brien and Caglar Keyder in 1978. Since then, their analysis, highly controversial when published, has garnered widespread support from French economists and historians. Despite this validation, their conclusions have been ignored in Anglo-American versions of how technological change affects economic growth and in comparative studies of the process of industrialization.[11] Viewed from the Continent or indeed from almost anywhere else, the dominant Anglo-American version of comparative industrialization, with its emphasis on the cultural aspects of the British model, appears terribly parochial.

As Peter Mathias pointed out so effectively, British industrialization shifted the context for those who followed. Following precisely in Britain's footsteps was impossible. Why would anyone expect France or any other country to industrialize on the same pattern as a pioneer, especially when that nation lacks the same mix of resources and expertise and has to contend with competition from the cradle in international markets?[12] This common-sense question gets us to the heart of the matter. As Martin Wiener, W. D. Rubinstein, Nicholas Crafts, and others have suggested in various ways, the underlying question or implicit challenge in this literature is to understand whether and how late-Victorian or post-1945 Britain lost its industrial edge, an issue that became more potent in the 1970s and the 1980s when a surging Japanese economy led many Americans to fear eclipse.[13] France's economic performance during the Industrial Revolution must be considered on its own appreciable merits, not through the lens of later eras. The French path to industrial society diverged from the British, but over the long term its seemingly tortuous route produced impressive results.

Since Landes, accounts of comparative industrialization focus on the divergent cultural outlooks about science and its application on the two flanks of the Channel. Adopting elements from Landes and from Max Weber, Margaret C. Jacob has provided the most coherent recent explanation for why the French were unable to take advantage of their opportunities while the English made the most of theirs. Her explication of national "scientific cultures" endowed an intellectual underpinning for the divergences within western Europe that was independent of more contested economic realities.[14] Following a similar line of argument, historians of the advent of consumerism have claimed that cultural and economic imperatives held back France's imitation of Britain's mass-market industrial approach. This approach has the added advantage of being able to justify France's "catch up" after 1850.[15]

Joel Mokyr's evocation of an "Industrial Enlightenment" took the cultural argument into new, more explicitly economic territory. His term referred to a particular way of thinking that emphasized how "useful knowledge" bridged the Scientific Revolution and the Industrial

Revolution. As for Landes and Jacob, this focus on the generation and use of knowledge, particularly scientific and technological knowledge, was both necessary and sufficient to explain Britain's accomplishments and France's retardation.[16] In his 2004 Presidential Address to the Economic History Association, Mokyr provocatively made the linkages implicit in the "cultural turn" explicit. In an attempt to have his cake and eat it too, the most recent version of Mokyr's Enlightenment added a "doctrine of *economic reasonableness*" to the Baconian program emphasized by Jacob.[17]

The consequence of these perspectives has been an identification of technical creativity with Britain and an assumption that—for cultural reasons, based on a certain view of science rooted in Bacon—Adam Smith was describing how the British people acted, not a model of economic behavior. As Marie Antoinette learned, however, breaking the cake of custom was far messier than the bloodless, straightforward industrial revolution depicted by Jacob and Mokyr. They talked about origins and results, but not about process. Incentives, profits, and labor relations do not get much attention in their versions of the Industrial Revolution. The "how" portrayed in my account challenges the "why" featured in recent cultural explanations of comparative industrialization.

In the still waters reflecting French backwardness and British triumphalism, the publication of Kenneth Pomeranz's *The Great Divergence: China, Europe, and the Making of the Modern World Economy* cast widening ripples. Pomeranz asserted that *all* national experiences must be considered as deviations and that none should be elevated to the status of the normative, no matter its timing or the relative strength of the economy in question.[18] This profoundly timely warning about interpretations of the British Industrial Revolution was seconded by R. Bin Wong's proposals for establishing reciprocal comparisons not only between countries but also between theories and their material basis.[19] As I began to investigate the sources and framed my research agenda, the threads of the Anglocentric argument—so patiently gathered by Landes on behalf of de Maupassant's peasant—unraveled.

Jan de Vries laid the groundwork for a reconsideration of Anglocentric accounts of comparative industrialization and technological change. De Vries' emphasis on an "industrious revolution" in much of western

Europe profoundly challenged Landes' version of industrialization and its link to economic growth during the age of revolution.[20] To de Vries, the more thorough use of resources and manpower and hence the growing integration of labor, commodity, and capital markets were the chief sources of economic growth before the onset of substantial mechanization in the second quarter of the nineteenth century.[21] His argument has been buttressed by the findings of economic historians who keep pushing back the date for when mechanized production became the chief force in the British industrial economy. The 1820s are the earliest date currently advanced.[22] Taken together, these studies suggest the need to revisit accounts emphasizing convergence around a British model based on laissez-faire conceptions of the sources of industrial success.[23]

Based on more than a decade's archival research, I have fashioned an alternative vision of industrial development that crosses the traditional historiographical boundaries of the old regime, the Revolutionary decade, and the nineteenth century. The documentary record left little doubt about the uniqueness of French labor. Largely because of the emergence of Revolutionary politics, the relations of the laboring classes to entrepreneurs and to the state in France differed greatly from those in Britain. This divergence had profound repercussions both politically and on the shop floor. Secondly, as Landes asserted, a French style of entrepreneurialism existed, but it was rational, profit-seeking, and relatively successful despite the trials and tribulations of war and the upheavals of the Revolution. Finally, the French state developed idiosyncratic and effective means of mediating between these groups while seeking to accelerate scientific and technological innovation, to manage labor unruliness, and to encourage risk-taking. In short, I found that the state was more than just an obstacle to the operation of theoretically free markets—on both sides of the Channel. The French experience illustrates that there was more than one pathway to industrialization. Deviation from the British model of the interaction of state and society did not necessarily rule out long-term economic growth.

This book weaves together three arguments. The first is that the late-eighteenth-century French state attempted to emulate most of what policy makers understood as the English model of industrialization—which, I illustrate, was far removed from what "liberal" accounts would suggest. The second argument is that these attempts foundered because of the emergence of revolutionary politics in France. The possibility of a thoroughgoing social and economic revolution by the laboring classes ensured that neither the French state nor Continental entrepreneurs could safely maximize profits or innovate in response to labor militancy, as in Britain. The true divergence of industrial paths dated from the radical phase of the French Revolution. This view restores the prominence of the "political" component of political economy, which is absent from too many accounts of this turbulent era. It also builds on Donald MacKenzie's call to revisit and deepen our understanding of the technical consequences of relations between entrepreneurs and labor during industrialization.[24] Thus, this book embraces a different style of comparative history that considers not only what actually occurred, but also what avenues remained unexplored and why certain possibilities were conceivable in Great Britain but not in France.

A corollary of this second argument is that Great Britain could profit from its admitted advantages and achieve industrial ascendance because its entrepreneurs were able to control and exploit the working classes to a degree that was impossible in France because of the "threat from below." A commitment to allowing the market to set prices must not be permitted to mask the significance of British state action in industrial development. The ruthless repression of the British working classes, the powerful incentives provided to entrepreneurs, the acquisition and protection of an enormous empire, and the determined way that the Hanoverian state manipulated the ideology of liberalism produced concrete long-term economic advantages for Great Britain.[25] Landes to the contrary, the involvement of the state in British industrialization *never* incarnated the form of laissez-faire articulated by Adam Smith.[26]

Third, I argue that the French state, unable to emulate the "liberal" British route to industrialization, embarked on a search for a different path that was forged amidst the heat of war, revolutionary politics, and emerging dictatorship. A statist command economy formed to deal with the mortal threats of the Year II (1793–94) stoked the fires of revolutionary fervor and provided a potent alternative model of industrial development. The dependence of this model on the Reign of Terror made it anathema to French policy makers. With this initial attempt politically intolerable, the French state evolved a distinct longer-term institutional model of industrial development. This novel approach necessitated different means of accelerating scientific and technological innovation, managing labor unruliness and encouraging entrepreneurialism. These efforts yielded relative prosperity and, over the long term, permitted France to achieve levels of per capita income comparable to those of Britain. The path not taken traversed different thickets to emerge in the same place.

The Path Not Taken: Tracing France's Industrial *Sonderweg*

In chapter 2, "A Brave New World of Work: The Reform of the *Corporations* and the *Lettres-Patentes* of May 1779," I explore the complex relationship between the French royal government and the *corporate* economy of the ancien régime. Bounded by privilege, entrepreneurs, artisans, and workers did not have sufficient "liberty" to innovate technologically or in the manner of production, but too much license led to poor-quality goods that were difficult to export. A group of Enlightened thinkers known as the Physiocrats played the leading roles in developing this analysis of French "backwardness." Following their lead, several initiatives successively altered the legal framework governing the interaction of producers and laborers to allow both groups to be more creative, more innovative, and more efficient. By dissecting the meanings of liberty and license in French industrial regulation, this chapter delineates contemporary conceptions of international competitiveness.

After 1750, repeated attempts were made to increase the degree of economic "liberty" accorded to entrepreneurs while circumscribing the "license" of the laboring classes in a dense web of restraint that varied according to the political perspective of the sponsors of the reforms. During Anne-Robert-Jacques Turgot's tenure as Controller-General (1774–1776), a new legal framework to regulate the world of work became enmeshed in the effort to abolish the *corporations*. After his fall from power, Jacques Necker encountered the profound difficulties of regulating labor within the restored, but profoundly shaken corporate structure. *Lettres-patentes* issued in May 1779 and revised in September 1781 were intended to take France several furlongs down the path blazed by England. Chapter 2 traces the effects of these reforms of the *corporate* world on the shop floors and in the counting houses of France's most advanced industrial district, the province of Normandy. Resistance to reform in this model province demonstrated the limitations of top-down reform in the ancien régime, particularly in the face of divisive challenges stemming from privilege, liberty, and international economic competitiveness. The better-known reforms of the Revolutionary decade also stand revealed as part of a long-standing effort to reform the world of work and to enable French entrepreneurs to take better advantage of the productive environment.

Chapter 3, "Foreign Policy as Industrial Policy: The Anglo-French Commercial Treaty of 1786," traces the mentalité underlying the cross-Channel rivalry that dominated this period. In economic terms, and particularly in industrial terms, our understanding of French leads and British lags in productivity and in technology recasts the explanations of backwardness emphasized by Alexander Gerschenkron.[27] This chapter develops themes highlighted in Christine Macloud's revisionist article reminding us of Britain's technological dependence on the Continent and challenges Maxine Berg's assertion that fashion was at the heart of Britain's industrial dominance.[28] French attempts to imitate British economic practice are the jumping-off point for my attempt to plumb the depths of the discrepancy between the practice of laissez-faire envisioned by Adam Smith and the actions of the Hanoverian state.

The negotiation of the Anglo-French Commercial Treaty, that treaty's implementation, and (most important) the way French entrepreneurs responded to the heightened challenge of British competition demonstrate how issues of liberty and license, privilege, and profit worked out in practice. In an astonishing number of sectors, French entrepreneurs of the 1780s competed successfully with their English counterparts. Contemporary French competitiveness has been submerged by a deluge of references to the leading textile sector and by a deterministic emphasis on the impending collapse of Bourbon political authority. The fundamental presumption of state policy makers in signing the Treaty—that the French *could* compete and could beat the British at their own game—was not necessarily misguided. In areas where Britain was dominant, the French state's intervention sought to mitigate the Treaty's effects and stimulate competitiveness. This account demonstrates conclusively that the French policy makers' principal misconceptions were political, not economic or technological. French attempts at bolstering technological improvement and supercharging the competitive spirit of entrepreneurs foundered with the sinking ship of state. This chapter sets the stage for an analysis of the 1789 Revolution's effect on the course of the Industrial Revolution in France.

After 1789, labor relations differentiated British and French industrial conditions. Chapter 4, "The Other 'Great Fear': Labor Relations, Industrialization, and Revolution," examines the incidence of machine-breaking on both flanks of the Channel from the old regime until well into the nineteenth century. The effectiveness of government repression of machine-breaking was a sensitive gauge of the relative power of the British and French states. During the eighteenth century, machine-breaking by the restive English laboring classes was much more common than in France. The turning point in this domain, as in so many others, came in 1789. Although overshadowed by more spectacular events in Paris, the revolutionary moment embraced pervasive machine-breaking in several industrial centers. Because machine-breaking in 1789 was an aspect of the emergence of revolutionary politics, the supposedly assertive French state proved nearly powerless in clamping it down. Throughout the revolutionary decade (1789–1799), French industrial entrepreneurs could not rely on the state to repress working-class militancy.

In England, machine-breaking is usually associated with the Luddites of 1811–1817. If the weakness of French state support slowed the pace of mechanization and technological innovation, Great Britain, the archetypical "liberal" state, deployed impressive levels of coercion to repress labor militancy in general and the practice of machine-breaking in particular. Entrepreneurs in England exhibited a justified faith in the power and protection of the Hanoverian state against the "threat from below." E. P. Thompson and Adrian Randall emphasized the potential for political upheaval inherent in Luddism,[29] but events illustrated that the English working classes were rebellious not revolutionary. Sheltered from the peril of a political and social revolution, Britain could safely consolidate and extend its industrial advantages in the generation after 1789. By tracing the complex relationship of the state with the laboring classes during the revolutionary era, this chapter demonstrates how French machine-breaking shifted labor relations on the Continent to slow the pace of industrial transformation. Although far less well known, French machine-breaking in 1789 had a far greater effect than its English counterpart. This chapter chronicles the onset of a genuine divergence in the industrial pathways of these two nations.

Chapter 5, "*La patrie en danger*: The Industrial Policy of the Year II," is a case study of the effect of Revolutionary politics on industrial development. In 1793–94, the Committee of Public Safety faced not only the onslaught of the overwhelming coalition of states arrayed against the Revolutionary French government but also the Federalist and Vendéan rebellions. The French economy was in free fall, industrial production was collapsing, and the ports were cut off from the colonies. Only a deliberate policy of state-sponsored Terror enabled the Revolutionaries to enforce the wide-ranging economic measures needed to provide the weapons and food essential to victory.

For the French Revolutionaries, the key to increasing industrial production was the mobilization of scarce resources. The state mobilized human resources (including skilled laborers, entrepreneurs, and technological innovators), knowledge (consisting of both the best existing industrial practices and new processes and inventions), and raw materials. Against all odds, the Committee of Public Safety defeated both internal and external enemies and embarked on a crusade to bring the benefits of the French Revolution to the other peoples of Europe. Only through the Reign of Terror was French industry able to meet the extraordinary demands made upon it. Chapter 5 takes issue with how many historians of science and technology, most notably Ken Alder, have treated this period. Rather than focus on experiments of restricted application or generalizing about technological applications from limited data, I examine how agents of the Terror treated industrial problems and how entrepreneurs and laborers reacted to the draconian policies of the state, both in Paris and in the provinces. In the crucible of war and revolutionary politics, the economic relationship of state and society shifted fundamentally, but the reliance of the industrial policies of the Year II on revolutionary violence indelibly tainted this effective approach to managing the economy. Succeeding regimes groped for different means of mobilizing French resources for war. The link between the industrial policies of the Year II and the Terror also magnified the "threat from below" in the minds of entrepreneurs, thereby speeding up French divergence from the English model of industrialization.

Chapter 6, "From Allard to Chaptal: The Search for an Institutional Formula for French Industrialization (1791–1804)," investigates French administrative attempts to develop institutions capable of fostering industrialization in the context of Revolutionary politics. As early as 1791, classical liberalism was superseded by more cameralist views of how to stimulate technological innovation and support entrepreneurialism while keeping the laboring classes in check. Accelerated by the widening war and the crisis of the Year II, French policy makers experimented with novel institutional means of fostering competitiveness and innovation. Until the advent of the Consulate (1799–1804), these experiments generally foundered on the jagged rocks of uncertain French finances and political instability. In the fresh dawn of the Napoleonic era, Minister of the Interior Jean-Antoine Chaptal (1800–1804) masterminded a new approach to improving French industrial competitiveness that melded Physiocratic notions concerning resources and liberal attitudes about the rights of the individual with an activist vision about the necessity of state action in technological matters.

Chapter 6 delves deeply into the technological decisions made during the Revolutionary decade to illustrate the demise of the liberal paradigm and to explain how the economic and political situation constrained industrial and technical possibilities. Economic and political weaknesses limited the effectiveness of institutions, such as the first industrial exposition of 1798, that publicized and rewarded technological advance. Chaptal codified and extended the bureaucratic efforts of the Revolutionary decade. He emerged as the institutional architect of the nineteenth-century French industrial economy. Mokyr's description of Chaptal as a paradigmatic figure of the Industrial Enlightenment permits me to reflect on the limitations and utility of this term and the way that Mokyr related a particular way of thinking to the sustained technological innovation that took place after 1820.[30] Chaptal's institutional creations combined formal education and hands-on experience to permit France to find a middle way among the competing models of industrial development. He and his bureaucratic collaborators adopted and adapted elements of the British model, the hyper-centralized approach of the Year II, and classical laissez-faire to lead France down a different path to industrial development that laid the foundations for long-term economic success.

Chapter 7, "Facing Up to English Industrial Dominance: Industrial Policy from the Empire to the July Revolution (1805–1830)," traces the legacy of the Chaptalian framework. Chaptal and his approach to fostering industrial competitiveness were shunted aside temporarily during the era of Napoleon's ascendancy in favor of a mercantilist Continental System designed to ensure French economic hegemony by force of arms. The expansion of imperial borders to encompass Belgium, the Netherlands, the Rhineland, and northern Italy fortified expectations that an enlarged France verging on the dimensions of the original European Common Market could compete with Britain on its own terms. Tracing the vicissitudes of international trade in terms of the ups and downs of European-wide war reveals both the genuine possibilities and the mistaken assumptions inherent in the Napoleonic approach to economic competition. After Napoleon's two defeats, however, the institutions created early in the century resumed their place at the forefront of government industrial policy. The Restored Bourbons did not enjoy Napoleon's military successes or the bloated borders of the Empire, but they benefited from the education, technical training, and workplace experience his regime propagated to a host of the artisans, laborers, and tinkerers. During the Restoration (1815–1830), a period of diminished political expectations, state industrial policy focused on expanding the profitability and range of what France did well and reserving the home market for domestic manufactures. Only someone wearing the blinders of laissez-faire would expect the French to slavishly attempt to compete in international markets with Great Britain in the areas of its supremacy. If this led to slower growth than across the Channel, it did allow France to recover from its ordeals and to profit from its own competitive advantages.

Without the cover provided by war, competing with the paramount industrial power from an inferior position challenged French pride and hampered French exports. Royal policy makers recognized that, although basic capacity had to be developed in certain industries (such as steel), France had a unique mix of resources, skilled laborers, and product specialties that could form the basis of a solid prosperity despite English predominance. In a wide variety of sectors, the French state focused its efforts on providing what manufacturers lacked to enable them to take advantage of their opportunities. Within the confines of "domestic" laissez-faire, a uniquely French form of government intervention often entailed finding appropriate skilled labor, developing scientific knowledge, and furnishing technological expertise or start-up capital. The success of these efforts also bought time for the educational and training institutions created earlier in the century to work. Unsurprisingly in view of France's segmented markets and uneven transportation network, innovation was usually organized locally or regionally. A modified mercantilism dominated French industrial policy until the railroad inaugurated a new industrial era.

Chapter 8, "Coalitions and Competition: Entrepreneurs and Workers React to the New Industrial Environment," complements chapter 7 by revealing how entrepreneurs and laborers responded to state industrial policies in the first three decades of the nineteenth century. Both groups rejected passive acceptance of top-down policies and instead sought to develop tactics and strategies to improve their own situations. These maneuvers focused on limiting competition through either formal or informal coalitions or through some sort of fraud, very broadly defined. The prospect of competition either with domestic technological innovators or with favored international rivals made many French uneasy. Their response was to circumscribe and circumvent the market. Groups of entrepreneurs and laborers circumvented the law in ways that illustrate the limitations of state control and the continuing legacy of Revolutionary politics in the early industrial age, further complicating our understanding of the process of industrialization.

In the early nineteenth century, the collective power of the laboring classes was expressed intermittently, but the "threat from below" loomed large. The repression of worker militancy met with uneven success. When French manufacturers attempted to lower wages to reduce prices and stimulate sales rather than search for technological improvements or new outlets for their goods, their workers usually accepted it.

When workers banded together to restore the cuts, as they did in Normandy, at Le Houlme in 1825, entrepreneurs and the state reacted in ways that characterize the divergence of labor relations on both flanks of the Channel because of the lingering "threat from below" in France.

Producers also resisted the market discipline imposed by the state by forming coalitions and by counterfeiting, making knock-offs, and smuggling. The true scale and scope of these activities will never be known, but they peaked during Napoleon's Continental System (1806–1813). The willingness of entrepreneurs in a wide range of industries, market situations, and places to engage in illegal activities demonstrates the multiple sources of entrepreneurial profits missing from so many accounts of industrialization. Economic competition is fought out both over and under the table. The existence of so many opportunities for illicit profit-taking also attested to the weakness of the market mechanism in a country as large and diverse as France. The evidence presented in chapter 8 undermines optimistic Anglocentric visions of how industrial development ought to work by emphasizing how it actually functioned. The extent of these illicit sources of profit also helps to explain why many French industrial entrepreneurs did not wholeheartedly endorse mechanization and innovation, as their British rivals did: they did not yet need to. In France, many if not most entrepreneurs preferred to take advantage of alternative and fundamentally easier means of acquiring wealth. Economic rationality in an age of war and Revolution was not as straightforward or as mathematical as market theorists would have us believe.

Chapter 9, "Chaptal's Legacy in a Niche Industry," provides a glimpse of how the institutions founded earlier in the nineteenth century created a new industrial environment. As Chaptal envisioned decades earlier, entrepreneurs, tinkerers, and bureaucrats in the city of Troyes took full advantage of the institutional support that was available to them. By cultivating the niche market of *bonneterie* (the making of hats and stockings), Troyes rebounded from a deep industrial decline between 1810 and 1820 to become one of France's most dynamic centers of technical innovation. As de Maupassant's thrifty peasant well knew, the path not taken presented its own delights to those who knew where to look.

2

A Brave New World of Work: The Reform of the *Corporations* and the *Lettres-Patentes* of May 1779

Labor discipline stood at a nexus of French urban society. In the second half of the eighteenth century, the French royal government had two major, often competing, sets of concerns regarding the working classes. The public unruliness of French laborers was a continual problem of *police*—a power of oversight, regulation, and prevention of disturbances exercised by the courts, by municipal and clerical authorities, and by agents of the central state. In different times and different places, however, these same authorities could be even more anxious about the role of workplace agitation in retarding the technological progress and industrial development so essential to local prosperity and international political and economic competitiveness. Pressured by France's rivalry with England, both sets of concerns waxed in importance in the decades before 1789, leading every level of Bourbon administration to devote increasing attention to the knotty issue of regulating labor.

This conflicting set of problems was mirrored by the ambivalence of the laboring classes' attitudes toward the central state. French laborers were not hostile to, fearful of, or distant from all forms of government intervention, as much of the literature suggests. Although unruly laborers often resisted any attempt by the state to control their public actions, they also commonly solicited assistance from the courts, the *police des manufactures*, and the central government in other domains, particularly to resolve conflicts with their employers. We must complicate our historical understandings of the interaction among state action, the eighteenth-century world of work, and the process of industrialization in order to view the situation in France on its own merits rather than through the fractured lens of nineteenth-century British critics of French economic performance.

Nor was the French state monolithic in its approach to labor questions. Different levels of the administration had dissimilar goals and methods. This made the ancien régime a surprisingly complicated regulatory environment that is not at all consistent with the "absolutist" reputation of the Bourbon state. All French administrators wanted to remove the threat of public unruliness, but many understood that labor unrest was not necessarily a challenge to authority; in fact, sometimes it was a reaction to gaps in the state structure exploited by employers. (See chapters 4, 7, and 8.) However, the reformers, or, more accurately, the improvers, went far beyond this recognition—they did not look at labor discipline only in the negative, i.e. as actions or activities that had be stopped or prevented. In good Enlightened fashion, they hoped to build on the laboring classes' reliance on the French state to transform a "labor problem" into an economic opportunity. In other words, the Enlightened concept of *police* in France was not solely negative. As Foucault pointed out, there was a positive aspect. Foucault incorporated this aspect in his notion of social and individual improvement, which is fundamental to his concept of discipline and/or regulation. This vital and very French aspect of the oversight of labor and entrepreneurialism is often missing from a class-based Marxist reductionism and from much of the historical literature obsessed with the panopticon that deals with the interaction of labor and technological issues during the earliest phases of the Industrial Revolution.

By loosening or eliminating restrictions high and low, the improvers hoped to atomize both the masters and the men enmeshed in the world of work by destroying their ability to act collectively. Thus, an essential part of the improvers' program was to shift entrepreneurial ambitions of a thoroughgoing domination of the working classes toward a relationship more conducive to technological innovation capable of bettering France's competitive position. This chapter investigates how the French state linked the resolution of its labor issues, its plans for technological advance, and allowing entrepreneurs greater "liberty" in their decision making, all in the name of stimulating industry. For many important decision makers of the reign of Louis XVI, French industrial success depended on implementing a regulatory framework for laborers and the labor market that would allow entrepreneurs, officials, and workers greater freedom to experiment while also making labor more productive.[1]

"Liberty" was the justification for this rejection of the Colbertian notion that greater control over the standards of industrial output resulted in greater economic development. Such an attitude has often been associated with the nascent liberalism of the Physiocrats,[2] but those espousing this conception of how to improve the manufacturing sector encompassed a broad segment of French officialdom as well as a significant number of entrepreneurs. Strongly influenced by the English model of technological progress,[3] these improvers believed that the successful regulation of labor would eliminate certain public activities of the working classes while fostering liberty to experiment in the workplace. This liberty would permit technological development, allow the deployment of new machines, and boost French competitiveness throughout the industrial economy. Every echelon of the French state hierarchy sought to achieve these goals, because many if not most French administrators in this era of skill-based and tool-based manufacture looked at comparative economic competitiveness primarily in terms of labor productivity.

The distinctive role of the early-modern French state in regulating labor and in encouraging technological innovation has usually been understood to have been initiated by Jean-Baptiste Colbert, Louis XIV's long-serving Controller-General of Finances. Since 1989, an outstanding historical literature has explored the emergence and practice of the singular involvement of the French government in economic development and technological change.[4] With varying degrees of explicitness, the vast majority of these studies ask old questions in new ways, using the tools and perspectives of cultural history to enrich our understanding of the dynamics of the causal links between labor relations and the pace or timing of technological change. Such studies ensure that national, regional, or local choices about issues ranging from the role of the state in the economy, and dealing with restive laboring classes, to differing attitudes toward the maximization of profit are scrutinized on their own terms as potentially rational decisions. They do not dismiss contrary evidence in the rush to praise the developmental model deemed most efficient or productive in the short run.[5] Each of these pioneering works focuses on a different cultural aspect of the French economy or administrative structure while emphasizing the significance of various external political, social, and economic factors relating to war and rapid political change.

Their object, both collectively and individually, is to explain why France, despite its vast resources, its pre-eminence in basic science, and its active and informed bureaucracy, failed to keep pace in technology and manufacturing output with Great Britain at the dawn of the industrial era. Cultural history adds new dimensions to an exploration of how labor relations and technological choice interact to explain France's preference for certain technologies and not others, with important attendant consequences for industrialization.[6]

If the study of cultural practices and cultural issues has had a powerful influence on the history of technology, it has had just as great an effect on labor and economic history. For the purposes of this discussion, the most effective linkages drawn among economic practice, state action, and labor custom are William Sewell's examination of the language of labor, especially the corporate idiom, which helped bring the "linguistic turn" to the field; Steven Kaplan's and Michael Sonenscher's applications of cultural perspectives to ideology, which emphasize the issues of liberalism and natural law respectively; and William Reddy's revisionist look at the development of the market as a cultural construct.[7] Sewell's central argument that laborers organized to defend real interests that could be defined in economically rational terms, not to retain antiquated customs, is particularly important. Sewell provides a means of moving beyond English historian E. P. Thompson's groundbreaking vision of a "moral economy" articulated by the popular classes, a conception also championed by Natalie Zemon Davis. Ironically, this emphasis served to limit the historical exploration into the genuine economic rationality behind the actions of laborers.[8] In combination with the collapse of the Marxist paradigm, these major contributions have dramatically shifted the focus away from more traditional understandings of class development and the emergence of the factory system to a more nuanced and flexible conception of authority and of the interactions of entrepreneurs, workers, and administrators during the eighteenth century.[9] These cultural approaches to eighteenth-century economic, labor and technological history underpin my argument that the culture of the world of work, including the reactions of elites to pressures exerted by the laboring classes was central to the success of technological innovation and diffusion in the age of revolution.[10]

My investigation of the ambivalent nature of the supervision of French labor will focus on the reform of the fundamental late old-regime laws regulating labor. The *Lettres-patentes* of January 2, 1749, regulating the world of work and establishing the behavioral standards for masters and men, set the standard for government oversight of French labor in the second half of the eighteenth century; they imitated and referred directly to Colbert's 1667–68 reform of industrial relations. Although several half-hearted attempts were made to revise this standard, it remained largely intact until the wide-ranging deregulation of the world of work in 1776 under Turgot. After his fall, the administration led in this domain by Jacques Necker revived the 1749 ordinance as a stopgap measure to prevent chaos. However, the texts of these acts make it clear that circumstances had changed. The fall of Turgot and France's economic difficulties on the eve of the War for American Independence complicated the issues of regulatory control vs. liberty. Whereas the 1749 *Lettres-patentes* had been concerned fundamentally with discipline or controlling the actions of laborers, the abolition of the *corporations* had seemed to realize complete freedom in the world of work for those areas where it was implemented. Despite several attempts to resolve the situation under Necker, a new foundational set of *Lettres-patentes* which represented a compromise between the imperatives of police and economic liberty were issued only in September 1781 and required another three years to execute. Thus, the world of work was in turmoil for ten years after the death of Louis XV. Although inspired by the English model of industrial development and the Physiocrats' plans, no level of the administrative hierarchy could resolve the problem of labor discipline—at least in part—because of conflicting policy aims. The reform of the *Lettres-patentes* demonstrates the failure of the French central state to liberate the workplace from the empire of custom enough to allow entrepreneurs in key sectors to innovate while disciplining the public activities of French laborers.

Creating a New Legal Framework: Toward the *Lettres-Patentes* of September 1781

In the generation before 1789, French administrators concerned either with the economy or with public order had an overriding obsession: how to mix discipline, regulation, and "liberty" in the formulation of labor and industrial policy.[11] Despite constant innovation, which caused its own tensions, this obsession was never resolved definitively because of fluctuations in the leadership and political situation. This instability resulted in significant deviations in official policies toward masters and men and in what was considered the best approach to industrial development. The French were influenced by policies that have become known as "liberalism" when applied to the English economy, but the Bourbon administration's measures took greater account of local custom, international politics, and financial considerations. These conditions led royal officials to meld traditional approaches and fashionable economic doctrines to forge a "mixed" industrial regime which recognized that old-style discipline could not be adapted to new social, economic, and cultural conditions. The realities of international commercial competition with Great Britain helped ensure that, despite a distinct preference for a harsh brand of discipline on the part of most master artisans, entrepreneurs, and local officials, the laboring classes were not cowed and not without influence. Labor legislation under Louis XVI demonstrates that, for many influential government decision makers, both in theory and in practice, the laboring classes had to retain significant autonomy of action on the shop floor if technological advance on the English model was to occur.

Labor legislation per se did not exist under Louis XVI. The charter regulating the relations of masters and workers was the *Lettres-patentes* of January 2, 1749.[12] This measure placed workers under a legal subordination to their masters equivalent to the subservience of children to their parents; put simply, the state redefined the control of masters over men as an aspect of "domestic authority."[13] A major element of this control was the creation of the *billet de congé*, a notice of release. Now workers had to obtain the permission of their bosses to quit a job. Laborers had to give their employers sufficient notice (which varied according to the trade and the region, generally ranging from 8 to 15 days), and to have repaid all debts. A worker who did not acquire a *billet de congé* was liable to a 100-*livre* fine. Laborers were also banned from leaguing together. Serge Chassagne depicts this ordinance as a powerful weapon of a nascent manufacturing class, but the difficulties faced by *juges de manufactures* in enforcing this measure suggests that it must have been one of the most evaded or ignored measures on the books.[14] All over France, workers were able to overcome the repressive legal environment by uniting to secure wage hikes, to enhance piece rates, or to retain control over hiring, at least in certain shops and in certain trades. Thus, these *Lettres-patentes* were not as effective a tool as the masters and industrial entrepreneurs might have wished. For example, despite the legal prohibitions, the workers of Troyes "have not ceased to assemble in cabarets, cafes and other places. When they leave these places, they cause an uproar in the streets and commit various excesses against anyone they encounter there."[15] Masters in the textile trades complained that these "excesses" were intended to put pressure on them to give in to wage demands. Although the reality of worker unrest in the cities was far from the disciplined body of urban laborers envisaged by the 1749 statute, no major revision was undertaken until 1774, when Anne-Robert-Jacques de Turgot became Louis XVI's Controller-General.[16]

The saga of Turgot's ministry is well known. His attack on the *corporations* was part of a brief interlude of reform at the accession of Louis XVI that culminated, on February 22, 1776, in the promulgation of the infamous Six Edicts, in which the replacement of the *corvée* with a monetary tax was the other significant measure.[17] Sewell ascribes Turgot's assault on corporatism to a particular (generally landed) notion of property which placed greater emphasis on the "right to labor" than on ownership of material property. It was to "free our subjects from all infringements on this inalienable right of humanity" that Turgot wanted to abolish the *corporations*. He thought that the corporations limited trade opportunities by limiting access to industrial employment. This reduced competition resulted in higher prices and lower-quality goods. "All persons of whatever quality and condition, even all foreigners" should be permitted to "espouse and practice throughout the kingdom whatever sort of commerce or craft they may choose, and even to combine several [of them.]"[18]

The Edicts refer to masters as "entrepreneurs." Although Sewell discusses it in the context of property relations, this term also points to Turgot's recognition that *corporate* restrictions limited enterprise.[19] Breaking those shackles would lead to greater innovation and drive on the part of the most skilled agents in the industrial economy, thereby increasing France's international economic competitiveness. The conception of the manufacturing economy in these Edicts, therefore, mirrored Turgot's plans for agriculture, which was also to be "liberated" by freeing the grain trade. At the same time, however, Turgot himself, along with many of his contemporaries, seemed to recognize the danger that could result from granting laborers and masters "liberty" from *corporate* restrictions. By diminishing the supervisory and police powers of the state, liberty could be a threat to the privileged structure of the ancien régime.[20]

Turgot hoped to disarm the magistrates who feared the practical implications of "liberty" when the measure was promulgated in February 1776. The regulations on "crucial" professions such as the pharmacists, goldsmiths, booksellers, and printers were left in place, while specific guidelines were included for butchers, bakers, and apothecaries. Barbers and wig makers were exempted. Although all French subjects as well as foreigners could now exercise any trade or trades they wished, they had to register their occupation(s) with the police. All employers were required to keep written records on their employees. In addition, Turgot established non-*corporate* and non-judicial alternatives for the resolution of disputes among merchants, artisans, and workers. Lastly, article XIV extended the 1749 prohibition on "association" among laborers to "all masters, journeymen, workmen, and apprentices of the aforesaid guilds and companies to form any association among themselves under any pretext whatsoever." This measure demonstrates one of the most radical aspects of the Edicts: the fundamental equality of masters and men in Turgot's conception of the world of work.[21] However, this "equality" meant something quite different on the shop floor than it would in a salon, in a political pamphlet, or even in a judicial court. Turgot envisioned a productive environment in which every laborer had an *individual* relationship with each master. However, in market environments where workers functioned as individuals, deprived of the support and assistance of their fellows, masters dominated them. Equality of masters and men, for Turgot, must be understood as a means of ensuring the control of industrial entrepreneurs by eliminating associations of workers.

Turgot also initiated a new phase in what would become one of the central administrative controversies of the reign, namely the scope to be given to consumers to make individualized, rational decisions about the quality of goods for sale.[22] The ease of sale of goods produced in the relative freedom of the laboring environment of privileged enclaves such as the faubourg Saint-Antoine was proof, said Turgot, that *corporate* regulation and its protection of the consumer was unnecessary. Monopoly did not maintain the quality of the merchandise presented to the public. In a *mémoire* presented to Louis XVI early in 1776, Turgot stated: "All the world knows that the *police* system of the *jurandes* purporting to ensure competent work is entirely illusory."

The immediate response to Turgot's reforms split largely along what could be referred to as the relationship to the means of production. Near the summit of the social structure, the Parlement of Paris refused to register the Six Edicts, forcing the crown to resort to a *lit de justice* held on March 12, 1776.[23] In their remonstrance, the judges defended the *corporate* system as the guarantor of social control over the workers and of quality production by the masters. With regard to innovation, the judges worried that it would no longer be in the interest of either masters or workers to perfect their processes or improve their wares. They pointed out that England, the exemplar of economic freedom, also had guilds. Allowing anyone to practice a trade without an apprenticeship or a proof of competency, i.e. performing a masterpiece, would undermine consumer faith in French industry and harm sales. According to the First President, "the edict suppressing the *jurandes* would break the bonds of the established order created for tradesmen and artisans and leave turbulent and licentious youths without regulation and without bridle which would lead to all sorts of excess which are avoided through the interior discipline of the communities and by the domestic authority of the masters over the *compagnons*."[24]

Masters were troubled by many of the same concerns. In general, they defended the *corporations* on a number of disparate grounds, including economic utility, the maintenance of social order, that membership in a *corporation* was property, and the protection of privilege in general. Although masters and magistrates differed in some respects as to what was wrong with Turgot's proposals, they shared a distinct anxiety over the social consequences of any measure that would weaken the sacrosanct nature of membership in a *corporation* as a form of property. Equally important, both groups feared lessening the disciplinary control over unruly and undisciplined workers of masters who had a stake in society.[25]

While magistrates and masters sought to prevent the suppression of the *corporations*, workers, particularly the compagnons of Paris, but also some small-scale masters, were much more favorable to the measure.[26] *Compagnon* glazier Jacques Ménétra noted the gratification of the masters when "all these self-important juries and syndics were abolished." According to the bookseller Hardy, the common people were greatly excited by the prospect of the abolition of the *corporations*. In several quarters of Paris, illuminations set up by rejoicing *compagnons* accompanied popular dancing in the streets and brawls between certain trades. Unnerved by these incidents, the police called out troops to restore order.[27] It was, in part, the different responses by elites, masters, and men to this measure that prompted Steven Kaplan to describe Turgot's reforms as a "carnivalization of social relations."[28]

The Six Edicts also excited a storm of protests by nobles and clergy who feared a loss of social distinction if the *corvée* became a monetary payment, by magistrates who defended the privileged basis of ancien régime society, and by disgruntled masters who felt they had lost control over their workers.[29] These measures also elicited nervous responses from keen observers who noted that the continual reforms initiated by the central government, ranging from the reorganization of provincial administration by Controller-General Laverdy in 1765–1768 to Maupeou's refashioning of the judicial system in 1771–1774, were destabilizing French society. Nervousness about the boundaries of social discipline and political liberty led to the surfacing or new tensions concerning "improvement."[30] Numerous entrepreneurs and bureaucrats suspected that upsetting the applecart of government oversight might unleash the whirlwind of labor unrest. Turgot, staunch defender of "liberty" that he was, was so outraged by the volume and vehemence of protest that he solicited a royal edict forbidding the publication of *mémoires* in favor of the *corporations*.[31] This public protest, however, was superseded by Turgot's political isolation within the cabinet as a cause for his dismissal on May 12, 1776.[32]

Just as Louis XVI and his council revived the parlements in 1774, a new regulatory edict to govern the world of work in Paris was issued in August 1776.[33] However, the *corporate* structure was altered far more by the brief interlude of Turgot's reforms than was the legal system by Maupeou's revolution of 1770–1774. Led by the Count de Maurepas, the Royal Council revised the *corporations* in a new attempt to reconcile regulation and liberty. The Preamble reflected the Royal Council's hope that, with fewer rules to follow, the revived *communautés* would no longer "prevent progress in the arts."[34]

At the same time that the Edict sought to ensure that "liberty" reigned in the choice of what to produce, it also intended to return order to the labor market, also in the name of "liberty." The goal was a restoration of subordination in master-worker relations, "without which, commerce, talent and industry will be deprived of the advantages attached to that freedom [*liberté*] which ought to stimulate emulation without heightening fraud or licentiousness." Article XL restored the regime of worker surveillance established by the *Lettres-patentes* of January 2, 1749 as the most coherent set of restrictions on the books concerning the police of the *corporations* and journeymen.[35] Article XLIII forbade workers from setting up religious *confréries* under any circumstances to prevent them from becoming surrogate compagnonnages. Both masters and men were given three months to register with the police.[36]

If workers had to be disciplined, the new Edict widened access to membership in the *corporations* dramatically. Seigneurs regained the right to grant masterships in areas under their control, but the major reforms involved the elimination of the *chef d'oeuvre* and apprenticeship as requirements for becoming a master.[37] The Edict of August 1776 envisaged three means of entry into the restored *communautés*.[38] First, former masters could join if they paid a *droit de confirmation* fixed at one-fifth the regular entry fee of a revived *corporation* or either one-fourth or one-third of the jointure fee for a *communauté* that united several professions. Second, those who had set up a shop or begun a trade since February had to make an annual payment of one-tenth the entry fee or they could enroll in the new *corporation* at one-third the nominal cost. Third, former masters could become *agrégés* (incorporated or associated members) with the ability to exercise their profession, but without the other rights and privileges of a master, including, and, in particular, the ability to train apprentices. All of the entry fees were much reduced, and those without resident status [*forains*] no longer faced financial penalties if they became members of the *corporations*. Three-fourths of the funds collected as fees went to the royal treasury.

In addition to these loosened restrictions, the new Edict removed the gender bias of most trades.[39] After August 1776, upon making a formal declaration to the police, women could now exercise nearly all professions ruled by the *corporations*, although they could not participate in the political life of most *communautés*. However, for women, this measure was double-sided; men were also admitted to women's *corporations* on the same basis, which drew many female complaints. Widows could also acquire permanent authorization to follow the trade of their spouse at reasonable rates even after the typical one-year grace period. Religion endured as the only major discrimination within the *corporate* world, as was made explicit when the Six Corps of Paris used the August Edict to expel Jews who had entered their trades.[40]

In certain areas, some professions, such as the charcoal sellers and the wool combers of Rouen, were deliberately excluded from the realm of the *corporations*.[41] In Paris, 21 trades were no longer subject to corporate control, and only 50 corporations were authorized in August 1776 (versus 117 before February).[42] In many other cases, two or more professions were amalgamated into a single *communauté*, usually where the skills or raw materials were similar. For example, the *serruriers* (locksmiths) and *cloutiers* (nail makers) were soldered together in many cities on the theory that both trades worked with iron. *Fabricants de bas* (stocking makers) were associated with *merciers* (haberdashers) because they both dealt with clothing accessories. Yet the economic interests of storekeepers and merchants like the *merciers* were very different from that of a fabricant. Although *serruriers* and *cloutiers* had some skills in common, in practice they tended to compete for the right to make wares that lay between their specialties much more than they shared expertise in the use of coal as fuel.[43] Such amalgamations were to be a long-lasting, divisive legacy of the regulations of August 1776.

Paris' masters did not greet the August edict with rejoicing, as they had the restoration of the parlements. In fact, their overwhelming response was to hope that the ministry would change its mind yet again and modify the regulations during the three-month grace period in which they were supposed to make their decision about whether to join or become an *agrégé*.[44] Despite a lukewarm response to the new regulations, over the next four years, the Royal Council applied its provisions to most of the realm. Lyon led the way in January 1777.[45] Although in general the number of *corporations* was reduced, their control over the world of work simplified, and access to membership opened, in Lyon (France's second city) new *communautés* were established.[46] The 75 remaining cities in the Parlement of Paris' jurisdiction followed on April 25, 1777. Normandy received an analogous edict in February 1778, revised in April 1779. The areas overseen by the Parlements of Nancy and Metz were next, in May 1779. The *conseil souverain* of Roussillon brought up the rear in July 1780.[47] The measure was never applied in the resort of the Parlement of Guyenne although the tailors' *corporation* of Bordeaux reported that their workers felt themselves authorized by the law to act as if it had been applied in that city. Seventeen "*ouvriers non maîtrises*" in Bordeaux set themselves up in business as if "liberty" from *corporate* regulation existed there.[48] Thus, just as Turgot's attempt to abolish the *corporations* had not been applied in much of France because of local resistance, the re-establishment of the *corporate* regime was limited in execution.

Although the initial revision of Turgot's economic program began during Jean-Étienne-Bernard de Clugny's tenure as Controller-General, it picked up speed under his successors. Maurepas remained the leading figure of the Royal Council until his death in 1781, but Jacques Necker played an increasing role in both fiscal and administrative affairs.[49] He may not have enjoyed the title of Controller-General, but Necker and his coterie of supporters developed far-reaching plans for the regulation of commerce and industry that were implemented during this turbulent era of war and economic dislocation. These plans, which often appeared to be executed haphazardly in response to pressing short-term wartime financial considerations, were, in fact, generally based on relatively consistent principles. Necker embraced the need for governmental regulation, but believed that there were clear limits to what the central state could accomplish. As a result, Necker's reforms, ranging from the establishment of provincial assemblies to industrial regulation, invariably sought to mix consultation of local public opinion, a certain level of state involvement, and liberty. Whenever possible, Necker also wanted the legal situation to reflect actual practice, but this aim was frequently overshadowed by his hopes to increase the uniformity of the realm's economic and political institutions.[50] These often conflicting aims were reflected in the post-Turgot industrial policy of the French state under Louis XVI.[51]

Necker recognized that after Turgot's "experiment" the industrial situation could not be returned to the status quo ante.[52] At the same time, he understood that the central government needed more information if it was to reform the world of work, particularly for the textile industries which made up such a large percentage of French industrial production and manufactured exports. Although there had been four major inquiries into the *corporate* world in the 1760s and the 1770s, in 1777, a new set of orders went out to the inspectors of manufacturing to report on their areas of expertise. A renewed appeal followed in March 1778, accompanied by inquiries to intendants and deputies of commerce about a proposed "intermediate administration between the regulatory system and indefinite liberty."[53]

Necker deliberately co-opted such proponents of liberal reform as Inspector of Commerce Simon Clicquot de Blervache, Inspector of Manufacturing Nicolas Desmarest, and Pierre Samuel Dupont de Nemours (an assistant and protégé of Turgot) to help construct a new regulatory regime in textiles.[54] These improvers were all linked to the Physiocrats and were prominent in royal administration, in the salons, and in the academies. With their support, Necker could reject the notion that regulation undermined sales. Instead, he proposed a dual system in which producers could chose to follow the regulations that guaranteed quality or they could manufacture textiles by any means or in any form they wished, according to their individual sense of market demand.[55]

Necker's ideas resulted in the *Lettres-patentes* of May 5, 1779, which transformed the regulation of French textile production.[56] The creation of a dual regulatory system was strongly supported by Turgot's supporters and by the "experts" so essential to French administrative decision making, including the *intendants du commerce* and the inspectors of manufacturing.[57] This measure fundamentally shifted the state's regulatory apparatus relationship to France's largest industry. No longer would the officials of the central state involved in overseeing industrial production and their agents be limited to using inspection to enforce existing regulations; they could now focus on the encouragement of entrepreneurs.[58]

The manner of encouragement was particularly significant, because, in practice, the "intermediate system" meant "liberty" from regulation.[59] Thus, the *Lettres-patentes* of May 1779 trumpeted a new policy, namely that, although the quality production associated with the Colbertian legacy was officially preferred, the bulk of French textile production was to be unregulated—the fabric and the style were to be at the sole discretion of the producers. Just as Turgot's "liberty" was intended to release the bonds on the French inventive genius, Necker's policies allowed French textile entrepreneurs to concoct new types and designs of textiles as a means of winning over fickle consumers and keeping them loyal despite the subordination of wool, linen, and silk to King Cotton.[60] William Reddy's evocation of the emergence of "market culture" as a response to the position of the corporations as "anti-entrepreneurs" frames this discussion while reflecting Adam Smith's views on the matter.[61]

The unshackling of the French textile industries in May 1779 affected far more entrepreneurs, workers, and consumers than Turgot's reforms, and significantly loosened the regulatory chains yoking the rest of the world of work. Managing the dual system, given the multitude of local customs and the complicated requirements, proved a logistical nightmare.[62] Further *Lettres-patentes* issued on June 1, 4, and 28 of 1780 adjusted how the *bureaux de visite et de marque* (inspection offices) administered the textile trades since, beginning on July 1, 1780, the main task of these bureaux was no longer to enforce adherence to state-mandated norms, but to distinguish between types of production.[63] Towns were to establish bureaux to confirm whether the cloth had been made in France and whether it followed quality controls; the seals on the bales cost of 1 *sol* (later 1.5 *sols*) per bolt [*pièce*]. Other *généralités* received equivalent *Lettres-patentes* in 1780 and 1782. The new system was to be in full force by January 1, 1782.[64]

All these measures sharply diminished the role of the *corporations* in regulating the largest industry in France. No longer did individual entrepreneurs have to conform to *corporate* regulations on the size, shape, and composition of textiles. These measures also completed the emancipation of rural producers from the control of urban-based *corporations*, a process begun in the 1750s.[65] The exclusive right of certain places to manufacture particular types or styles of cloth, first successfully challenged in 1776, was now extinct.[66] Not surprisingly, formerly privileged urban centers, including Sedan and Lille, led the resistance to these reforms.[67] This version of "liberty" was, however, somewhat of a mirage. Although the creation and extension of the *bureaux de visite et de marque* meant that the degree of regulation was less, the responsibility was shifted from the locality to representatives of the central state. Thus, what was intended as liberation was not always experienced that way by entrepreneurs.[68]

This reform agenda concluded with the promulgation of the *Lettres-patentes* of September 12, 1781, which regulated the relations of masters and men both inside and outside the corporate world. Although it was issued after Necker's fall from power, this ordinance was the logical culmination of the regulatory trend inaugurated in August 1776. The *Lettres-patentes* of 1749 remained the basis of labor relations, but the Royal Council (led by Charles Gravier and Jean-François Joly de Fleury) thought it needful to "add precautions which they thought capable of maintaining police and the subordination of the workers."[69]

This subordination was meant to be thorough but not unlimited. The deck, however, was stacked heavily in favor of employers. In return for concrete advantages demanded by masters for decades, both large-scale and small-scale entrepreneurs had to fulfill certain obligations toward their workers. The advantages included the requirement that "every worker" who wanted a job had "to register his name with the police." Workers could not leave their masters without finishing jobs that had been started, reimbursing any advances, and giving 8 days' notice. The *billet de congé* was issued only when these conditions were met. A model certificate was included in the circular announcing the measure. Work could not be offered to a laborer lacking a *billet*. Entrepreneurs were also forbidden from pirating workers, either directly or indirectly, from another entrepreneur. The fine for both actions was fixed at 100 *livres* plus any damages to the worker(s) involved. All disputes between entrepreneur and worker, including conflicts concerning the theft of raw materials or tools or regarding the *billet de congé*, were to be resolved without delay by the *juge de police* instead of the *juge de manufactures*. Workers were also forbidden to assemble or meet under any pretext whatsoever. Stipulated prohibitions included membership in a religious *confrérie*, influencing the placement of workers, planning retribution against employers, and forcing the discharge of workers, whether French or foreign. The subordination of the laboring classes was deemed necessary to social control and shop-floor discipline as a means of improving France's industrial competitiveness.

From the standpoints of industrial policy and labor discipline, the major goal behind these measures was simple: keep workers in one place. Although the fundamental ancien régime assumption that police control over the lower classes was desirable played a role, this measure went beyond the traditional distrust of workers on the tramp.[70] A major lesson drawn by Enlightened reformers, in part from the English example, was that, in order to make workers (and the lower classes more generally[71]) more productive cogs in the industrial economy, it was necessary to tie them down geographically. Between February 1776 and 1789, the precedent from across the Channel was buttressed convincingly by contemporary anxiety about growing militancy among *compagnons*, textile workers, and unskilled laborers.[72] The constant itinerant movement of workers, particularly *compagnons*, undermined labor discipline and provided easy means for recalcitrant employees to escape the mechanisms deployed for their "improvement."[73] The fact that most recent studies have found that workers roamed in search of better conditions or control over access to employment rather than higher wages per se confirms the matter from the perspective of the workers.[74] Thus, through the medium of the *Lettres-patentes* of September 12, 1781, the Royal Council realized an element of the project of Enlightenment. Those individuals of the laboring classes who were tied to the locality could be educated, disciplined, and/or transformed into more economically productive "new men" capable of meeting the challenges of international industrial competition.[75]

Although these provisions would permit masters a greater degree of control over their workers than had existed previously (at least in practice),

these *Lettres-patentes* were not completely one-sided. Such even-handed-
ness might have been the result of some sort of Enlightened view (as pos-
tulated by Minard and Kaplan), or it could have been a recognition of
the natural-law tradition (as emphasized by Sonenscher), but at a very
pragmatic level this ordinance acknowledged the boundaries of administratively imposed labor discipline.[76] Government officials recognized the
limited success of the disciplinary regulations that had been passed at
regular intervals since the seventeenth century. This time, the Royal
Council wanted results; as a consequence, these Enlightened policy makers were willing to accept a somewhat more equitable industrial regime.
Such a regime would maintain worker discipline and social subordination by diminishing labor's geographical mobility. Yet entrepreneurs had
to fulfill their bargains as well, especially the one sought by the state: to
make their laborers more productive by any means available to them. It
was in this context rather than any sort of spirit of generosity or sense of
equity that workers were able to achieve some of their ambitions in the
Lettres-patentes of September 1781.

Workers acquired access to faster, cheaper, and probably more impartial justice in industrial disputes or against employer persecution through
the substitution of the *juge de police* for the *juge de manufactures*.
Employer culpability was spelled out in the regulations, particularly with
regard to the non-delivery of the *billet de congé* or pirating workers,
providing laborers with explicit additional legal protections. In such
instances, workers could obtain a favorable judgment "without payment," a rare situation under the ancien régime. Furthermore, in some
places, laborers had been subjected to registration for decades; now it
was stripped of its most oppressive elements, particularly the payment
of variable clerk fees. Finally, while workers were required to fulfill their
contracts with employers "faithfully," so too were masters prohibited
"from dismissing their workers . . . without legitimate cause before the
term fixed by contract." These limitations on the subordination of
workers may seem minimal to contemporary sensibilities, but they
excited a storm of protest from masters in the 1780s and emboldened
some laborers to demand even more such protections, particularly for
collective action, despite the Royal Council's explicit restrictions on
their activities.

The government hoped to nurture enterprise, in part through the use of market forces. However, since a true market economy does not exist unless all the component elements can be bought openly and freely, this "market culture" remained fractured and limited.[77] What the administration desired was to generate both more and more daring entrepreneurs dedicated to a certain "industrial" approach to development without empowering labor enough to undermine economic progress.[78] In this context, the reforms of Turgot, the restoration of the *corporations* in August 1776, and the *Lettres-patentes* of May 5, 1779 and especially those of September 12, 1781 take on new meaning. The *Lettres-patentes* of 1781 went beyond those of 1749 by legislating the domination not only of masters over *compagnons* and apprentices but also, more generally, of industrial entrepreneurs over workers. Subordination was to be enforced differently however; what was being proposed was a new form of paternalism in which producers were to be responsible not just for the conduct but also for the education of their employees, who would become more resourceful, more efficient workers,[79] thereby foreshadowing the emphasis on creating a "new man" so essential to the revolutionaries of 1789.[80] That after 1779 this paternalism could be and was exercised by a significant number of women is a sign of the new possibilities of the productive environment. But as with so many of the policies devised in this tumultuous period, this approach could not bridge conflicting social and economic goals. As we shall see in subsequent chapters, the French industrial experience from the 1780s well into the nineteenth century revealed the impossibility of liberating the entrepreneur while subordinating the laborer.

The New Industrial Regime in Normandy

Normandy was the usual testing ground for ancien régime industrial policies The *corporate* system had deep roots in the area, but rural outworking and subcontracting in the textile trades had undermined much of its economic power, particularly over labor.[81] The proximity of the English Channel and the density of transplanted British inventors and entrepreneurs meant that the English example had particular resonance.[82] The far-reaching tentacles of Rouen's textile trades were the result of and an encouragement to commercial entrepreneurialism.[83] A sympathetic Parlement and a succession of progressive intendants ensured that in the province of Upper Normandy the industrial reforms sown by the Bourbon administration under Louis XVI fell on fertile soil.

Not until February 1778 did Normandy receive an edict analogous to that of August 1776 for the resort of Paris. The delay was prolonged by the Parlement, which did not register the edict until March 1779. In response to vociferous local complaints, a revised edict followed in April and was quickly registered. The *Lettres-patentes* of May 5, 1779, however, were registered swiftly on July 16, but the *Lettres-patentes* of September 12, 1781 were hampered in their application to Normandy; the Royal Council did not even promulgate an edict until February 6, 1783. The parlement registered it two months later.[84] These long deferments point to two major problems in applying edicts written for Paris to another place. First, as the repeated requests for data in the 1760s and the 1770s illustrated, the central government recognized that it needed local expertise to prescribe an effective industrial regime. Second, this dependence on their cooperation meant that local authorities, from the intendant to the inspector of manufacturing to the parlement, could delay the drafting of central government legislation by various tactics, ranging from withholding information on provincial conditions to stalling to drafting mémoires, pétitions, or remonstrances.[85] The delays, pleas of exceptional circumstances, and personal conflicts at work in Upper Normandy were multiplied in attempts to standardize French industrial legislation.

In the new industrial era, the *corporations* of Rouen were frustrated in their attempts to control the labor force, but, at the same time, the masters were unwilling to abide by the revised rules of the game.[86] As the registers of the deliberations of the *communautés* make abundantly clear, the precepts restricting worker movement, whether established by royal *Lettres-patentes*, by seigneurial ordinance, by municipal statute, or by corporate regulation, had not been followed before 1776. Issuing *billets*, establishing placement offices and forbidding masters from recruiting workers remained problematic issues up to 1789.[87] The master locksmiths systematically evaded the regulation requiring them to register their *compagnons* and apprentices with a central placement office in order to minimize their liability for the head tax.[88] On their side, the *compagnons* refused to get the compulsory *billet* or keep their work histories up to date. The masters also complained repeatedly that, despite all their efforts, the *compagnons* continually "assembled and intrigued amongst themselves to subtract themselves from the dispositions of the regulations and to convince others to do the same." The masters threatened to "take the necessary measures to identify the intriguing *compagnons* and to impose penalties proportionate to their disobedience."[89] In retaliation, the masters petitioned to increase the notice that workers had to provide to 15 days. Refused by the Parlement, the lieutenant-general of police granted their request in early November 1782.[90]

Masters also sought to use the *billets de congé* from the placement office as a proto-*livret* that the employer could withhold or annotate with comments about the labor actions or work history of the individual. In September 1784, *compagnon* locksmith Gabriel Parel successfully petitioned the lieutenant-general of police to force his employer Carbin to give him a billet so he could take a new job. The unwillingness of the *communauté* in general and this master in particular to conform to the decision led the lieutenant-general of police to impose a penalty of 50 *livres* for non-compliance. This controversy continued for more than six months because Carbin, seconded strongly by the syndics of the *corporation*, refused to pay the fine or deliver the *billet*. The locksmiths seem to have chosen to take their stand on this case. The syndics even destroyed Parel's papers, claiming that he had subtracted "six notations from his work history, no doubt unfavorable to him." The Parlement ruled in the *compagnon*'s favor; the masters had to pay Parel 72 *livres* 12 *sols,* and 3 *deniers*, give him new papers, and allow him to start his new job. A few years later, Parel became a master and joined the others in evading tax payments by not registering his *compagnons*. This particular troublesome worker was co-opted, but the vast majority of *compagnon* locksmiths of Rouen continued to circumvent the requirements of the *Lettres-patentes* of 1781.[91]

Rouen's locksmiths had conspicuous difficulty controlling their workers, but they were not alone. In Nantes and Lyon, master locksmiths had similar difficulties.[92] In a completely unrelated trade, in 1784, the *toiliers* (cloth manufacturers) of Rouen complained bitterly about the "collective

demands of their agitated workers," not only for an increase in piece
rates, but also to avoid the requirements of the regulations. After receiving a favorable judgment from the lieutenant-general of police, the
masters continued their offensive in May 1785 by trying to force the
"laborers who work illegally at home to return to the shops of the masters of the community." The objective was to "prevent all the daily
abuses . . . by weavers who work for several masters and deceive them
by exchanging qualities of cotton or by slowing down work for one particular master." Their solution was to "force weavers to work for only
one shop and to ensure that the same rules for *billets de congé* used for
the weavers applied for all other workers."[93] This solution was not
implemented, but the master cloth makers repeatedly pushed for more
stringent enforcement of the regulations against workers as a means of
protecting their own privileges, which they claimed was essential to
ensuring their economic survival.[94] More than 15 professions, ranging
from the coffee sellers to the perfumers to the starch makers, shared this
view. The *cahier de doléances* [list of grievances] of the cider and beer
merchants demanded that "the statutes of each community of the arts
and crafts be conserved along with formal rules to maintain the quality
of merchandise by stopping or preventing the abuses which have been
introduced in all types of manufactures."[95]

For the *corporations* of Rouen, access to the privileges of the *communauté* was a major concern under the new industrial system. There were
two aspects to the issue of access. The first concerned the rights of
agrégés, which was linked to uncertainty concerning the prerogatives of
the sons and widows of masters and *agrégés*.[96] Continuous bickering
erupted about the proper cost of mastership, the peripheral fees to be
charged, and the *corporate* customs that could be enforced, and over
how long reduced entry rates would apply. On February 6, 1783, the
Royal Council promulgated new *Lettres-patentes*, which were registered
in Normandy on April 9, to resolve the questions of access until the various communautés had an opportunity to write new regulations. *Agrégés*
were required to pay only one-fourth the full cost of acquiring this status, but only if they joined within six months. Widows and sons continued to pay half the nominal fees. The detailed provisions of these
Lettres-patentes had a clear intent, namely to ensure the widest possible

access to the remaining privileges of the *corporate* system. At the same time, the masters' local financial obligations were severely limited, as were the potential sources of revenue of the revived *corporations* (in order to deprive these institutions of their ability to obstruct the state).[97]

The Royal Council could not heal breaches between trades joined against their will or between the masters of the pre-1776 *corps* and the masters of the "new communities." The locksmiths suffered from both these difficulties. The competition between the *serruriers* and *cloutiers* remained a distraction in the affairs of the *corporation*, particularly in the inspection of goods. The syndics of the "former masters" delayed handing over the papers and ritual trappings of the *corporation* until May 1783 as part of a "cabal" by the "old masters." The former syndics shifted the bulk of *corporate* taxes to the "new masters" by refusing to pay their fair share of the taxes assessed by the new syndics.[98] Divisions between "old" and "new" groups of masters continued through 1789. The *cahier de doléances* of the shoe makers included sections listing the grievances of "the former community" and "the new community."[99] Such divisions sapped the vitality of Rouen's *corporate* world in the decade before the Revolution.

The second major issue regarding access involved the incorporation of masters from the 14 nearby *seigneuries* into Rouen's *communautés*. Incorporation would eliminate troublesome privileged areas of exception to *corporate* regulation, but had the potential to increase significantly the size of the affected *corporations*. *Corporations* in the textile trades, the locksmiths, and the perfumers continually obstructed the efforts of masters, especially widows from other areas, to join their community, particularly since they only paid one-fourth the nominal fee if they joined within three months. The syndics had several effective delaying tactics at their disposal. They could refuse to give the necessary examinations, they could seize the tools or goods of aspirants by claiming that they had engaged in trade prematurely, or they could interpret the residency requirements stringently. However, such maneuvers could not stand up on appeal; masters from the *seigneuries* were admitted repeatedly because of orders from higher authorities. Identical tactics were used against *compagnons* who wanted to become masters, particularly those who took advantage of the opportunities of the *Lettres-patentes* of

February 6, 1783.[100] The masters of Rouen successfully enlisted the assistance of their long-time patrons, the parlement, to try to cut off the areas of haut justice and force those masters not only to join the city's *corporations* but also to pay the full price. When the Royal Council nullified this effort, many syndics used their powers of inspection to force the masters from privileged areas to join the *communautés* of Rouen or risk repeated seizure of their goods. These tactics were especially effective in trades facing heavy competition, such as the *bonnetiers* (hat and stocking makers) and the lace makers.[101]

Standardized regulations imposed by the Royal Council caused strife within and among the *corporations* of Rouen and increased tension between the principles of regulation and liberty. Inspection was usually the bone of contention. In most trades, two syndics and two adjuncts chosen from among the wealthiest masters were charged with inspecting workshops and stores at least four times a year to ensure fulfillment of the regulations. They were paid 20 *sols* a visit.[102] Several Rouen locksmiths complained bitterly about the costs of such frequent inspection.[103] The lace makers and the cloth manufacturers reported that their syndics ignored the new regulations and did not inspect the surrounding areas now under their jurisdiction.[104] Another type of conflict occurred in the *corporation* uniting *merciers* and *fabricants de bas*. Before 1776, it was the merchant hat makers and stocking makers rather than the manufacturers who inspected hats and stockings entering the city. Now they oversaw the city's manufacturers too. Intendant of commerce Jean-François Tolozan refused to sanction a separate means of inspection. He stated that a new office was unnecessary and that the haberdashers should continue at this task.[105] Unwilling to allow merchants rather than producers to have this degree of control over their livelihoods, the manufacturers twice offered to pay a government inspector to undertake this chore. Not only did Tolozan refuse the requests, he chastised the new *communauté* for contravening the new regulations by naming a deputy to plead their case. The syndics were compelled to pay the costs of the delegation out of their own pocket because this sort of deputation was no longer permitted.[106]

Such incidents illustrate not only the cleavages within the *corporations* but also how the state disciplined the masters. Here the issue was inspection, but divisions inside the *communautés* concerned a wide range of issues, including the regulation of the labor force, the implementation of new technologies, and the proper mix of quality vs. price in manufacturing. The divisions within the *corporations* were a microcosm of the divergences of opinion between French policy makers and among those administrators most directly responsible for overseeing the industrial economy. The difficulties experienced in the 1780s during the attempt to impose a new regulatory regime through the *Lettres-patentes* of May 5, 1779 illustrate the complications that a conflicted and deeply divided administration of Bourbon France had in implementing sweeping change that cut across geographical, legal, and economic boundaries.

Philippe Minard's study of the nationwide network of inspectors of manufacturing revealed that an impressive number of masters rejected the quality specifications imposed by these regulations.[107] Unwillingness to use a lead seal (*plomb*) that correctly designated the quality of the goods was a particular problem often uncovered by inspection. Goods in contravention of the rules or merchandise produced illegally were seized and sold. Half the profit went to the crown (it was usually dedicated to paying the inspectors of manufacturing and their subordinates who maintained the *bureaux de visite*) and half to the person who discovered the contravention.[108] Throughout the 1780s, deciding who to visit, what to do about what was found, where the site of inspection ought to be, and what precisely was against the rules were among the most contentious issues facing Rouen's *corporations*, especially those in the textile trades.[109] The situation was complicated further by the intendant who demanded authorization to make inspections in trades such as metalworking as had been the case before 1776.[110]

The reorganization of the *corporations* begun in 1776 gutted the institution, weakening long-standing *corporate* bonds and paving the way for the formation of new ties and new loyalties. The best way to illustrate the degeneration of the *corporations* as institutions is through an example. The dyers' *communauté* of Rouen united specialists in wool or silk and those who did other types of dyeing, particularly of cotton goods. Already fragmented by specialty and placed at a remove from *corporate* affairs by the system which mandated that 25 deputies manage all the business of the *communauté*, the 80 master dyers showed no group loyalty.

It even became necessary to impose fines of 3 *livres* on each of the 25 deputies for skipping meetings. In November 1781, the few deputies who did show up asked the lieutenant-general of police to force all the deputies to attend meetings at which important matters were to be discussed.[111] From this moment, in stark contrast to the situation before the implementation of the reforms, meetings witnessed only the division of the *corporation's* tax burden.

Despite the lack of enthusiasm for the new *communauté*, it appears that most of the masters of the pre-1778 group joined the restructured *corporation*. However, an important clique who included several important dyers opted to become *agrégés*. They retained that status for three years, but these independent-minded dyers reconsidered because of new decisions taken in Paris instigated by other *corporations* dissatisfied with the quality of the current work being done by the dyers. Cheating the remaining regulations by marking goods dyed with *petit teint* as *bon teint* (figuratively, light vs. thorough dyeing) was so widespread that, on the request of the lace sellers, the deputies of commerce decided to eliminate many of the various categories of dyeing and replace them with a new classification dubbed "*impression de Rouen*."[112] As a result of pressure from merchants to improve quality and the increased latitude now afforded to masters, twenty *agrégés* decided to join the new *communauté* in 1784–85. Despite their earlier violent attacks on the integrity of the institution, these formerly recalcitrant masters were asked only a few seemingly pro forma technical questions before being allowed to become members. Despite the influx of independent-minded members, the register of its deliberations reveals that the group did not become more vital. At the meeting to write the *corporation's cahier de doléances* in March 1789, less than half the members attended. They dashed off only two articles, one attacking the Commercial Treaty and the other demanding proportional taxation.[113] This profit-oriented and technologically savvy group of guild masters in Rouen had been rendered mute and politically impotent by the reform of the world of work, but these disciplinary changes did not eliminate or even lessen the number of complaints about the quality of their product.[114]

Enjoying a privileged status represented both prestige and influence in the *corporate* world of the provincial capital. In Rouen, the *corporations* diminished in number from 60 to 37, but this was a less drastic reduction than in many other cities.[115] In Rouen, the numerous makers of earthenware who exported all over France and the impoverished wicker makers (*pannetiers*) had been joined together in a *corporation*. After being assured by intendant Louis Thiroux de Crosne that wicker making was "a poor profession which deserved no official attention," Necker and the *bureau des arts et métiers* decided that this occupation "should be emancipated to become a resource for poor workers."[116] Many of the professions rendered *libre* when the *corporations* were reestablished petitioned for recertification, including brewers, wool combers, charcoal sellers, glove makers, exchange brokers, skinners, parchment makers, and leather dressers. The state rejected each request except in situations where public health was at risk: Tolozan mandated that the police must authorize new brewing establishments.[117]

From the thriving woollens centers of Elbeuf, Louviers, and Yvetot, *corporate* privileges were defended as the only means of maintaining the quality of production and preventing the theft of raw material by workers. The merchants and manufacturers of Yvetot petitioned for a new regulation on the number of threads in the warp for woollen cloth. Skimping on thread count had become so widespread that "very few manufacturers were not lax in this regard." The manufacturers claimed that it was the merchants who fostered "this type of contravention," but de Crosne believed that "it is the manufacturers themselves who scrimp in order to profit from the fraud that they perpetrate." Because of such swindles, the quality of many different sorts of fabrics had been diminishing and sales had fallen.[118] The central government's response was to establish stricter guidelines for inspection by the *gardes* of the new *communauté* and to renew prohibitions against worker theft and embezzlement. The Royal Council empowered the intendant to judge all cases relating to woollens manufacture for the next three years, bypassing the lax *haut justice* of seigneurs and the delays of the *parlement*.[119]

The manufacturers of Elbeuf actively demanded state intervention to protect their privileges.[120] They stated that quality was all-important in selling goods abroad. Allowing *liberté*, they argued, would lead to "the decline in the reputation of the product of Elbeuf because of greed." The manufacturers also resented losing control over membership in their corporation. The less experienced syndics of the new *corporation* did not require any demonstration of competency; the petitioners wanted to restore requiring a masterpiece. The 71 current masters claimed that they could supply all existing demands if they had enough high-grade wool and if they could get enough trained workers. Increasing the number of masters only increased wages for qualified workers, which emboldened them, leading to greater labor intractability: "Everyday experience proves that the more workers are paid, the less they work and the less productively they work." This claim was the basis of their hope to refuse admission to anyone who had not exercised the profession consecutively for more than six years unless he was the son of a master. In addition, they wanted to prevent anyone who had not been a master for six years from taking apprentices.[121] De Crosne stated that "the manufacturers of Elbeuf intend to retain the available work of production within their families and close off entry to the community to everyone but their children." The manufacturers' complaints of problems with work and wages were a smokescreen masking their desire to maintain their *corporate* privileges.[122]

The woollens districts were not alone among the *corporations* of Normandy in their vehement defense of privilege. This defense was conducted at a variety of legal, economic, and political levels and by a vast array of tactics. The masters of the small cities of Caudebec, Montvilliers, and especially Neufchatel strongly resented the loss of their *corporate* status and petitioned to get their privileges restored. Only Caudebec was successful.[123] In 1780, 22 merchants in Neufchatel followed up their initial complaint with a more strongly worded grievance which stated that *jurandes* were needed to protect the consumer from "defective" merchandise being sold by peddlers and merchants from outside the city. They also complained that if the similarly sized neighboring municipalities of Gournay and Aumale could have *corporations*, they should regain their former status, since the rights of those masters were authorized only by seigneurs, while their *jurandes* had been legitimated by the king and by the parlement. An assembly of the inhabitants of the county of Neufchatel supported this petition as a means of protecting the consumer. De Crosne responded that the city was too small, that it "had no commerce other than cheese," and that it "could not support a *corps de métiers en jurande* without doing [financial] harm to the inhabitants." Necker accepted this logic without comment.[124]

Other contentious *corporate* issues of the 1780s included recruitment and the central government's distrust of the regulatory oversight of manufacturing exercised by parlements and seigneurs. Local resistance to central government initiative was widespread. For example, in June 1781, when a Laigle tool maker named Boucher pirated a *compagnon* named Julien from a pin maker named Pottier, the latter complained to the seigneur as *juge des manufactures*. Pottier groused that his remaining workers could not complete their tasks. He had tried to convince Julien to return because it would be "the honest thing to do." Only when Julien refused had Pottier gone to the seigneur, who rebuffed his complaint as did the Parlement. In July 1782, the deputies of commerce overturned the Parlement's denial of the appeal, stating that even if the 1781 prohibition against filching workers was not yet in force, the regulation dating from 1749 was. The Parlement was criticized for not consulting the appropriate statute. The deputies found that "a worker is always a worker. He does not lose that quality when he changes employers. He is still subject to the regulations and to the *police des manufactures*." The deputies of commerce concluded their opinion with their fear that "if workers could leave their masters when they please, insubordination and anarchy will result and ruin manufacturing." Julien left Boucher's employ and was fined 300 *livres*.[125]

Although this type of legal override did not take place frequently, it was not uncommon. In January 1784, the Parlement of Rouen unilaterally prolonged by five months how long the former masters had to decide on their *corporate* status. The judges also lowered the fee charged a widow for reception in the new *communautés* from one-half to one-fourth of the nominal fee. The Parlement's defense of widows continued even after the Royal Council voided this initial decision.[126] Thirty months later, the Royal Council had to explicitly reconfirm this particular article of the *Lettres-patentes* of 1781.[127] The Royal Council was determined not to permit the lingering patron-client relationship between the Parlement and the masters, particularly of the "old" communities, to undermine its effort to create uniformity of productive conditions.

Laying the Foundations: Technological Innovation and Entrepreneurialism under the New Industrial Regime

The preceding section investigated the political and economic underpinnings of technological innovation. The experiences of the province of Normandy illustrate a conclusion reached both by the French Royal Council and by present-day historians: that the *corporate* structure of the ancien régime impeded technological innovation. Based on the English model of industrial development as they imagined it, dedicated and knowledgeable French policy makers attempted to evade *corporate* controls in manufacturing sectors that depended on new machinery, such as textiles, while preserving their success in many traditional industries. This evasion was also intended to protect individuals capable of developing new technologies and to foster commercial entrepreneurialism. This pragmatic two-tier strategy, which took advantage of the strengths and weaknesses of the late-eighteenth-century French economy, characterized industrial policy during the reign of Louis XVI.

The regulatory reform of the textile sector in May 1779 and the continuous attempts of the central state to help French textile entrepreneurs during the 1780s generally foundered on the rocks of contemptuous entrepreneurial attitudes toward workers. French entrepreneurs were hampered by an acute sense of social and even physical fear of what laborers would do if machines were installed in workshops or distributed to workers for home-based production. The concerns about the organized power of *compagnons* illustrated so powerfully by Cynthia Truant must be placed partly in this context, as should the emphasis discussed in the preceding section on the masters' desire to maintain and even extend their control over the workers. The limits of their success and of patriarchy became starkly apparent in the summer of 1789, when workers in Normandy destroyed hundreds of advanced textile machines.

Even the most Enlightened philosophe hoped to educate French laborers in mechanics, mathematics, and drawing, so as to enable them to tinker as they believed British artisans routinely did, yet everyone feared the potential social dislocation that might follow.[128] Some policy makers identified a major French weakness in the use of coal as fuel or other specific craft techniques.[129] British mechanics who crossed the Channel with hopes of transplanting Samuel Crompton's mule jenny or James Watt's steam engine complained bitterly about the inability of French workers to comprehend, build, or utilize these critical artifacts of industrial advance, as did their French colleagues.[130] Many entrepreneurs were hesitant to invest money or time in mechanization or technological innovation because of their misgivings about the ability of their laborers to use new machinery properly and economically; they knew how difficult it was to naturalize all types of productive innovations in a new environment unless accompanied by a skilled practitioner to demonstrate the operation—this was true of transfers to England and of technologies and practices conveyed in other directions.[131]

Despite their fear of unemployed workers and their concerns about the ability of laborers to comprehend mechanical innovations, the French entrepreneurial elite had a somewhat complacent attitude toward their employees, both potential and actual. It was generally recognized that British workers earned significantly more than their French counterparts. The merchants who dominated the textile trades, particularly with regard to goods sent long distances (whether inside France or exported) had little incentive to innovate. Confident in the power and stability of the Bourbon state, these entrepreneurs believed that French workers could produce saleable goods in the face of British competition by slashing wages.[132] If present workers would not accept reduced wages, the poverty and swelling numbers of adults in the French population ensured that others would fill the breach. Although this attitude was based on an incorrect understanding of the extent of the economic advantages provided by mechanical spinning, it was nonetheless widespread, especially in areas with extensive "putting out."[133] The relative lack of emphasis on wage demands either by masters or laborers noted in previous sections should also be understood as support for this argument. This perception that traditional methods could continue to be successful was bolstered by the fact that a significant number of French regions, sectors, and individuals maintained or even improved their economic position during the 1780s.[134]

Entrepreneurs persistently applied traditional means to new conditions instead of resorting to technological innovation. In Picardy, Languedoc, and Lorraine, French entrepreneurs responded to competitive challenges by committing fraud, by smuggling, or by reducing the wages of their workers in order to maintain sales and profits.[135] In practice, these practices could mean skimping on the number of threads in a bolt of fabric, thereby cheating both middlemen and consumers. It could also mean putting English labels on French-made goods or vice-versa as a means of evading tariffs. Smuggling also moved from the margins of the economy to being a major opportunity for entrepreneurialism in the eighteenth century because of competitive differences in the quality/price relationship of Europe's manufacturers. Yet when hard times hit, the first measure applied by entrepreneurs from Rouen to Strasbourg and from Lille to Clermont-de-Lodève was to reduce wages as a means of lowering prices to maintain both sales and profits. Reliance on this tactic illustrates a major difference between England and France during the early industrial revolution: such a competitive strategy would not be possible in Britain until the outbreak of the French Revolution.

Policy makers recognized that a productive environment featuring a patchwork of rules and jurisdictions inhibited the use of economies of scale and undermined investment in new technologies. Although both Turgot and Necker had sought to create a uniform economic environment, neither had pushed the position to its limits. Their successors as directors of French economic policy, notably Charles-Alexandre de Calonne, Controller-General from 1783 to 1786, learned from their difficulties in implementing industrial reform. Calonne shifted this Sisyphian task to another domain as a means of creating uniform economic conditions. He focused on the "single duty project" of pushing the customs barriers to the national frontiers.[136] Such a measure would undermine the mishmash of local exceptions and create a truly national market that French entrepreneurs could use as a base to compete not only with their British archrivals but also with their competitors in the Netherlands for high-end paper, with those in Switzerland for ribbons, watches, and some luxury items, with those in the German lands for certain woollens, with those in Piedmont for silks, and so on. But these efforts were scuttled by impending bankruptcy. Only in the period 1789–1791 did the National Assembly successfully construct a French national market with uniform economic and legal conditions.

The example of the reform of the world of work in Normandy delineates a number of issues. The resistance of masters, merchants, and manufacturers to reform of the *corporate* structure and industrial framework in this model province plainly frustrated those most interested in mechanization, new technology, and industrial competitiveness, ranging from ministers of state to local administrators to inventors. The doughty defense of privilege to escape or minimize competition demonstrated to improvers that more radical measures had to be undertaken. Yet the obstinacy of both masters and men Normandy must not overshadow the fact that this province was at the forefront of French technological development. A nearly inexhaustible supply of innovating entrepreneurs developed and applied new machines and methods, not only in the textile trades, but also in earthenware, chemicals, and copper smelting. This creativity was stimulated by the presence of numerous English migrants and by the looming threat of Albion across the Channel.[137] These accomplishments suggested to attentive policy makers that if the *corporate* chains fettering French industry were shattered, the "liberated" manufacturing economy would experience explosive growth much like that taking place in Great Britain.

French improvers beginning with Turgot chipped away at the tottering edifice of privilege, but could not topple it until 1791.[138] The labor legislation of the reign of Louis XVI sought to free the entrepreneur enough to allow them to develop and utilize new technology while maintaining the discipline of the industrial work force. Although the National Assembly implemented most of the ideas advanced by the improving ministries of Turgot, Necker, and Calonne, the French industrial economy did not blossom as had been hoped, in large measure because of the political restiveness of the laboring classes. As a result, initiatives begun by French improvers bore fruit long after the improvers had fallen from power or had moved on to other areas of endeavor. However, in the nineteenth century, the seeds sown during the reign of Louis XVI yielded a rich harvest.

3

Foreign Policy as Industrial Policy: The Anglo-French Commercial Treaty of 1786

Informed contemporaries in France knew that changes of consequence were taking place in the late-eighteenth-century English industrial economy. But they did not despair. Thanks to a flock of spies to complement close cultural and deepening economic ties, the French developed a distinctive and somewhat surprising conception of the political, economic, and technological components of British industrial success. Looking at the situation through their own prism shaped by differing cultural and political perspectives, French policy makers thought that, with only slight modifications, they could adapt the British model of economic and technological development and beat the British at their own industrial game. It is worth remembering that it was a French envoy to Berlin who coined the term "Industrial Revolution" in 1799, stating that it was "under way" in France.[1]

In hindsight, these French pretensions not just to economic competitiveness, but to industrial dominance appear to be just another example of overweening Gallic pride. Or do they? Reconstructing how contemporaries understood the technological and economic developments known as the Industrial Revolution and comparing these views to what historians have identified as the crucial strides in industrial organization and technology is revealing. Without considering contemporary understandings on their own terms and in relation to what seemed possible to these individuals,[2] it is impossible to appreciate the era's competitive industrial environment.

Events on the other flank of the Channel fascinated many Enlightened French, especially the Physiocrats. Both the British and the French were essential components of what Margaret Jacob,[3] Joel Mokyr,[4] and Larry Stewart[5] have described as an "Industrial Enlightenment" that "placed applied science at the service of commercial and manufacturing interests."[6] Following up on the model of the seventeenth-century provincial agricultural academies that brought together government officials and landed proprietors in order to spread knowledge of improved agricultural techniques, an important wing of the Physiocrats hoped to fashion a new sort of economic partnership between state and society.[7] Charles Gillispie has identified a distinct cluster who embraced this goal, including Vincent de Gournay and his circle (which included Daniel de Trudaine, his son Charles-Philibert Trudaine de Montigny, the Abbé Morellet, the royal censor Chrétien-Guillaume Lamoignon de Malesherbes, the Controller-General Anne-Robert-Jacques de Turgot, and Pierre-Samuel Dupont de Nemours).[8] As opposed to most of the other Physiocrats, who tended to be writers and mercantilists to a greater or lesser degree, these men were all high-ranking state servants whose administrative experience led them to advocate lifting the burden of government restraint on economic activities. Removing state impediments did not imply a "hands-off" attitude. These activists saw the economic role of the state as positive; they wanted to help people to compete, and they abhorred regulations that prevented things from being done. This cluster of Physiocrats wanted to inform people of their opportunities and guide them into making productive decisions rather than relying on constraint to manage the economy. Thus, their conception of how to foster economic growth explicitly rejected Colbertian mercantilism.

Proponents of the concept of an "Industrial Enlightenment" have dismissed the Physiocrats as obsessed with agriculture. These officials, however, did not restrict their conception of the basis of national wealth; they recognized the necessary contributions of commerce and manufacturing to economic prosperity and included these sectors in their plans. Albert O. Hirschman illuminated how this group of Physiocrats came to understand that, although the wares distributed or produced by the commercial and industrial sectors were not strictly essential in the way that agriculture was, by offering useful employment and by using the products of agriculture, commerce and industry provided revenue that was fundamental to French national prosperity.[9] Turgot's stewardship of the *généralité* of Limoges was the model for how this strain of Physiocracy conceived of the benefits an activist state could bring to economic relations.[10] These Physiocrats were in dialogue with Adam Smith, and their ideas and disciples played a major role in the French Revolution.[11] Thus, Physiocracy was both more industrially minded and far more influential in early industrialization than its reputation suggests. These major participants in the "Industrial Enlightenment" understood industrialization in England as liberal, but certainly not as laissez-faire. This understanding determined the French approach to economic development and industrial policy at the end of the eighteenth century.

This chapter explores two significant questions: What did French policy makers think were the essential components of British industrial success? What steps, if any, did they take along that same path? The first question requires the tool kit of the cultural historian of mentalité to establish how contemporaries construed the economic situation. French perceptions must then be understood in their political and social context to reconstruct the potential range of policy objectives available to French officials. Only on this basis can the lengthy strides taken by late-eighteenth-century French officials toward the implementation of their conception of the British industrial model through the passage of the Anglo-French Commercial Treaty in 1786 be fully appreciated. This approach also avoids looking backward for signs of British industrial domination or for the conditions that led to political and social revolution in 1789.

French Perceptions of English Industrial Success in the Late Eighteenth Century

British technological change and the early phases of industrialization did not take place in isolation. Even in wartime, the French knew a great deal about developments in eighteenth-century England.[12] There were thousands of British emigrants in France, most of whom worked in some aspect of production. The French could also take advantage of newspaper accounts of scientific or technological advances and the published reports of learned societies or leading scientists. Lively correspondence developed among industrially minded people on both sides of the Channel who faced similar commercial, technological, or scientific challenges and

constraints. Industrial spies made regular reports, which were complemented by the reports of experts sent openly by entrepreneurs, regional
administrative bodies, or the central government to investigate English
industrial organization and technology. Diplomatic dispatches also played
a part in fleshing out impressions of industrial progress in England.[13] This
steadily deepening stream of information inspired many of the policy decisions of the French government in industrial and technological matters
during the age of revolution.

Present-day historians have distorted the eighteenth-century impres-
sion of industrialization. This historiographical lack of attention to the
evidence has been based on an overemphasis on the singularity of the
British industrial experience or model and a tendency to mistake Adam
Smith's prescriptions as descriptions of reality.[14] As was discussed in
chapter 1, the main controversy revolves around the role played by the
British state in fostering industrialization. In a manner clearly linked to
Smith's emphasis on the importance of laissez-faire in fostering economic
growth,[15] economic historians join historians of science and technology
to systematically downplay the significance of British state intervention
in industrial matters in favor of celebrating individual initiative, especially the gifted amateur toiling alone in an obscure shed.[16] Even those
who recognize the important role of the English government in the mobilization of capital and the creation of a national bank insist that, as Peter
Mathias puts it, "Industrialization in Britain after the mid-eighteenth
century, in distinction from the expansion of trade, was not the result of
deliberate government policy sponsoring industrial progress."[17] Such an
interpretation is bolstered by contemporary estimates that during this
period a mere 0.5 percent of public revenue was spent on "development."[18]

No perception could be further from most contemporary French (or,
for that matter, British) understandings of the role of the Hanoverian
state in industrialization. In part, this difference stems from the fact
that, in the eighteenth century, trade policy and industrial policy could
not be separated.[19] French *commissaire général du commerce* Édouard
Boyetet, one of the chief negotiators of the Commercial Treaty, noted the
link in 1785:

England has convinced itself that France will certainly succeed in its project [of
improving hardware production] and this has decided it to make every effort and

to use every means possible to prevent it. The admission of English hardware will assure the complete success of their plans because that Power will not spare the money or avoid any sacrifice when it is a question of procuring some advantage for its commerce. It is capable of providing goods at a loss until it has destroyed the manufactures of France which have arisen in imitation of their own. This is how England has conducted itself on all occasions and in every country where its interests are affected. England will certainly employ the same means in France because it has reason to fear the activity, the genius and all the resources of our Nation.

Before embarking on an analysis of the benefits of relevant British state activities, such as providing incentives to exporters, patent protection, cheap credit, and a foreign policy designed to increase British commercial advantages, Boyetet observed that English policies deliberately aimed to destroy French industries "before they can acquire the degree of perfection they are capable of attaining."[20]

Jean-Marie Roland de la Platière, an inspector of manufacturing and a noted author on industrial topics, went even further. In the early 1780s, after he had visited Great Britain twice to investigate industrial practice, Roland concluded that the actual source of British superiority in the production of cotton goods was not cheaper prices or higher quality. Rather, the true wellspring of British success was a more effective, state-directed political economy.[21] A report written by two members of the Chamber of Commerce of Normandy after a visit to England came to many of the same conclusions about the importance of state action in industrial success. Meanwhile, on the prompting of the Chamber of Commerce of Bordeaux, the deputies of commerce also lauded British commercial policies that facilitated industrial expansion.[22] Perhaps the most important of these policies was the bounty on exports of cotton goods to France known as the "drawback." Enacted in 1784, this policy underlay the booming English trade with France by stimulating production and thus economies of scale.[23] However, there were other factors that contemporaries emphasized as signs of the depth of British state involvement in industry, such as assistance in developing markets, accumulating credit, patent protection, successful repression of worker activism, the power of various authorities to set wage rates (little used after 1773),[24] and the illicit acquisition of needed workers or technologies. Here was a markedly different perspective on the nature of British industrialization than the one provided by most historians of British industrialization. These contemporaries support the re-evaluation of the eighteenth-century British state begun by John Brewer and Patrick O'Brien and followed up by Martin Daunton, who restated the Smithian problematic by describing British state involvement in economic matters as "The Visible Hand."[25] Eighteenth-century perceptions sustain a profound revision of historical understandings of the nature and prospects of cross-Channel industrial competition.[26]

Jean-François de Tolozan, director of the Bureau of Commerce, was responsible for evaluating and ameliorating French industrial production. With reports, both statistical and individual, coming in from all over the kingdom, as well as from abroad, Tolozan was perhaps the best-informed French official regarding the competitive balance between England and France. In a report on the state of French commerce published in 1789, Tolozan adopted a defensive tone. He painted a vivid picture of an economic slump made worse by rational consumer behavior. Cheaper English goods had flooded the French market, displacing domestic wares. Yet the situation was not thoroughly negative. In fact, Tolozan stated that French and British foreign trade were roughly equivalent in volume in 1789.[27] While recognizing French inferiority in techniques and technology in agriculture and in a wide range of important industries, including metallurgy and textiles, Tolozan suspected that government policies helped to undermine French competitiveness by discouraging inventors and investors. He believed that the French method of giving privileges to inventors had "held back, though its excessive caution, the progress in our understanding of the sciences, damaging national industries."[28] If present government policies relating to patents and the lack of a systematic application of French scientific prowess to industrial technology were investigated, he suggested, "we could construct a parallel between France and England that would not be to our advantage."[29] Tolozan believed that a thorough revision of the means and methods of state oversight concerning the development of new technologies and their application to industry would permit rapid technological advance and bolster France's already competitive economic position.[30]

English-born John Holker, *inspecteur général des Manufactures étrangères* since 1755, was also singularly well informed about the relative technological positions of France and England. After having worked in the nascent Manchester cotton industry, he came to France in 1749 because of his Jacobite political leanings. For nearly 40 years, he played an important and unique role in the Anglo-French economic duel. Holker conducted regular inspection tours of industrial districts and had a large number of English correspondents. He and his son were instrumental in importing a host of English inventions and techniques, including James Hargreaves' spinning jenny and Richard Arkwright's system of production, primarily by convincing workers and entrepreneurs to emigrate to France. The elder Holker was also an entrepreneur in Normandy, directing large-scale enterprises in cotton spinning and chemical manufacture. Consequently, he had strong, well-founded opinions about France's ability to compete with England.

In 1785, Holker wrote a letter to the chief French negotiator of the Commercial Treaty. Printed four years later, this letter surveyed the competitive situation in a number of industries, particularly textiles, iron making, and porcelain. Presciently, Holker concluded that the English enjoyed a cost advantage of up to 30 percent in cotton spinning, especially for finer-grade threads. This advantage was based chiefly on greater mechanization in England, a circumstance rooted in the exceptional personal wealth of most English entrepreneurs and amplified by the greater availability and cheapness of investment capital in England. But there were other reasons for British preeminence, said Holker, namely superior access to water transport, higher-quality raw wool, and greater supplies of cheaper and better-burning coal. The granting of exclusive privileges by the state and the many regulations that curbed the creativity of entrepreneurs also hindered technological advancement in France.

Yet Holker believed that France could compete, and in the very near future, if its Enlightened administration implemented the proper policies. Having made both his administrative and entrepreneurial careers by importing English workers, machines, and techniques, Holker believed that the French government had to follow an even more interventionist path than its cross-Channel rival. Indeed, he argued, the French state must act quickly and decisively to dam the flood of contraband goods deluging the country, just as the British government had done through tariffs. To assist the textile industries in his adopted home in Normandy, the immediate task was to mechanize the spinning process on the Arkwright model. A specialist in personal technology transfer, Holker argued that eliminating exclusive privileges would entice "the numerous [British] artists in this genre who seek to expatriate hoping to better their position." The relative lack of fast-running streams could be bypassed by substituting horses. Continuing to eradicate the many regulations that hemmed in the production and distribution of textiles would enable French entrepreneurs to adapt more quickly to market conditions. The state must help by remedying shortages of capital and expertise. Through government sponsorship, qualified mechanics and entrepreneurs from England could erect an additional 8–10 establishments using the Arkwright system, powered by water, thereby allowing France to compete with Great Britain on an equal basis: "We could then make our own cotton fabrics at just as low a price and even sell them in England."[31] Holker thus sketched a series of measures calculated to drastically improve French competitiveness which were independent of but parallel to those conceived at the highest levels of government.

Three roving inspectors of manufacturing tasked to survey the critical cotton textile sector agreed that state action must play a central role in the French economy. Their report disclosed that at the end of 1787 there was widespread unemployment among French textile workers and concluded that Tolozan's plan for improving French economic competitiveness was well suited to its objectives, that the decline in sales of French textiles was only "momentary," and that "the spirit of emulation" would lead manufacturers and the government to undertake the "proper remedies."[32] According to Gérard de Rayneval (*premier commis* of the foreign ministry) and Pierre-Samuel Dupont de Nemours (*Intendant de commerce*), this modernization of French manufacturing techniques was to be expedited by the accelerated acquisition of English machines and workers capable either of utilizing them and building more. This policy was greatly facilitated by the establishment of close commercial relations and the reciprocal elimination of passport requirements.[33] With such powerful faith in the ability of France to compete with Great Britain coming from informed officials (particularly Dupont de Nemours, who compiled reams of industrial statistics in support of the plan), the French decision to sign a commercial treaty becomes more understandable.[34]

A detailed understanding of the French trade position and the sources of British success was in no way restricted to high government officials or to areas with close ties to England. French entrepreneurs knew about the phenomenal rate of growth in the English cotton industry. They believed that England began to experience rapid growth around 1760 and then detected a "take-off" in the mid 1780s. According to a set of figures gathered by a Rouennais merchant and published by the National Assembly, England produced about 5 million *livres*' worth of cotton goods in 1760, but by 1789 that figure had reached about 180 million, with nearly half produced in and around Manchester. English cotton manufacturing employed 159,000 men, 93,000 women, and 101,000 children, for a total of 353,000. Two years later, in reference to this report, the merchant-manufacturer Louis-Ézechias Pouchet, remarked that "what was extraordinary was that in 1784 this industry, which had already grown so astonishingly during the previous 20 years, doubled its production in the four following years."[35] Pouchet observed this growth first-hand by visiting England in 1786, but was not discouraged from competing. Upon his return, he immediately established a workshop to build English-style textile machinery in Rouen.[36]

Pouchet identified two major causes of rapid English industrial expansion. One was that the British did not suffer from the excessive *corporate* controls that continued to hamper the French. Pouchet emphasized the "freedom" enjoyed by British manufacturers and compared it to the "slavery" of French producers. A second English advantage was the widespread use of Arkwright's water frame. Only three large-scale French establishments currently employed this "ingenious discovery," which permitted not only the mechanization of cotton spinning but also the production of higher-quality thread at a lower price.

These regulatory and technological advantages could be remedied, said Pouchet. France could achieve a superior competitive position because of three major advantages. First, French manufacturers consistently displayed superior creativity in new designs.[37] Second, the cost of labor was decidedly lower in France. And finally, thanks to the Revolution begun in

1789, French industry would "prove to the universe" that "although held back for centuries, it will develop brilliantly now that it is being encouraged [by the government.]" While he acknowledged the efforts of the royal administration before 1789, Pouchet believed that the Revolutionary government had embarked on more effective policies to boost industry. Thus, Pouchet recognized the scope of English success, but did not think that any of the major elements of British competitive advantage—commercial, technological, or social and political—were irreversible. His response to current British preeminence was not as resigned as Tolozan or as optimistic as Holker's, but like them he understood vigorous state action to be a major source of English industrial success.

Contemporary French views of the English model revolved around the vital role of the state in industrialization. Policy makers and entrepreneurs emphasized the fundamental connection between foreign and commercial policy and their link to the interests of industrialists and inventors. In French eyes, the Hanoverian state—which opened markets, utilized tax-based export incentives, and provided more successful protection of patents and access to cheaper capital while effectively funneling private support for invention—was essential to British industrial success.[38] Yet the very success of the British state pointed to how the French could catch up and pull ahead.

That some English saw themselves as irrevocably superior to the French in technological matters is unquestioned. The oft-cited Arthur Young personified this self-image during his travels in France. Visiting the renowned French chemist Antoine-Laurent Lavoisier, Young witnessed an experiment featuring a machine that distilled water from air; it impressed him mightily: "If accurate, (of which I must confess I have little conception), it is a noble machine. Mons. Lavoisier, when the structure of it was commended, said, *Mais oui monsieur, et même par un artiste Français!* with an accent of voice that admitted their general inferiority to ours." In the same entry, Young lauded the inventions of noted Parisian mechanic and watch maker Nicholas Lhomond, including a prototype electrical telegraph and an improved vertical spinning jenny.[39] The yawning gap between Young's dismissive and condescending tone and the high level of scientific achievement and mechanical dexterity displayed by Lavoisier and Lhomond parallels a chasm in the analyses of generations of historians concerned with the industrial revolution and shows a similar insensitivity to irony. There is an overwhelming tendency to portray English industrial superiority as inevitable, much like the outbreak of revolution in France.[40]

Somewhat surprisingly in view of the recent attention paid to the pace and timing of the economic transformation of the Industrial Revolution, no major present-day account appears to consider what British contemporaries thought about their relative technological position or about the technological basis of the economic changes taking place.[41] This applies even to those who make assertions that would be supported by such contemporary evidence.[42] This lack of attention should not be taken to mean that the British were silent either in public or in private about their international technological and industrial standing. Investigations of the cultural worlds of individual innovators such as James Watt and Josiah Wedgwood[43] and discussions of the old canard that early breakthroughs in industrial technology were invented in France and perfected in Britain have long made use of contemporary views.[44] On the other hand, historians focusing on the *longue durée* of economic change have downplayed the depth or breadth of the British technological advantage in the late eighteenth century, at least as the dominant factor in British predominance during the initial stages of the Industrial Revolution.[45] Across a broad spectrum of industries, including metallurgy, chemicals, and textiles, on the eve of the French Revolution, British scientists, administrators and entrepreneurs would not have assumed their scientific or technological superiority. According to Harold Parker, the most important technologies imported by Britain from France in the eighteenth century included casting plate glass, chemical bleaching, improved dyeing procedures, and the production of synthetic alkali and soda. Britain also imported the theories needed for important advances in the utilization of water power and in civil engineering.[46]

James Watt repeatedly recognized the advanced state of French chemistry, and Matthew Boulton regularly sought to acquire for his machine shops French (or other foreign) workers who were conversant with superior methods of gilding, silvering, and engraving.[47] English silk producers assumed the superiority of French means and methods and claimed that they could not compete.[48] The entrepreneurs of that dynamic center of technological and industrial innovation, Manchester, knew that the French had technological knowledge they lacked. They coveted Rouen's method of utilizing scarlet dye on cotton goods and sent "travelers" to convince French dyers to bring the secret to England.[49] In plate glass, the French enjoyed a 30–40 percent price advantage.[50] Much of the high-quality paper used in England during the second half of the eighteenth century came from the Netherlands or was smuggled in from France.[51] Even in areas where British technology was superior, such as in the making of a heavy cotton fabric known as fustian, a group of Manchester trade representatives claimed that their advantage was temporary. The French, they said, "follow us very fast," and they estimated that they could be overtaken in a mere three years.[52] Josiah Wedgwood commented: ". . . our manufacture [of pottery and earthenware] may now [have an advantage] . . . but how long has that been the case? It is within my memory that the earthenwares of France were superior to ours; the revolution, therefore, has been sudden, and its effects may be temporary; the same circumstances whatever they have been, that turned the scale in our favour in this age, may, in the next, vary as much in the favour of France."[53] Nor was the mighty English Navy too complacent to adapt or adopt metallurgical advances in arms making developed abroad, particular after defeat in the War of American Independence.[54] In 1786, a newspaper usually allied with the Whigs reported fears that "England will be a better customer to France, than France will be to England."[55] Finally, the outcry in 1786–87 from entrepreneurs in a host of industries demanding renewed or even increased protection from French competition during the discussion over the terms of the proposed commercial treaty is a useful reminder that British industrial dominance was not universal.[56]

The previous overview should not be understood as an attempt to present a counterfactual argument. In most sectors, French industrial technology was not ahead of or even equal to that of Britain. Instead, my goal is to refocus attention on investigation of the historical record rather than arguing backwards from theoretical or nationalistic assumptions.[57] Influential and knowledgeable contemporaries believed that the industrial gap between Great Britain and France was not as wide or as deep as most recent commentators have argued; nor did they see British technological dominance as inevitable. Nor can it be argued that these observers simply underestimated the scope of the changes taking place in England. Rather these informed contemporaries credited France's willingness and capacity to learn from their island neighbors and to apply those lessons productively, particularly if the limitations imposed by the system of privilege, the relative unproductiveness of French labor and political economy could be reduced. With these observations about relative competitiveness in mind, French industrial policy on the eve of the French Revolution comes into much clearer focus. To fine-tune that vision, the next section will focus on the most significant state policy of the 1780s regarding international competitiveness: the Anglo-French Commercial Treaty of 1786.

Peace and Prosperity: French Industrial Policy after the War of American Independence

The Treaty of Paris ending the War of American Independence contained a clause calling for the negotiation of a new commercial arrangement to replace one dating from the 1713 Treaty of Utrecht. This agreement was to be concluded by the end of 1786. After two years of British stonewalling, Vergennes, the French principal minister, forced the British to negotiate seriously by prohibiting some British goods in July 1785 and more the following October. The British sent William Eden, later Lord Auckland, to bargain in earnest in September 1785. On September 26, 1786, Eden and the chief French negotiator, de Rayneval, signed an agreement often known as the Eden Treaty. Ratified two weeks later, it was to go into effect on May 10, 1787. An "Additional Convention" signed in January 1787 settled the customs procedures and preserved certain local privileges while establishing supplemental means of reining in smuggling. Far from being a "free trade" agreement, the Treaty of 1786 significantly lowered customs rates in a roughly reciprocal manner based on the concept of "most favored nation" status, but left most duties at 10–12 percent. Some goods remained prohibited by each nation. The Treaty also specified rules of conduct governing such thorny issues as smuggling, privateering, traveling without passports, and wartime commercial behavior. The Eden Treaty established the greatest degree of official cooperation between the two rivals in more than 100 years.[58]

There are a number of traditional explanations for French sponsorship of a commercial treaty with England, running the gamut from sheer stupidity to miscalculation to a misguided attempt to forge a lasting peace. All such explanations share one major element: they investigate the Treaty from the perspective of a foreordained British industrial ascendancy. Historians interested in the age of Enlightenment or in the origins of the French Revolution tend to gloss over the Treaty as an example of the incompetence of the Bourbon administration in economic matters. At most, they see the Treaty as a cause of unemployment and unrest justifying the fall of the monarchy.[59] For those who focus on the Industrial Revolution, the Treaty, if mentioned at all, is portrayed as either a curious oddity or a grievous French miscalculation that hastened British industrial dominance.[60] David Landes mentions it in passing in his chapter "Continental Emulations" as a factor speeding the mechanization of the cotton industry in France. Musson and Robinson, Berg, Crafts, Mokyr, Mathias, and Floud and McCloskey do not mention the Treaty, even when reporting burgeoning trade statistics which stemmed directly from it.[61] Historians concerned with politics or diplomacy do refer to the Treaty, but generally in reference to the political genius and/or great good fortune of the government of William Pitt the Younger. These commentators recognize that the French were the ones who demanded the Treaty despite enormous misgivings by the English.[62]

This resume raises important questions. Why would the English hesitate to sign a commercial treaty with a rival over whom they had a clear technological edge? It certainly could be attributed to nationalism or Francophobia, but might not there have been more material reasons that gave the English cabinet pause? I contend that the Treaty deserves a more nuanced and thorough investigation because of its central place in French industrial history, not to mention the Treaty's important effects on French public opinion in the late 1780s. If the Treaty is not placed in its proper context, French state action during the last 15 years of the century is largely unintelligible.

The argument that the Anglo-French Commercial Treaty led to the economic crisis of the late 1780s seems finally to have been consigned to the dustbin of historiography.[63] Nor can it be claimed that contemporaries blamed the Treaty exclusively for the economic situation.[64] According to Shapiro and Markoff, the *cahiers de doléances* of the Third Estate were far more likely to call for increased commercial freedom than for trade protection.[65] Yet all commentators agree that, at the very least, the Treaty deepened the economic crisis and contributed to declining employment, which reinforced the growing political ferment of the era. Why then did the French—the Treaty's instigators—want to sign it? What were their goals? Why did they fail so badly?

Under Vergennes, a fundamental goal of French foreign policy was to initiate "a thaw in the Anglo-French cold war."[66] He believed that a commercial treaty would prevent future wars between the two long-standing rivals and ensure the stability of Europe, thereby allowing the French to act as the arbiters of the Continent.[67] Louis XVI seems to have either agreed with this policy or to have helped to elaborate it. Louis XVI, according to the *premier commis* for foreign affairs, "desired only to consolidate the fine harmony that exists between Himself and the King of England. . . . His Majesty is persuaded that the surest means of fulfilling this object of uniting these two nations is through commercial relations."[68] Thus, for Louis XVI as for Vergennes, politics and economics were firmly linked in policy planning.[69]

Yet both Louis XVI and his principal minister had additional reasons to support the notion of a commercial treaty with England. At a practical level, both men hoped that a formal commercial relationship would lessen or even eliminate the vast smuggling traffic crisscrossing the Channel. A crackdown on illicit activity would diminish the speculation on commodities that distorted regional and international trade while increasing state revenues.

Vergennes and the other ministers were influenced by the Physiocrats' political economy.[70] This influence was exercised partly by Dupont de Nemours, but also by Tolozan, Holker, Boyetet, deputy of commerce Guillaume-François Mahy de Cormeré, and the dominant member of the Chamber of Commerce of Normandy, Jean-Barthélemy Lecouteulx de Canteleu. According to Quesnay, Mirabeau père, and Le Trosne, trade barriers only caused economic dislocation by stifling agriculture. Since industry was only a secondary consideration, they argued that increased agricultural opportunity would produce more wealth than industrial expansion. These analysts saw the ascendancy of agriculture in the French economy as a fundamental advantage against the English; an exchange of agricultural goods for industrial products benefited France.[71] A Physiocratic conception of wealth fortified the cultural mindset of important officials, strengthening their faith in France's ability to compete.[72]

Vergennes consistently linked commercial agreements and France's international position. Trade agreements were to solidify France diplomatic influence, limit smuggling and provide export opportunities for French agriculture and industry. According to Orville Murphy, Vergennes negotiated at least 16 treaties, conventions or diplomatic talks between 1774 and 1787. On his urging, Louis XVI issued innumerable edicts and Lettres-patentes to regulate or implement trade agreements or to clarify customs regulations or consular procedures. Vergennes successfully pursued commercial treaties with the United States, Russia, Spain, Portugal, and the Netherlands. This diplomatic offensive under Vergennes' leadership permits a re-evaluation of the position of the Eden Treaty within French policy. Far from being an isolated effort, French decision makers under Louis XVI appear to have pursued a consistent and coherent diplomatic policy designed to facilitate the continuation of peace and provide opportunities for France to improve its economic competitiveness.[73] In this context, Vergennes' prohibition of British goods in July and October 1785 and the numerous concessions he made to ensure that the projected commercial treaty would become reality make much more sense.[74]

The other major French player in the push for a commercial treaty was Charles-Alexander de Calonne, who served as Controller-General from November 1783 to April 1787. Recognizing that important factions such as the Physiocrats wanted an agreement, Calonne sought to get the best treaty possible. Since he had no overarching objective other than competitiveness, Calonne focused on using the proposed commercial compact to promote different policy goals such as increasing agricultural exports and lessening smuggling. Calonne intended to use the enhanced tax revenue to fund additional encouragements for favored technological and industrial projects.[75]

As the treaty negotiations drew to a close, Calonne and other high officials escalated or initiated important programs to support French industry. The creation of additional free ports and the attempt to recreate an East India Company were precursors of this trend. The French accelerated their campaign to entice English workers, inventors, and entrepreneurs to bring their expertise and discoveries across the Channel, ranging from Boulton and Watt[76] to Philemon Pickford (who sought to naturalize the Arkwright system[77]) to the mechanic William Hall (who built carding machines that improved on John Kay's).[78] Through the twin avenues of espionage and sponsored commercial acquisition, the French state acquired all sorts of English industrial machinery.[79] The royal administration also eased restrictions on foreigners establishing industrial enterprises in France.[80] In one sector, the central state imitated the English export bounty by returning a portion of the taxes collected on imported raw sugar to enable French refined sugar to compete in European markets.[81] New duties were levied on certain industrial raw materials leaving the country through Alsace and on English textiles entering through any of the provinces reputed "foreign," thereby closing an important loophole in the customs.[82] Rates were lowered on such essential imports as the glazed cardboard used in the finishing of cloth.[83]

French policy makers were encouraged by reports from Normandy that entrepreneurs were mechanizing faster there than in other regions of the country, spurred on by the proximity of English competition.[84] Additional funds were given to flagship French industrial firms utilizing advanced technology such as Le Creusot, but Calonne spread the wealth much more widely than any of his predecessors in hopes that a widened web would attract more attention from potential investors, industrial entrepreneurs and inventors.[85] Thus, the Treaty was part of a general industrial policy aimed at improving French industrial and commercial competitiveness, strengthening the finances of the central state and bolstering France's international position.

For Calonne, as for many others, these programs and the Treaty itself were useful only if they led to the creation of a unified French market, in other words, pushing the customs to the national frontiers. Dubbed the "Single Duty Project," this plan had been under consideration for generations, but became the subject of considerable study under Louis XVI.[86]

Despite the support of Vergennes, all the Deputies of Commerce, nearly all the Inspectors of Manufacturing, and numerous other reform-minded high-ranking administrators, not to mention much of the commercial and industrial elite, the project fizzled in the Assembly of Notables when Calonne fell from power in April 1787.[87]

Since it was supposed to precede the Eden Treaty,[88] the Assembly of Notables' failure to approve the Single Duty Project burdened French entrepreneurs once the Anglo-French commercial agreement went into effect.[89] This unforeseen penalty stemmed from the fact that British goods paid a single duty of 10–12 percent on entering France. The numerous internal customs barriers applied to *French* goods meant that the duties on items arriving from England were often less than those on domestic items traveling down major internal communication arteries. For example, over the course of the approximately 150 miles between the *grandes villes* of Lyon and Marseille, the Farmers General levied nine different duties—and there were tolls, too! As a result, iron mined in the Franche Comté paid 35 percent of its value to be shipped via the Rhône to Marseille while Russian iron arriving from the Baltic paid less than 20 percent of its face value in transport and taxes.[90] From the petition of the Chamber of Commerce of Lille which made implementing a Single Duty Project the focus of its complaints in 1788 to the lampoon of the customs by Louis-Sébastien Mercier to the *cahiers de doléances*, a host of evidence illustrates that this customs problem was well known to contemporaries.[91] Thus, British productive advantages were accentuated by the French tax structure despite the best intentions and efforts of the reform-minded administrators who negotiated the Treaty.

Yet the same voices within French administration that claimed France could compete with Great Britain feared the effects of a commercial agreement. From Rouen, Holker père warned that the 30 percent British cost advantage in cotton textiles would lead to widespread unemployment. Still, this firm opponent of the Treaty did not favor increasing or even maintaining a high tariff; that would only increase smuggling. Holker stated: "The only means of succeeding is to diminish as much as possible, our [France's] costs of production." As was noted above, Holker suggested a raft of state policies to enable France to mechanize production swiftly. Thus, Holker's chief concern was the timing of the

Treaty.[92] Such anxieties were understood and acted on at the highest levels. Plans had been made to minimize the adjustment to the lowered duties on English imports. The French government subordinated certain industrial interests to other concerns, but the Eden Treaty was not signed through ineptitude or from ignorance of the potential effects, or without planning for those consequences.

The French policy makers who negotiated and approved the Treaty may have recognized that a difficult period of adjustment loomed, but they could not have predicted the multitude of structural and accidental factors that contributed to turn the 1786–1789 period into a deep economic crisis that facilitated the outbreak of revolution in France.[93] These well-intentioned improvers also could not have foreseen the freak storm of July 13, 1788, which destroyed much of the cereal crop in northern France, aggravating an already dire subsistence crisis. Nor could the French officials have expected Spain to shut her borders in 1785, the death of Prussian King Frederick II in 1786, the depth of the Dutch crisis of the 1780s, or the disastrous effects of the Russo-Turkish War of 1787 for the Levantine and Baltic trades. This series of international events limited French exports just as the domestic market faced an increased level of penetration by the British, undermining the carefully laid plans for export substitution by Vergennes and his colleagues.[94]

Nor could French policy makers have predicted the extent of the political morass that began with the formation of the Assembly of Notables in 1786, continued with sweeping reforms of provincial administration and the legal system in 1787–88 only to culminate (or so it was hoped) in the meeting of the Estates-General in May 1789. The effects of the political imbroglio were readily apparent to contemporaries. When assessing the rapid decline of Troyes' cotton industry in 1790, Jacques-Claude Beugnot, *procureur-général-syndic* of the newly formed *département* of the Aube, reported: "The first and most powerful cause of this decline is the host of events that have taken place over the past two or three years with a rapidity capable of fatiguing human thought. The scarcity of currency, the interruption of diplomatic relations, caution, hope, fear and the frenzy that has held France in its grip ought to be and have been disastrous for our industry which can only flourish when it is calm and under the shade of a protective government. To this first cause which did not affect Troyes less strongly than the rest of France, must be added the effects of the Commercial Treaty between France and England."[95] The Eden Treaty played a part in the economic, political, and diplomatic crisis, but, as was illustrated above, the Treaty often has unjustly received the bulk of the historiographical blame.

The French Response to the 1786 Commercial Treaty

The Treaty was to take effect in May 1787. Yet as soon as the draft treaty was announced, English wares flooded into France through the Atlantic free ports and in the hulls of uncounted smugglers.[96] Once the duties were officially lowered, the tide flowed in more rapidly. From 16 million *livres* in 1784, legal French imports from Britain reached 23 million *livres* in 1786 and 27 million the following year.[97] This deluge greatly deepened the industrial crisis begun two years earlier, plunging several textile-producing regions (notably Normandy, Picardy, Flanders, and Champagne) into deep depression. This conjunction led public opinion to associate the industrial crash with the Treaty. French complaints about the Treaty, however, preceded the effects. Concerned that their interests had not been consulted during the negotiation process, and dreading the effects, the Chambers of Commerce of Rouen and Amiens led the attack on the Treaty.[98]

Beginning in 1786, the entrepreneurs of Abbeville and especially Amiens complained bitterly that they could not compete with the English in either the price or quality of their woollens. These complaints were received with little sympathy or were even downplayed by informed observers. A January 1787 report to the Quai d'Orsay stated: "Amiens seems to be the only city that the Treaty has disconcerted, but its commerce began to slacken more than a year earlier." This report blamed the loss of sales on the systematic fraud practiced by the Amienois through misrepresentation of the quality of both the dyeing and the fabric. Skimping on quality had become so widespread that faith in the products of Amiens had fallen both domestically and abroad. In addition, the report suggested that merchants and manufacturers in Picardy were overly dependent on profits from smuggling.[99] Upon being told that the local woollens "cannot in any one instance . . . stand the competition" and that "Amiens would be ruined," Arthur Young gave an uncharacteristically optimistic reply: "The manufacturers of all countries are full of these apprehensions, which usually prove extremely groundless. In all probability the effect would be as expected if a counter stream of emulation and industry did not work against it. The introduction of English fabrics may be hurtful for a time, but in the long run may be beneficial, by spurring up the French manufacturers to greater exertions and to a keener industry."[100] Stopping fraudulent productive practices and getting the wealthy and knowledgeable entrepreneurs of Picardy and other provinces to shift their attention to making goods capable of withstanding English competition was one of the stated French goals in signing the Commercial Treaty. Contemporaries and recent historians both consider the crisis in Amiens to have been the result of structural causes in operation long before the signing of the Commercial Treaty, including a widespread decline in wool production stemming from sheep distemper.[101]

Although the causes of Picardy's economic downturn excited little goodwill, the effects of the Treaty were genuinely devastating. Merchants from Bordeaux who handled the bulk of Amiens' trade with the south of France and with the Spanish colonial market reported that "Amiens' woollens could not be sold even at dirt cheap prices."[102] Louis Villard, inspector of manufacturing, reported that 30,000 were out of work in Picardy and that most of the *fabricants* could only sell their goods at a loss, which would lead to a raft of bankruptcies. He noted that the great popularity of cheap English goods in France was not mirrored by demand for fine French products in England. Without less expensive and higher-quality raw wool or mechanization, the Picards could not compete no matter how low the cost of labor. Working examples of the key inventions existed, but production methods were so archaic that the government must take an active role and distribute models freely and widely to facilitate French production. Lastly, Villard proposed that the French government force the Spanish to reopen their market.[103]

"*Anglomanie*" and lower-quality raw materials were complaints also evoked by merchant-manufacturers from Troyes, capital of Champagne. Their complaints centered on the barriers concocted by the English to retard French exports and the English purchasing of cotton from the French colonies to supply the needs of Lancashire, depriving French industry of essential raw cotton.[104] Tellingly, however, the Troyens expressed their concerns in terms of the growth of unemployment much more concretely than the Picards, pointing to the potential for labor unrest from the 9,000 people in surrounding districts looking for work. Yet the official response to Champenois protests was only slightly more supportive than it was for the Picards. *Intendant de commerce* Simon Clicquot de Blervache wrote to the deputies of the province that since they understood that the English price advantage came from their machines to spin and card cotton and the Troyens could purchase these machines directly, they should do so. He suggested that mechanization along with the purchase of fabric by concerned elites to be sold abroad was the proper response to their fears about the social effects of growing unemployment.[105]

Little public outcry greeted the Commercial Treaty of 1786. Ultimately, however, the Chambers of Commerce of the provinces of Normandy, Artois, Flanders, Champagne, and Picardy came to oppose the Treaty, pushed by Boyetet, who had withdrawn his support when he saw the effects.[106] As discussed above, the implementation of the provisions of the Treaty proved both difficult and complex for France in part because of the defeat of the Single Duty Project.[107] Additional difficulties in implementation included the ongoing collapse of the French state as it struggled to deal with diplomatic humiliation over the Dutch situation, massive administrative reorganization involving extensive decentralization, ministerial instability, and a deepening economic crisis. Observable in almost every field of administrative endeavor, this internal decay was nowhere more evident than in the oversight of internal and external customs. Since some duties varied according to transport costs or local privileges, French manufacturers could only take advantage of the terms of the treaty if these rates were calculated precisely.

The situation worsened when Pierre-Charles Laurent de Villedeuil, Controller-General for part of 1787, unilaterally raised the tariff rates on most British cotton goods on May 31 while opening every port to British goods (from nine). Most ports were unprepared for the complicated process of accepting British goods for entry into France. The free ports continued to avoid collecting any and all duties, allowing at least 3 million *livres'* worth of English goods to enter France with a major price advantage.[108] These developments deprived the nation of significant revenue intended to be used to protect French industry and to acquire English technology.

The Treaty established clear means of preventing "dumping" or other unfair competitive practices.[109] Other than a brief interlude in the spring of 1788 under Claude-Guillaume Lambert, Controller-General from September 1787 to August 1788, hampered by ministerial changeover, the French government, in stark contrast to the British, proved unwilling to use its treaty rights aggressively to protect the interests of its citizenry. For example, the provincial assembly of Champagne complained: "The English have established an additional tax of 5 percent they call a stamping duty on our fabrics. Without examining the Treaty to see if they have this right, it would seem that a similar tax on their goods would diminish the violent damage done by their commerce to our manufactures. . . ." The state's response was to do nothing; instead, the Champenois were told to remedy the situation as best they could.[110] Rayneval went so far as to blame the inactivity of Controller-General Archbishop Loménie de Brienne for the negative effects of the Treaty. Although this charge made in hindsight is clearly self-justificatory, the French government did recognize that the Treaty was causing widespread unemployment—particularly among textile workers, a reported 185,000 of whom had lost their jobs by Christmas 1787.[111]

It was in this context that the Chamber of Commerce of Normandy published a series of pamphlets attacking the terms of the Commercial Treaty.[112] The resonance of the objections raised by the Normans led Dupont de Nemours to defend the Treaty publicly.[113] This debate has framed much of what attention has been paid by historians to the treaty.[114] Beyond the polemics about prior consultation, an informed debate about the ability of France to compete with Great Britain was at the heart of this controversy. The Normans admitted that there were product areas, notably cotton goods and anything where coal could be used as fuel or that utilized metals other than iron, where the English had a significant price advantage. They attributed the bulk of that advantage to mechanized productive techniques that relied on lower prices for raw materials and fuel.

The central government came in for sustained criticism, mostly vis-à-vis British practice. According to the Chamber of Commerce, the French state's lack of proper protection for invention, burdensome regulations, and high taxes on manufacturing accentuated the most obvious failing of their government: it did not exclude goods where there was a clear national interest, as the British did with silks. The Normans also applied a cultural comparison to blame the French themselves for the economic situation. They claimed that if French machine builders applied themselves to doing quality work, there would be an adequate supply of the necessary machines. The Normans also devoted considerable attention to the preference of the English for English-manufactured goods, which contrasted with the unceasing French demand for novelty. These well-founded and oft-cited criticisms have often been used as a summary of French reactions to the Commercial Treaty. But they only tell a part of the story.

What has received little attention is that the Normans also discussed at length the goods where the French had the competitive edge. As might be expected, these products included silks, high-quality woollens, and linen. One of their chief complaints about the Treaty was that the English would be able to steal French industrial secrets; they feared the emigration of their most skilled workers. The complaints of the Normans must not mask the fact that these pamphlets must be understood as extended "white papers" that recommended a bevy of new policies designed to help them compete. Nor did the Normans call for the abrogation of the Treaty; what they demanded was the mitigation of those provisions that gave the English advantages that were not counterbalanced by benefits to the French.

The Chamber of Commerce judged that mechanization and market access were the twin keys to British industrial success, particularly in cotton goods. They wanted the central government to equalize the rules of the game by enacting a Single Duty Project and establishing shipping regulations analogous to the British Navigation Acts. These measures would force the French colonies to utilize French ships, thereby restoring first crack at that important market to domestic manufactures.[115] They also urged the government to induce the Spanish to restore access to this important market. The Normans were concerned that William Eden had gone to Madrid to negotiate a commercial agreement. To facilitate the changeover to mechanized production, the Chamber of Commerce suggested that the government offer bounties for exports. Another bounty was proposed for those who shifted to linen manufacture. Finally, they urged the government to support a research effort to discover a better means of bleaching linens—a discovery which would give the French pre-eminence in that branch of the textile trade. These demands were realistic and reasonable responses to the sales downturn and resultant unemployment that came in the wake of the Treaty. The Chamber of Commerce of Rouen undertook a serious and constructive criticism of the Treaty and suggested practical means of bolstering French economic performance in Normandy and in other areas hard hit by British competition.[116]

Yet not every area or every industry had such a negative experience in 1787–88. The primary sector of the French economy boomed. Despite poor grape harvests in Bordeaux, exports of wine, especially to Great Britain, more than doubled. Brandy sales increased even more rapidly, becoming the largest single element of French trade with England. Large-scale brandy refining in Picardy, in Flanders, and in the Aunis-Saintonge region around La Rochelle spread the benefits of access to the English market. The mining of coal and the smelting of iron also reached all-time highs in 1787–1789. Rives-le-Gier (in the Forez region) and Anzin (in the North) received their first Watt-type steam engines from the Périer workshops at Chaillot in 1789.[117]

Nor was the picture for manufactured goods as bleak as the portrait drawn by scholars seeking to provide an economic explanation for the outbreak of the French Revolution. Although English cotton goods displaced French manufactures, many textile sectors remained surprisingly robust. Despite fears to the contrary, fine French woollens from Abbeville, Lodève, Bédarieux, Elbeuf, Louviers, and Sedan continued to find buyers at home and abroad. (The lower-quality and lower-priced goods of these regions, and especially of Champagne and Languedoc, did not fare quite so well.) Silk and ribbon production suffered from the recession, but did not lose markets to the English. Linen production, especially around Saint-Quentin and Cholet, flourished, albeit less than expected, with production reaching all-time highs in 1787–88. Finally, trade in lace burgeoned, becoming the largest single French textile export to England. Beyond textiles, several glass-making establishments using coal as fuel were founded, most notably at Saint-Étienne.[118]

The data suggest a more nuanced regional image of the effects of the Commercial Treaty. If much of northern and northwestern France suffered greatly, it was not universal. Although textile production fell by at least one-third in Amiens and one-fourth in Beauvais, Rouen, Reims, Troyes, and the Lille-Roubaix-Tourcoing region between 1787 and 1789, it increased in Saint-Quentin and held steady in Abbeville, Elbeuf, Louviers, and Sedan, at least until France sunk deeper into recession. More solidly on the plus side, eastern France enjoyed rapidly expanding trade with the German lands and Poland, especially in manufactured goods. Lyon and Dauphiné were largely untouched by the Commercial Treaty.[119] The other silk-working regions around Nîmes and in the Forez boomed. If Carcassonne and some of the other woollens centers in Languedoc were hit hard, other parts of the Midi prospered. The Atlantic ports and Marseille flourished as French overseas trade reached unprecedented levels thanks to sharp increases in trade with the colonies, Africa, the Netherlands, the Baltic region, and, most notably, England. French exports to Great Britain grew from 20 million *livres* in 1784 to 34 million in 1787 and 35 million in 1789. If there was still a massive deficit in France's trade with England (25 million *livres* in 1787) it came from minerals (especially lead, copper, iron, tin, and coal) and foodstuffs (butter, salted meat, grain). The French trade deficit did *not* come from manufactured goods, which made up only 40 percent of imports from Britain. France needed these raw materials to support industrial development. British coal was essential to the extensive glass-making industry of the Bordelais. British iron and copper were indispensable to the construction of textile machines and the perfection of techniques for rolling lead in Normandy. The government's strategy of facilitating access to higher-quality and less expensive raw materials was successful—at least according to Smithian, if not Physiocratic, principles.[120]

With France's total foreign trade (although not per capita) surpassing that of England by a considerable margin at 1.02 billion *livres* versus 775 million, on the eve of the Revolution, there was cause for long-term optimism despite the harsh immediate effects of the Treaty in some industries and regions. French trade performance ran counter to most of the economic trends of the post-1775 period. Agriculture, including the large wine sector, suffered from poor harvests in 1787–88, but that was clearly temporary. Colonial re-exports were booming. Given the conjunction of structural and short-term depressive factors at work in 1787–1789, French trade performance was particularly impressive and illustrated the strength and competitive resilience of the national economy. In short, the Eden Treaty had drastic repercussions, but they were not solely negative; this survey illustrates that amidst the devastation of the English flood tide of imports, there were justified reasons for cautious optimism about France's long-term ability to compete.[121]

Countering a "British Invasion"

Improving the French economy was the centerpiece of the embedded industrial policy implicit in the desire to negotiate a commercial treaty with England. Reminiscent of Jean-Jacques Rousseau's adage about "forcing men to be free," the Eden Treaty was intended to coerce the French into being more competitive. An impressive number of entrepreneurs and officials throughout France answered the state's call for technological and productive modernization on the English model. Despite the formidable squeeze on available revenues stemming from administrative centralization, negligent customs collection, and deadlocked tax reform, various governmental bodies mobilized substantial funds to combat the effects of the economic depression. A significant portion of these moneys supported projects that sought to develop technical expertise or assisted industrialists by spreading scientific knowledge or access to English-type machinery. The positive responses of bureaucrats and entrepreneurs to the deepening crisis illustrate the modernizing potential of France on the eve of 1789.

Despite the contrary claims of the critics cited above, the policy makers of the Bourbon state responded swiftly to complaints of unfair commercial practices and, when possible, attempted to even the industrial playing field by eliminating trade inequities through legislative action. For example, urged by the deputies of commerce, the royal council enacted measures to increase the tariffs on certain English textiles and to enforce more stringently the anti-smuggling measures included in the treaty.[122] These measures continued long-standing policies.

The central government also stepped up its efforts to develop French technological expertise. After Calonne's fall, that effort pivoted on acquiring

English machines and spreading access to them as widely as possible. As is well known, the French state concentrated on copying Arkwright's water frame and the mule jenny, both of which were crucial to England's competitive edge.[123] Industrial spies/entrepreneurs including Bonaventure-Joseph Le Turc and Thomas LeClerc were commissioned to acquire these technologies.[124] British machine builders, including James Milne and family, William Douglas, Edward and George Garnett, Philemon Pickford, Nicolas Brisout de Barneville, Thomas Foxlow, William Hall, William Clarck [*sic*], and John Macloud, were rewarded for coming to France and given subsidies for each set of machines they sold.[125] Other machines were distributed free to worthy areas and organizations. Qualified French mechanics, such as Adrien Delarche and Nicolas Lhomond, received government patronage to construct machines approved by the Academy of Science.[126] The Bourbon government also paid the wages of at least 100 foreign workers in machine building and provided large subsidies to innovative French entrepreneurs who financed the construction of advanced textile machinery. Before the adjudication of Arkwright's second patent in 1785, no less than three French mechanics were building roller-spinning machines.[127] Doggedly, if haphazardly, government action enabled hundreds of English-style (if not always functionally equivalent) carding and spinning machines to be put into operation in nearly every major industrial district in France between 1786 and 1789.[128]

To further technological development, the French government sought to develop and standardize "best practice" in the manufacture of crucible steel. Three members of the Academy of Science—Alexandre-Théophile Vandermonde, Claude Berthollet, and Gaspard Monge—were ordered to investigate the process of making steel "fired by the English method" and to publish their findings.[129] The *Mémoire sur le fer considéré dans ses différens états métalliques* (1787) was the result; it was then printed (in 1788) and distributed without charge to local governments and prominent forge masters all over France. Qualified workers were offered bounties for each French apprentice they trained in English techniques. Huge cash prizes were offered to anyone who could imitate Henry Cort's puddling process in a French foundry.[130] Two gunsmiths enticed to Saint-Étienne from Liège in the Austrian Netherlands brought more advanced forging and boring techniques to France's largest weapons-making complex in 1784–85.[131]

The French government's attempts to develop native technology and science were complemented by efforts to acquire machines and expertise from abroad. The French state provided financial support for the diffusion of existing mechanical models and craft knowledge as widely as possible. Where practicable, it deployed legislative and police measures to facilitate French competitiveness. In large measure, these well-known policies were the mirror image of methods on the other side of the Channel, where the English state used its prodigious capacities to encourage technological development and transfer and to provide competitive advantages in key industries. The assertion that British development was largely indigenous[132] has been challenged by an avalanche of recent research.[133] The British state capitalized on the administrative and financial abilities developed for the purpose of waging war to follow a consistent policy of industrial and commercial encouragement. For example, Britain lagged notably in plate glass, in paper, and even in copper smelting. In the late eighteenth century, British agents were active all over Western Europe, hoping to acquire needed machines or to convince knowledgeable individuals to emigrate. If the British preferred to support science, technology, industry, and commerce through the application of police and judicial power or, less directly, through tax policies or diplomatic actions or by providing "drawbacks" to exporters, such a strategy differs only in means from the methods used in France.

The French state made use of existing institutions, such as the Academy of Science, which judged new inventions and published technological treatises. The navy supported metallurgical improvements particularly in smelting and rolling copper.[134] State-sponsored professional schools devoted to scientific research included the École des Ponts et chaussées (school of bridges and highways), founded in 1747; the École du génie (school of engineering) at Mézières, founded in 1748; and the École des Mines, founded in 1783. Supported by the Mint (1777), these establishments produced notable scientific breakthroughs especially in chemis-try, metallurgy, mineralogy, and civil engineering. According to Patrice Bret, the most recent historian of these institutions, the impressive outputs both in hands-on knowledge and in trained human practitioners of improved methods suggest that a practical French scientific culture capable of competing with Britain in technological matters was emerging in the 1780s.[135]

Advances in French science and technology made during the 1780s support this view. Although the mathematical and chemical discoveries of Jean-Antoine Chaptal, Pierre-Simon de Laplace, Louis-Bernard Guyton de Morveau, Lavoisier, and Monge have been well documented and given their just credit, French technology should not be dismissed. The notable developments of the decade included Berthollet's discovery of chlorine bleaching, the pioneering ballooning successes of the Montgolfier brothers, Nicolas Leblanc's soda-making process and the Marquis de Jouffroy's functional steamboat of 1783.[136] French advances in paper-making machinery demonstrated the possibilities of incremental technological improvement.[137] Despite the well-worn adages about the relative success of British technology and French science, the distinction between "pure" science and its "practical application" in technology does not seem to have been a major French disadvantage vis-à-vis Great Britain before 1789. The economic weight of the sectors in which this distinction might be said to exist, namely textiles and metallurgy, should not camouflage the limited conception of technological advance being utilized in many historical discussions of the early industrial revolution.

The central government attempted to build on these successes and to limit French liabilities. When the scope of the "English invasion" became manifest, the Bourbon regime, despite profound financial problems, took additional action to cope with the economic crisis. The central state allocated more than half a million *livres* to local governments that took proactive stances to assist unemployed (mostly textile) workers and to mitigate the industrial crisis.[138] In April 1788, to encourage entrepreneurs and to maintain employment, Controller-General Lambert authorized the clandestine purchase of cloth through intermediaries in the hardest-hit areas. Lambert authorized the expenditure of up to 60,000 *livres* a week.[139] Nor should such expenditures be seen solely as emergency measures to cope with a deepening crisis. They were part of a sustained state commitment to provide the resources to enable French entrepreneurs and workers to recover from the rapid shift in the competitive environment. In 1791, the Constituent Assembly set aside 3.85

million *livres* to encourage commerce and industry. That sum included support for science and for the spreading of advanced technologies. The same sum was budgeted for 1792, but war intervened and it was not disbursed.

The policy makers of the central state were not the only French subjects to respond constructively to the challenges of the Eden Treaty. Local elites enmeshed in the institutions of provincial administration and scientific advance exploited their ties to the central government to exact discretionary funds to combat the Treaty's effects. Rouen and Amiens led the way by establishing a Bureau of Encouragement gathering together merchants, manufacturers, royal administrators, local savants and municipal and provincial authorities who provided the kind of interaction deemed essential to effectively assisting technological improvement.[140] The depth of the economic difficulties experienced by the capitals of Normandy and Picardy correlated to the impressive willingness of local elites to unite to deal productively with competition. The royal council rewarded Rouen with a promise of 300,000 *livres* (of which about half was ultimately received) and Amiens with 180,000. Beauvais sought only to maintain employment among textile workers. These more limited goals received far less support. Beauvais garnered a loan of 26,000 *livres* to purchase textiles to help keep weavers at their looms.[141] Depressed areas (including Troyes, Reims, Orléans, Carcassonne, and Lyon) that sought money only to support the newly unemployed were told to rely on local charity.[142]

The Bureaus of Encouragement of Amiens and Rouen adopted policies pioneered by the central state modeled on Physiocratic notions of the British model of industrial development. To fund these initiatives, capital funds received from the central government were supplemented by local institutions and numerous individuals to wage this economic "war" of industrial competition. Cardinal de La Rochefoucauld donated 300,000 *livres* and the Chamber of Commerce of Normandy set aside 29,000 *livres* for the purchase of jennies (at 300 or 120 *livres* each, depending on size) to be distributed free to interested entrepreneurs.[143] Like the Beauvaisais, the Bureaus of Amiens and Rouen purchased locally produced cottons and woollens to maintain employment and then tried to find new markets to sell their mushrooming stockpiles. A new wave of agents were sent to England to entice workers with useful skills to emigrate and to convince technologically advanced British entrepreneurs to shift their operations to France where they would receive subsidies and could take advantage of lower wages. Entrepreneurs willing to shift to the manufacture of linens also received support. Macloud, Barneville, the Garnett brothers, and other mechanics were given impressive sums to come to the province in question and were given bonuses for each of their devices employed in local industry. Nor did any of these institutions simply throw money at the problem of English competition. Machines were tested both by reputable scientific experts and by workers. Both groups prevented a number of large-scale purchases of flawed machines deemed either unsuitable for local needs or unacceptable in quality.[144]

The local elite of Normandy were most active in responding to the technological challenge of English competition. In addition to importing English machines, the Rouennais sent men to acquire English sheep to naturalize in France hoping to eliminate the British advantage in this natural resource. Experienced miners prospected for local coal deposits to offset that British edge. Other funds were devoted to buying English-style machines made in France, such as the cards of Wood and Hill employed at Louviers. Considerable cash prizes were established for the construction of better models of English machines as well as for new technologies. In Rouen, a charity workshop was established to make use of the advanced English-style machines built by Barneville donated by the central government. Additional machines were purchased by local elites, who formed a prototype joint-stock company capitalized at 96,000 *livres* that was given use of a large workshop owned by the parish church of Saint-Maclou. Weekly prizes were awarded for the most productive female workers, and model workers were brought in from successful establishments in other areas to show local workers the most efficient techniques.[145] These measures were investments in the future. Contemporaries saw them as a first step toward sustaining English competition. The Rouennais, in particular, were explicit in asserting that France's most pressing problems were not technological. Rather, they claimed that with thorough implementation of the policies pioneered by the central government France would be able to compete with Great Britain in just a few years.[146]

The views expressed and the policies adopted by members of state-sponsored institutions would mean nothing if entrepreneurs would not implement them. The French central government had *always* counted on the commercial, industrial, and scientific elites to respond in kind to the multifaceted challenge posed by Britain. For policy makers, the capacities of these elites would enable France not only to compete, but to triumph. Although far from uniformly constructive, the reaction of regionally based economic elites to increased competition from Great Britain was impressive in its quantity and quality. Far from believing that the race to industrial predominance had already been won, a significant number of entrepreneurs decided to fight fire with fire. Dousing the fire meant adopting or adapting as many of the elements of British economic success as was possible.

Perhaps the most impressive response to British competition came from entrepreneurs already involved in technology transfer. In Amiens, the leader was the president of the Chamber of Commerce, Jean-Baptiste Morgan. He had brought workers from England in 1765 and had acquired spinning jennies from William Hall in Sens. "Best practice" spinning techniques taught by "spinning mistresses" sent by Holker enabled him to establish a vertically integrated manufacture of cotton velvets that was swiftly recognized as a royal manufacture. Morgan sent new agents to recruit more English workers. Arriving in yearly batches from 1788 to 1790, they provided Morgan with a detailed and precise knowledge of English techniques, and with the mechanical expertise to construct the needed machines and instruct workers in their use. As a result, in 1788, Morgan and his English partner Massey were the first French manufacturers to possess an English-style mule jenny capable of making thread as good as that produced in Manchester. The following year, they installed a water-powered flying shuttle built by Macloud. For the princely wage of 48 *louis d'or* a week, Macloud came to Amiens and constructed dozens of these machines in 1788. The Bureau of Encouragement at Amiens provided capital for these endeavors in exchange for the firm's commitment to share their techniques and technical know-how. Entrepreneurs from Abbeville and Amiens came to Morgan and Massey to examine and copy their mule jennies. Development of new fabrics and styles also was encouraged. Workers were invited to tinker with models or to build sample machines of new design. As a result of such "heroic" efforts, the mechanization of textiles in Picardy was greatly advanced while the unmechanized elements of the industry closed shop or cut jobs.[147]

A similar pattern—where those who had already utilized or had access to advanced technology sought to acquire additional infusions—can be traced in many other areas. In Elbeuf, innovative entrepreneurs like the Grandin family and Jean-Baptiste de Flavigny purchased a great deal of new machinery after 1787, just as they had bought the Anglo-French Commercial Treaty. Jean-Baptiste Décretot's initial woollens manufacture at Louviers was established using English skilled labor and machinery and was constantly fortified by new arrivals. Décretot employed two Irishmen named Malloy to build the jennies. In October 1788, Arthur Young described this product as "the first in the world." He then reported on a cotton mill run by four Englishmen from Ark-wright's enterprise and established at a cost of 400,000 *livres*. As a means of spreading his expertise even more widely, Décretot also formed partnerships in other textile centers with individuals skilled in managing workers or building machines. From a relatively humble beginning, through the inspired use of English technology and workers and his own abilities in marketing, design, and labor management, Décretot became one of the largest textile entrepreneurs in one of the most productive and competitive regions of France. His business expanded so rapidly that he could not fill all his orders. Décretot stepped up his "modernizing" activities because of the Commercial Treaty. He seized the moment. Recognizing the opportunity for himself and for French industry, he did not regard the Treaty as a death knell.[148] In fact, an impressive number of enterprises ranging from "small establishments for spinning cottons by jennys" to limited partnerships with impressive capital resources, were founded in Rouen as a result of such resourceful leadership.[149] In addition to the pioneering efforts of innovators, new or advanced textile machinery was installed in Elbeuf, Limoges, Metz, Montpellier, Noyon, Rochefort, Saint-Étienne, Soissons, Troyes, and Toulouse in 1788–89, usually for the first time.[150] Thus, the *Anglomanie* so evident in French consumer taste in the 1780s was reflected in a widespread hunger for English machinery.

The high nobility took part in the efforts to bolster French competitiveness. The Duke of Orléans and the Duke of La Rochefoucauld-Liancourt continued to invest in textile machinery and established new enterprises to spin and card cotton as part of their interest in progress, in order to help their tenants and dependents, and, not coincidently to profit. The Duke of Orléans purchased so many machines from Milne for his establishment at Montargis that one informed observer wondered if he sought a monopoly. He also helped Leblanc establish a soda-making facility near Saint-Denis in 1789.[151] Other nobles invested in mining enterprises especially in coal, or in traditional French strengths in luxury goods.[152]

Not all the French enterprises founded to combat English competition were in textiles. The noted chemist François-Antoine Descroizilles set up a new manufacture in Rouen to provide the sulphuric acid that was necessary to the bleaching process invented by Berthollet. In association with local textile entrepreneurs, he also founded a chlorine-bleaching plant; it soon had to turn away business. In Montpellier, Chaptal followed a remarkably similar trajectory in 1788. A new partnership to manufacture porcelain using English coal-firing techniques was formed in Lille. Coal was first applied to many different industrial endeavors all over France, but particularly in Normandy, with its easy access to English sources. At the same time, new pits were dug in many places, and the output of existing mines was stepped up. Two new foundries utilizing copper-rolling techniques equivalent to England's were opened along the lower Seine and immediately began large-scale production. Several other copper foundries were established in the Lyonnais in 1788–89.[153] Nor should the various Parisian enterprises devoted to technological development or machine building, such as the entrepreneurial haven of the Quinze-Vingts or the depository at the Hôtel de Mortagne, be excluded or forgotten.[154] This survey is not meant to be exhaustive; it provides only a sense of the range of entrepreneurial responses geographically—and in various industries—to heightened British competition after 1787.

The goal of this discussion has not been to argue that France was on the verge of catching up to English textile technology or metallurgical skill. Although I have some reservations about the counting methods used, I do not challenge the assertion made by the French government in 1790 that there were 900 spinning jennies in operation compared to 20,000 across the Channel, not counting 7,000–8,000 mule jennies.[155] My purpose in this section is not to downplay the depth of the economic crisis of the late 1780s in the paradigmatic industrial sectors that was magnified by the "British invasion" unleashed by the Eden Treaty. Nor do I seek to minimize the economic realities that underlay the demands for help voiced by the Chamber of Commerce of Normandy and found in many of the *cahiers de doléances* written in the spring of 1789.[156] These historical "facts" are not in question here.

My intent is to illustrate the significance and comprehensiveness of attempts by the late ancien-régime French state to follow what it understood as the English model of industrialization. The positive and practical responses of an impressive number of French scientists, inventors, entrepreneurs, and bureaucrats at all levels illustrate that these innovators did not believe that English dominance was either foreordained or inevitable.[157] They recognized that the French were willing and able to learn from their rivals across the Channel, just as the British themselves had done and continued to do. Technological leadership in a few key sectors did not yet signify industrial dominance. Inspired by the Physiocrats, they believed that France could and would compete with the "superior" British if the state provided the proper assistance and direction. Realistic well-informed contemporaries pointed to the relative lack of textile machines as a sign of how much France had to do. They also used this example to show how swiftly change could come. "Improvers" influenced by the English example knew that France had acquired its machines in just a few years; they had faith that the model would enable them to close the gap on their cross-Channel rivals. The manner of the nineteenth-century French industrial "catch up" illustrated how well founded these views were.

French producers suffered from the *Anglomanie* of the consumer, but the Eden Treaty stimulated domestic willingness to innovate, to experiment, and to compete. Just as a great number of French from Louis XVI on down were willing to learn some political lessons from North Americans, this chapter illustrates that many Enlightened French, particularly those influenced by the Physiocrats, were ready to implement a linked commercial/industrial policy that would allow them to follow the English model of fostering greater technological creativity and industrial productivity. The next chapter will explore the response of the working classes to the challenge of English-style industrialization and suggest that the positive reaction of many Enlightened French administrators and entrepreneurs to English competition did not survive the emergence of greater working-class militancy. The outbreak of the French Revolution dramatically affected the course of French industrialization and explains why the British model, despite such a promising start under the late ancien régime, became the path not taken in France.

4

The Other "Great Fear": Labor Relations, Industrialization, and Revolution

Timing, in some cases, is everything. "In the eighteenth century," Maxine Berg asserted flatly, "there was no Machinery Question. The machine was then simply a material contrivance which demonstrated the culmination and success of the division of labour."[1] That bold statement may have held true for Great Britain, but it certainly did not hold for France. During the pivotal decade of the 1780s, a significant number of French laborers, entrepreneurs, and government officials recognized that mechanization was a double-edged sword that required economic and social sacrifices. For fundamentally political reasons, wringing real concessions from a wide variety of interests was not always possible. The difference in timing between the emergence of a "machinery question" on either flank of the Channel demonstrates the divergence of industrial paths followed by the two nations. Many economic historians believe that industrial growth in Britain resulted from endogenous sources.[2] But sometimes, from a British perspective, the world is turned upside down: France encountered a deep-seated and troublesome machine question a full generation ahead of England, a fact that led the French to develop a divergent approach to industrialization and technological change.

The voluminous literatures examining workers on the shop floor, the emergence of trade unionism, and the eighteenth-century "moral economy" in Britain emphasize the autonomy, importance, and widespread existence of labor organization, especially important in considerations of mechanization. These literatures highlight the effective measures undertaken both by the government and by industrial entrepreneurs to control and/or discipline the laboring classes and depict a British experience that roughly paralleled developments in Revolutionary France.[3] The similarities should not, however, overshadow the differences. Organized labor and discernible elite fears about working-class militancy existed in Britain, but not on a comparable scale. The essential point is that in Britain, despite assertions to the contrary by E. P. Thompson and Adrian Randall, the independence and revolutionary potential of the working classes was several orders of magnitude less than it was across the Channel.[4] The Revolution undermined France's ability to follow industrial policies pioneered by Britain. Instead, the emergence of a new form of politics in France necessitated the adoption of a more defensive strategy that was less dependent on labor discipline or mechanization for productive success. Over the course of a generation, and through the super-heated crucible of war, Revolutionary politics, and dictatorship, social fear transformed the relationship among entrepreneurs, the laboring classes and the state.

This chapter explores the responses of French workers to mechanization during the early stages of industrialization and compares them to their English counterparts. As the two previous chapters demonstrate, machines represented more than a more efficient means of production. On the shop floor, workers encountered an artifact whose meaning went beyond mere utility. Machines incarnated the efforts of state officials and entrepreneurs to follow the English model of industrialization. Working-class violence as part of an emerging revolutionary political culture did not simply destroy industrial capacity. More fundamentally, French labor militancy rejected a form of industrialization that entailed their domination by the owners of the means of production.[5] No revolutionary government was able to eliminate the *menace d'en bas* (threat from below); elite fear about its revival lasted well into the next century.

In the 1790s, the British government kept the lid on social tensions, avoiding the possibility of revolution and greatly facilitating a spurt in industrial development. Some segments of the British elite were not always convinced that the lid remained firmly in place. Their anxieties generated a great deal of chatter. Perhaps, it was this discourse of fear that convinced Thompson and Randall of the reality of the English working classes' revolutionary pretensions. But when the focus shifts from rhetoric to practice, in the main, British industrial entrepreneurs operated on the assumption of domestic stability. In France, industrial entrepreneurs had no comparable faith in the willingness of the French people to be led by the economic elite, nor could they rely on the state to protect their workshops and factories from the wrath of disgruntled workers. This lack of faith undermined entrepreneurial willingness to invest in mechanization, constrained the choice of appropriate technologies and left the economic elite reliant on state initiative. The depth of the "threat from below" crushed French attempts to imitate the English economic model and created a decisively different industrial environment. Thus, the commonly used depiction of "dual revolutions" described an authentic and underappreciated difference between the industrial experiences of Britain and France during a critical time of rapid technological change.[6]

Luddism, Labor Militancy, and the State in England

Labor militancy in eighteenth-century England was more widespread, more deeply rooted, and more violent than its French counterpart.[7] The culmination of this militancy was the Luddite movement of 1811–12, in which formidable crowds destroyed thousands of machines in the West Country, the Midlands, Lancashire, and Yorkshire, leading to massive repression by the British state to curb popular exuberance. Pioneering historians of popular action, the organization of the laboring classes, and crowd behavior, animated by Eric Hobsbawm, George Rudé, and E. P. Thompson, chronicled the labor militancy that culminated in Luddism. They identified the sources of discontent and located the actions of the crowd within an ideological nexus generally termed, following Thompson, "the moral economy of the English crowd."[8] A central concern of this distinguished cohort of historians was to eliminate the common misconceptions that the Luddites' machine-breaking was the last resort of a desperate and unorganized mob and that popular violence was ineffective because it could not survive a confrontation with the waxing power of capitalism.

Hobsbawm distinguished two forms of machine-breaking. The first, "collective bargaining by riot," has had the most historiographical influence. Hobsbawm argued that, beginning in the late seventeenth century and accelerating over the course of the eighteenth, machine-breaking, and "riot and wrecking" more generally, was the most effective tactic available to workers to pressure exploitative employers and ensure their own solidarity; it was not about hostility to machines per se. In this context, "wrecking was simply a technique of trade unionism in the period, before, and during the early phases of the Industrial Revolution." Secondly, in those instances where machine-breaking was motivated by antipathy to mechanization, Hobsbawm insisted on the planned nature of the demolition and on the high level of popular support enjoyed by the wreckers. Hobsbawm emphasized the successes of labor militancy to provide a kind of prehistory for twentieth-century trade unionism. Moreover, he conceived of a labor movement that did not simply react to the depredations of capitalist entrepreneurs.[9]

Thompson's discussion of machine-breaking and the Luddites was the beating heart of his book *The Making of the English Working Class*. Luddism was associated with radicalism and therefore enjoyed revolutionary potential. This outbreak of machine-breaking was, according to Thompson, a turning point in the relationships among labor, entrepreneurs, and the state. The abrogation of paternalist legislation dating from the Elizabethan period during the Napoleonic era and the passage of the Combination Acts in 1799–1800 completed the state's withdrawal from the complex equation of shop-floor relations between employer and employee and undermined a customary moral economy. This deliberate abdication marked the growing identification between the interests of industrial capitalists and the actions of the British state in the early industrial era.[10]

George Rudé took up the theme of "collective bargaining by riot" and applied it in expressly comparative fashion to crowd behavior from 1730 to 1848. He created a typology of the various forms of riot, categorized the instances of popular action and demonstrated the focus on violence against property, not persons, even in direct popular action. Rudé argued that the outbreak of Revolutions in France (1789–1795, 1830, 1848) politicized the crowd in new ways. From that perspective and within the confines of his Fabian predilection for non-violent trade unionist forms of association, Rudé grouped Luddism with the Captain Swing riots in 1830–31 as the most impressive instances of popular organization and activity in Britain until a more politicized, more French movement arose: Chartism.[11]

Hobsbawm, Thompson, and Rudé inspired a flood of outstanding work by a distinguished group of historians. Led by Maxine Berg, Adrian Randall, and John Rule, these historians broadened and deepened our understanding of Luddism, its context, and its effects on the development of industrial capitalism so that the literature now goes beyond the ideological confines that hemmed in earlier interpretations. Their work also points toward new ways of understanding the role of machine-breaking and Luddism within the confines of English industrial and technological exceptionalism that go beyond the misguided view, expressed by David Landes, that "the workers, especially those bypassed by machine industry, said little but were undoubtedly of another mind."[12]

The British laboring classes were restive in the eighteenth century.[13] Market sensitive, they organized extensively, usually by trade and region. At the same time, however, they developed a moral economy in their dealings with entrepreneurs and the state. This moral economy was based on custom, but its foundations rested on established legal protections and on the power of local officials, notably the county justices of the peace, to set wages.[14] When their moral economy was violated through what usually was called an "innovation," either in the manner of payment, in the mode of work, or in the division of labor, or through the introduction of technology, and when the state did not intervene, English laborers had recourse to various tactics, including petition, various forms of intimidation, "combination," the strike, and machine-breaking. In many (perhaps most) cases, protest, intimidation, and the threat of direct action led state officials to impose concessions in favor of custom on innovating entrepreneurs in the name of the public good.[15]

Throughout the British Isles, laborers defended their interests with recurrent machine-breaking. During the eighteenth century, however, some areas and industries were particularly prone to resistance to mechanization; this resistance included, but was not restricted to, violence. In his investigation of the West Country and Yorkshire woollens industries, Randall argues convincingly that it was the nature of the local community that determined the reception of the machine and the possible range of popular responses.[16] The colliers of Northumberland were likely to resort to violent action; they destroyed machinery at the pit head in the 1740s and again in 1765.[17] Berg notes suggestively that, in the 1730s and again in the 1770s, the displacement of female labor was a crucial source of anti-machinery sentiment. Rule asserts that resistance to machinery in provincial England was intimately linked to the issue of apprenticeship.

The Spitalfields silk weavers rioted against the introduction of machines in 1675, in 1719, in 1736, and in the 1760s.[18] In the course of the "Wilkes and Liberty" campaign, Charles Dingley's new mechanical saw mill was attacked and taken apart by a crowd of 500 sawyers in May 1768.[19] James Hargreaves' first spinning jenny was dismantled in 1767; two years later more of his machines were destroyed. In 1776, the West Country experienced widespread popular sabotage of almost every form of machinery used by the woollens industry.[20] Three years later, a mob around Blackburn demolished every carding engine and every jenny that used more than 24 spindles, and other machines utilizing water or horse power. The same year, the water frames at Richard Arkwright's works at Chorley were destroyed, as were several recently established cotton mills. Machine-breaking in Lancashire and the Midlands punctuated the years 1778–1780. In the West Country, the introduction of the flying shuttle sparked riots at Trowbridge in 1785, in 1792, and several times between 1810 and 1813.[21] In Leicester, Joseph Brookhouse's attempt to utilize Arkwright's techniques to mechanize the spinning of worsted yarn provoked a violent response and was abandoned.[22] In 1792, Manchester was the scene of an attack on a factory containing 24 of Edmund Cartwright's power looms; ultimately, the factory was burned by outraged handloom weavers.[23] Recent accounts emphasize that these events were an element of a wide-ranging industrial protest rather than knee-jerk reactions to industrialization and economic modernization.[24]

If anything, resistance to the machine led to violence more often in the early nineteenth century than in the previous century. According to Randall's definitive account, in Wiltshire and Somerset, the elite workers in the woollens industry, the shearmen, formed a powerful union. Associated with their counterparts in the West Riding district of Yorkshire, beginning in 1799, West Country workers led a major campaign against the introduction of the gig mill and the shearing frame. Petitions to Parliament and legal action asserting the illegality of introducing this type of machinery were attempted: only when these tactics failed did the shearmen resort to a strike, which is usually referred to as the "Wiltshire Outages." Three strikes broke out in 1802. The shearmen enjoyed impressive public support including the backing of some sympathetic local magistrates, but they failed to halt the introduction of new technologies. Only then did the shearmen turn to violence, destroying not only the offending machines, but also considerable private property belonging to the owners of the machines and numerous bolts of cloth produced with them. After a short period of direct action in which a few dozen machines were destroyed, the Wiltshire shearmen returned to petitions and the courts to combat the introduction of machines. The Wiltshire Outages marked a transition between forms of industrial protest and primed the pump for more organized, more widespread resistance to machines a decade later.[25]

Machine-breaking had been an important, customary aspect of British industrial relations throughout the eighteenth century, but it assumed a darker and more tragic place in the folklore of industrialization with the Luddite movement of 1811–1817. Named after a supposed Leicester stockinger's apprentice named Ned Ludham who responded to his master's reprimand by taking a hammer to a stocking frame,[26] the followers of "Ned Ludd," "Captain Ludd," or sometimes "General Ludd" targeted this machine for destruction. The movement began in the lace and hosiery trades early in February 1811 in the Midlands triangle formed by Nottingham, Leicester, and Derby. Protected by exceptional public support within their communities, Luddite bands conducted at least 100 separate attacks that destroyed about 1,000 frames (out of 25,000), valued at £6,000–10,000.[27] Luddism in the Midlands died down in February 1812, but it had already inspired the woollens workers of Yorkshire to take action, beginning in January. A third outbreak took place in April among the cotton weavers of Lancashire. Factories were attacked by armed crowds. Thousands participated in these activities, including many whose livelihoods were not threatened directly by mechanization. Despite the diversity of the crowds, the Luddites generally destroyed only machines that were "innovations" or that threatened employment. They left other machines alone. The specific causes of these outbreaks varied not only according to region but also by sector. Collectively, these initial episodes of Luddism caused perhaps £100,000 of damage.[28] Further waves of machine-breaking, in which a few hundred

additional stocking frames were destroyed, came in the winter of
1812–13, in the summer and fall of 1814, and in the summer and fall of
1816 and the beginning of 1817.[29]

Machine-breaking did not disappear with the followers of Ned Ludd.
Incidents of it accompanied extensive rural rioting in East Anglia in 1816
and again in 1822. The targets there were the mole plough and the threshing machine. In 1826, Lancashire endured a wave of machine-breaking
more extensive than that of 1811–12. Twenty-one factories were assaulted,
and 1,000 looms, valued at £30,000, were smashed.[30] Three years later,
power looms were the target of Manchester's working classes. Repeated
recourse to machine-breaking culminated with the Captain Swing Riots.
Named after the swinging stick of the flail used in threshing,[31] this campaign ran from 1829 until 1832, peaking in late August 1830. The agricultural laborers relied on arson, but machine-breaking was an important
means of expressing popular anger. Although blackened with the term
"riots," the Captain Swing movement can best be characterized as a peaceful series of mass demonstrations by the poor and laboring classes across
a broad swath of southern England into the Midlands. The goals of the
demonstrators varied by region and were quite localized. The general
theme, however, was the redress of popular grievances, although political
concerns played a role in some areas. The grievances had to do with
wages, the use of Irish labor, the implementation of the Poor Law, the tithe
rates, and the introduction of threshing machines. In more than 1,500 separate incidents, an impressive proportion of England's threshers were
destroyed. A fair tally of industrial machinery was also wrecked in
1829–1832 as part of Captain Swing.[32] A separate attack on Beck's steam
factory at Coventry in 1831 concluded the era of machine-breaking in
Britain. Recent commentators, most notably Joel Mokyr, have portrayed
these events as last-ditch efforts with little chance of success.[33] After 1831,
the British popular classes abandoned machine-breaking as a major means
of resolving industrial or work-related disputes, perhaps because of the
success of other forms of worker activism in conditioning the "very nature
of production."[34] The general pattern of militancy contributed greatly—
though indirectly—to the rise of Chartism and the attendant campaigns
for the ten-hour day, for limitations on child labor in mines and factories,
and for the establishment of public health standards.[35]

If machine-breaking's longevity, geographical scope, and popular support in Great Britain impress, the magnitude of government repression astonishes. The Duke of Wellington began the Peninsular Campaign in 1808 with fewer than 10,000 men, but the British state deployed 12,000 to stop Luddism in 1812. On February 14, 1812, Parliament made frame-breaking a capital crime. Rudé provided a first approximation of the virulent response of the English state and courts to all popular riots and disturbances. Against a grand total of two fatal victims of the Luddites and the Captain Swing movement combined, British courts hanged more than 30 Luddites in 1812–13, and in 1830 nine swung for machine-breaking among the 19 executed in the aftermath of Captain Swing. These figures do not include the casualties involved in the attacks themselves. In repulsing the Luddite attack on Daniel Burton's steam-loom factory at Middleton in Lancashire on April 18, 1812, five were killed and 18 wounded. Later that day, a crowd of colliers returned to finish the job. In addition to the dead and maimed, dozens more Luddites and 200-plus machine-breakers involved in Captain Swing were sent to Australia. Nearly 650 were imprisoned.[36] More generally, Rudé found that in the course of more than 20 major riots and demonstrations between 1736 and 1848, the English "crowd" killed no more than a dozen. The courts hanged 118, and 630 died from military action.[37] These figures include the "Wilkes and Liberty" movement, which saw 11 demonstrators killed in London in 1768; the Gordon Riots of 1780, which saw 285 killed or wounded; the 110 casualties at the Bristol toll gates in 1793; and the 11 killed and 420 wounded at Peterloo in 1819.[38]

Rudé concluded that machine-breaking was only one aspect—if the most spectacular—of the popular restiveness of the early industrial period. Just as machine-breaking was an important customary aspect of industrial disputes and an important stage in the development of the possibility of revolution among the English working classes, so too, the emergence of other elements of British state repression was partly related to machine-breaking. To mention only those measures directly pertaining to the industrial work environment, new facets of state repression included Pitt's Two Acts restricting individual liberties in 1795; the suspension of the Habeas Corpus Act; the 1797 Administering Unlawful Oaths Act; the Combination Acts of 1799–1800; the final abrogation of paternalist industrial legislation in woollens in 1809; and the repeal of the Elizabethan apprenticeship statutes in 1814, eliminating the power of officials to regulate wages.[39] To such legislative action could be added the enrollment of property owners in a "patriotic" militia that could be and was used to confront direct action by the popular classes. An army of government spies blanketed the most restive districts. The regular armed forces assumed a new mission; 155 military barracks were constructed in industrial districts between 1792 and 1815. E. P. Thompson summed up the effects of these measures: "England, in 1792, had been governed by consent and deference, supplemented by the gallows and the 'Church-and-King' mob. In 1816 the English people were held down by force."[40] This wistful conclusion expressed Thompson's regret that no revolution broke out in England. This missed opportunity stemmed from the amazing skill of the English government at averting insurrection and revolution during the period 1792–1820.[41]

The possibility of revolution has been assessed from a variety of perspectives. According to Roger Wells (the most notable recent proponent of the view that the outbreak of revolution was a genuine possibility in England), the threat of invasion, food shortages, a generalized economic crisis, Jacobin and/or Paine-ite agitation, the consequences of the Irish uprising, navy revolts, millenarianism, and war-weariness could have triggered a revolution in 1799–1801.[42] A generalized historiographical consensus rejects this argument, asserting that Britain was too well governed and too thoroughly dominated by the elite for revolution to have broken out, especially once war with the Napoleonic regime resumed in 1803.[43] Thus, in the era of industrial unrest that is of such great interest to historians of labor, the crowd and industrialization has *not* been associated with the threat of revolution.[44]

Why is that consensus significant? The essential element is the effective, even vigorous action of the British state—in conciliation, in the mobilization of nationalism, and in repression—which even Thompson and Rudé acknowledged as having forestalled the emergence of a revolutionary moment until 1831–32.[45] Such activity hardly conforms to standard accounts of the laissez-faire nature of the British state after the publication of *The Wealth of Nations*. Thompson and Randall, among others, asserted that Luddism punctuated the transition from

intermittent paternalist protection of the laboring classes to the imposition of a laissez-faire political economy upon and against the will of the
working classes.

Rule's argument and the provocative new interpretation of Leonard
Rosenband provide another way of understanding government action
during this period. They believe that the primary purpose of the Combination Acts was not to destroy unions or to prevent the spread of political radicalism; rather, they depict the difficulties encountered by
employers determined to replace customary practice either with their
own discipline or with another form of discipline from above as the chief
motivation of these infamous measures.[46] In the aftermath of Luddism,
according to Randall, the English state increasingly identified its interests
with those of the large-scale "innovating" manufacturers. This identification led to a more systematic administrative implementation of laissez-
faire ideas at the expense of customary protections. Moreover, this policy
flourished despite considerable support among segments of the elite and
many small producers in favor of retaining these protections.[47] What
these findings suggest is the need for a fundamental reconsideration of
the part laissez-faire ideas played in technological change during the
early Industrial Revolution in Britain. That reconsideration must also
reassess the role of industrial protest in stimulating post-1815 government espousal of laissez-faire policies.[48]

That machine-breaking, among other popular actions, evoked a dis-
proportional *state* response should frame any consideration of the effects
of machine-breaking in England. Decoding the timing of the adoption of
machines is tenuous at best, but in the wake of extensive machine-breaking,
the task becomes even more uncertain. Landes and Mokyr head the list of
those who dismiss the possibility that any brakes on the process of mechanization could stem from the direct action of the laboring classes.[49] Yet
machine-breaking had some limited temporary successes in Great Britain.
The woollens industry in the West Country was most successful in resisting mechanization through direct action. After the "Wiltshire Outages"
of 1802, the gig frame was not restored until after Waterloo.[50] A 1787
attack on machinery in Leicester discouraged the introduction of mechanized spinning until after 1815.[51] The other major triumph of the
machine-breakers was registered by agricultural laborers who destroyed

thousands of threshing machines during the Captain Swing outbreak; these machines did not return in anything like the same numbers to most of southern England for at least 20 years. Short-lived successes included higher wages, the stoppage of the practice of making "cut-ups" in the Nottingham hosiery industry, the suspension of shearing frames in Yorkshire in 1812, and a wage increase after Captain Swing in 1830.[52] Thus, beyond the confines of the city of Leicester, from the standpoint of industrial technology, the only relatively unequivocal success by English machine-breakers occurred in an area that was rapidly becoming marginalized by the West Riding in an industry that was steadily being superseded by cotton.

What was the relationship between the energetic, even excessive, response of the English state to machine-breaking and the lack of concern among innovating entrepreneurs interested in mechanization? My answer is that, despite the well-documented militancy, widespread organization, and politicization of the English laboring classes, the British peoples—unlike their French counterparts—were generally willing to follow the lead of the elites.[53] The relative lack of violence surrounding British political action and the predilection for attacks on property rather than persons in industrial protest were signs of this willingness.[54] As a result, determined British entrepreneurs overcame customary industrial practices antithetical to mechanization and/or the imposition of the factory system through a variety of methods which astound those more familiar with Continental conditions.[55] First, the dependence of early factory masters on unfree labor was nothing less than extraordinary. Approximately one-third of factory workers were unable to leave their jobs due to apprenticeship restrictions or parish appointment. Second, entrepreneurs took advantage of variations in regional wealth and competition for employment in the four realms to attract the truly desperate; generally, they were the only ones willing to enter the factory and remain there. Third, the stagnant or even declining wages of the era forced families to send large numbers of boys and girls into the factory. British entrepreneurs took advantage of these conditions to impose a measure of industrial discipline that the laboring classes had resisted successfully in the eighteenth century.[56] This achievement bore abundant fruit after 1830, when the economic benefits of mechanization had spread to enough trades, thereby encouraging a growing number of entrepreneurs to follow in the footsteps of the pioneers.[57] Not coincidently, it was at this time that recent accounts situate the rapid acceleration of economic growth associated with an industrial "take-off."[58]

What cannot be emphasized strongly enough is that the reason why British entrepreneurs were able to embark on and ultimately complete this generation-long project was the absence of a genuine revolutionary threat to their position. This situation can be attributed to more effective administration or to the existence of political outlets or even to the greater willingness of the English elite to accommodate power sharing; the cause of British stability is not the essential issue here. The fact that industrial protest in Britain tended to occur during the upswing of a boom also pointed to how entrepreneurs could minimize the possibility of revolution.[59] Innovative manufacturers in Great Britain could rely on the state to endorse their interests and assist them in the task of "breaking" the British working classes.[60] This faith in the state was justified; more than 60 legislative measures were enacted during the crucial period 1793–1820 to prohibit working-class collective action.[61] Although such frequent intervention also illustrated the doggedness of resistance, it surely was not a coincidence that political reform from above came in the 1830s—only *after* a generation raised under the new discipline was at work.

Evidence for this interpretation of the British entrepreneurs' mentalité can best taken from how they acted. If the activities of the "heroic" British industrialist are well known, the ability of these entrepreneurs to overcome determined labor resistance with the support of a powerful, repressive state apparatus must be seen in comparative terms as unique.[62] For example, in the Continental context, James Hargreaves' actions were incomprehensible. After his first spinning jenny was destroyed by a mob in 1767, and after a crowd dismantled others in 1769, Hargreaves moved to Nottinghamshire to set up a new establishment.[63] Nottinghamshire had a well-deserved reputation for industrial protest and machine-breaking. The next section demonstrates the difference between this behavior and that of French entrepreneurs. Laissez-faire ideology justified the actions of both entrepreneurs and officials, but the activities of the British state during the early industrial era have little resemblance to the distant, limited role for government advanced by present-day proponents of Smithian economics.[64]

Machine-Breaking and *la menace d'en bas* in France

In England, machine-breaking was a consistent and persistent element of industrial work relations from the late seventeenth century until well into the nineteenth. France experienced a drastically different trajectory; in the eighteenth century, both English and French commentators recognized that French industrial laborers were less organized and less prone to violence than their English counterparts.[65] After the Gordon Riots of 1780 rocked London, Louis-Sébastien Mercier made the well-known observation that such "terror and alarms" on the part of the popular classes could never occur in Paris.[66] Yet machine-breaking had a greater effect on the course of industrialization in France. An important element of the national cultural difference is revealed in the same passage from Mercier, who believed that the Gordon Riots "took a course unimaginable by Parisians; for it appears that even in disorder the crowds were under some kind of control. For instance, a thing which a Frenchman can hardly credit; the houses of certain unpopular men were fired, but their neighbours not touched; our people in the like circumstances would show no such restraint."[67] Presciently, Mercier pointed out that the Parisians' lack of self-discipline meant that any disturbance would spiral rapidly out of control. This did not occur before 1789. But that fateful summer, popular action throughout the country transformed the nexus among the laboring classes, entrepreneurs, and the state, fundamentally altering the course of French industrialization.[68]

The contribution of machine-breaking to the revolutionary events of 1789 has consistently been undervalued by historians of labor relations and of industrialization.[69] Machine-breaking in eighteenth-century France did not have the same deep roots as in England. Whereas the silk weavers of Spitalfields were among the most dogged British workers in resisting mechanization,[70] labor relations in Lyon's *grande fabrique* suggested that this issue was less important than the relative positions of merchants, masters, and men, the role of municipal oversight, the imposition of work rules affecting employment, and opportunities for female employment. Despite notable worker hostility to the introduction of Jacques Vaucanson's cord-pulling machine (they believed it would eliminate the position of the "draw girl"), only a few machines were broken when a riot broke out in 1744. This minor damage was remarkable in view of the retreat of royal troops in the face of angry crowds of silk workers.[71] However, in the early 1780s such restraint diminished, overwhelmed by a swelling crescendo of labor agitation. This tumult surged during the recession that began in the mid 1780s: for instance, in Paris, workers in the building trades struck to resist a wage cut imposed by employers in July 1785, and in January 1786 carriers and porters struck to prevent the creation of a rival transport monopoly. The following March, the carpenters struck and several other trades joined the "fermentation."[72] Similar patterns of increasing labor militancy can be traced for Bordeaux, Lille, Lyon, Rouen, and Troyes. This spike in labor agitation paralleled that of Great Britain, but in France there was significantly less recourse to machine-breaking or others form of violence against persons or property.[73]

Yet across a variety of trades in diverse regions, this situation began to change on the eve of the French Revolution. Perhaps the most notable outbreak of resistance to the machine before 1789 took place in Saint-Étienne. Beginning in 1785, labor agitation in the region exploded as workers defended customary practice when faced with innovations in mechanization, the division of labor, and the adoption of manufacturing technique brought from abroad. Motivated partly by a xenophobic protection of French industrial custom, agitation began in the metallurgical trades when two workers from Liège brought new methods to forge musket barrels using trip hammers that would eliminate one step—and thus one job—in the local productive routine, but increasing the productivity of others. Metal workers responded violently to this innovation that threatened their livelihoods. The Belgians were driven from the city after being beaten; the municipality supported the workers and explicitly defended local customs. In late July 1787, a coalition of silk ribbon makers called a strike. They were concerned that, thanks to the high price of raw silk, there would not be enough work to carry them through the winter. Their first goal was to force all non-French workers to leave. Then the hand workers sought to use threats and peer pressure to convince workers in small mechanized workshops to stop using machines in order to maintain employment. The mounted police easily dispersed the ribbon makers; four were arrested. Worker violence against "foreigners" bringing new skills was renewed in March and April of 1789, with "Germans" employed in the coal mines now the target of attacks. Between 1785 and the spring of 1789, ribbon makers, miners, and metal workers in the arms industry intervened publicly on at least seven occasions to prevent the introduction of advanced machinery and to cast out Swiss, Belgian, and German workers who had brought new industrial techniques that might lead to short-term unemployment by improving the productivity of labor. While the ancien régime lasted, the violent tactics of the workers of Saint-Étienne conserved local customs.[74]

The deepening economic recession was accentuated in 1788 by the onset of a subsistence emergency. The pathetic harvest led to drastically higher bread prices, just as industrial unemployment plummeted. The political impasse stemming from the inability of the Bourbon government to reform the tax structure in order to pay its bills contributed to the looming sense of crisis. Harbingers of the coming storm took place all over France as popular unrest skyrocketed in response to the economic situation.[75] The increasing restiveness of the textile workers of Troyes led the municipal authorities to demand patrols by the royal garrison in April. They hoped to intimidate the workers enough to prevent them from molesting their employers. Throughout the summer, the *compagnons* of Bordeaux demonstrated in the streets on behalf of the Parlement of Guyenne, clashing several times with the garrison and the city watch. Emboldened by these successes, the *garçons cordonniers* (journeymen shoe makers) struck to obtain higher salaries in September. In lower Normandy, the cotton spinners of Falaise wrecked their own machines on November 11 after being laid off. Beginning in December, bread shortages regularly turned into food riots. Several customs posts were attacked in northern France.[76] When abbé Emmanuel-Joseph Sieyès published *What Is the Third Estate?* in January 1789, for many French any possible answer must have begun with the word 'turbulent'.

The situation remained uncertain throughout the spring of 1789. Widespread politicization stemming from meeting in assemblies and drawing up *cahiers de doléances* kept the kettle at full boil, particularly in the provinces. The laboring classes voiced their grievances concerning encroaching mechanization and their fears of losing their status as skilled labor if they used a machine whenever they could. In both rural and urban cahiers, among a wide variety of occupational groupings, complaints about machines were made most commonly in places hard hit by the Commercial Treaty such as (both Upper and Lower) Normandy and Champagne. Economic elites in some of these areas shared popular misgivings about the machine, particularly as English-style machines were constructed and diffused at record rates despite high unemployment.[77]

The backlash against the machine and fears of "deskilling" seemed to culminate in the Réveillon Riots in Paris' Faubourg Saint-Antoine on April 27–28, 1789. The riots countered independent, but related, comments by two important manufacturers—Dominique Henriot and Jean-Baptiste Réveillon—that apparently called for the reduction of the daily wage despite the dramatic rise in bread prices. The following day, impressive crowds of up to 2,000 sacked the homes first of Henriot and then of Réveillon, shouting patriotic slogans. The *Gardes françaises* opened fire, killing an undetermined number, with 900 the highest estimate. Three "ringleaders" were hanged after a massive military crackdown on the Faubourg the following day. The causes and meaning of this event have been hotly debated. For the purposes of comprehending the course of French industrialization, however, Réveillon must be understood as an entrepreneur who utilized a variety of innovative productive means, including machines, to manufacture high-quality wallpaper in a complex of shops located in and around his home at Titonville in the Faubourg. Réveillon was popular with his workers, but his innovative methods infringed on the customs of many important trades. Although Réveillon's workshops suffered the crowd's fury, it was peripheral damage because, as was common not only in industrial disputes but also in food riots, Réveillon's home was the central target. Because his shop and manufactory suffered relatively slight damage, the 40,000-*livre* recompense provided by the central government enabled Réveillon to resume production in the fall; he appeared to have weathered this outbreak of crowd anger and the attendant machine-breaking. However, in May 1791, as a new wave of labor militancy began to crest, Réveillon left the trade and his prestigious status as a royal manufacturer for good, never to return. Thus, Réveillon's fate as an entrepreneur interested in innovation who, with state support, manufactured goods capable of competing successfully with England had greater significance than the usual depiction of the event as a signpost warning of Parisian working-class violence to come in the spring of 1789.[78]

The *cahiers'* attack on machines and their evident culmination in the Réveillon Riots should not be permitted to dominate our understanding of French attitudes toward machinery as the Estates-General met at Versailles in May. Without question, objections to mechanization and to other innovations were heard from many quarters and violence against them had increased from previous (low) levels. But, as shown in the previous chapter, mechanization in France accelerated rapidly in 1788–89, which reveals the danger of concentrating on a few famous incidents (such as the Réveillon Riots) or on a readily available set of documents (such as the *cahiers*).

The emergence of a revolutionary situation in France in 1789 is too well known for more than a brief outline here. Food riots linked not only to high prices and uncertain supplies of bread, but also to hoarding and bread quality broke out all over the country. Popular unrest tied to political events as well as to the economic situation boiled over on July 12–14, peaking in the fall of the Bastille as Paris "saved" the Revolution from royal countermeasures. In that uncertain atmosphere, the enormous unease of many French when faced with "brigands"—often desperate people on the tramp looking for food—combined with deep concern about the possibility of an aristocratic reaction to the formation of the National Assembly to spark a "Great Fear." Rural folk all over France sacked noble chateaux and fired the property and debt records. The frightened deputies in Versailles renounced most of the privileges that typified the ancien régime and set the keystones for the Declaration of the Rights of Man and Citizen on the frenzied night of August 4–5. In urban areas, popular agitation led to municipal revolutions in which new individuals and social groups came to power. "National guards" were formed in imitation of Paris to protect property and the propertied from the increasingly militant popular classes. On October 5–6, a crowd led by women "captured" the king and his family and brought them from Versailles to Paris. This transfer was intended to ensure the delivery of sufficient bread to the new capital and to forestall the possibility of coun-terrevolution. A new political environment resulted from these events that laid the groundwork for modern democracy while spawning a viru-lent conservative response.[79]

If the outline of popular activities in 1789 is well known, machine-breaking is mentioned only in passing, if at all. Machine-breaking unfolded as part of the revolutionary moment. Appropriately enough, the first and largest incidence of machine-breaking took place in the heart-land of French attempts to imitate English industry, in Normandy.[80] In Rouen, the food rioting on July 11–13, which included attacks on grain convoys, mills, and markets, required the intervention of not only the city's bourgeois militia, but also the local garrison. At the same time that Parisian crowds stormed the Bastille, infuriated woollens workers from the nearby textile town of Darnetal estimated at 200–300 strong broke through the picket of royal troops charged with guarding the bridges over the Seine River. Arriving in the manufacturing faubourg of Saint-Sever, these hand workers destroyed or burned English and English-style machines in the district's warren of workshops and proto-factories. The large and newly formed establishment of Debourges and Calonne & Company, which made cotton velours, was invaded by 300–400, who had to break down the heavy wooden front door with paving stones to get at the machines. Thirty machines were demolished and the cording section of the enterprise sacked before the firm's own workers repelled the mob with weapons distributed by the owners. Just a few feet away, led by the manager, the workers at the concern (owned by Holker's son Jean) fired on the crowd, saving their stock of English machines from the flames. Despite such spirited defense of new machinery, hundreds of spinning jennies and a number of recently constructed carding machines were wrecked before the city's militia arrived to confront the crowds. Five were killed in the clash. On the other side of the river, in Rouen proper, another crowd ravaged the home of *procureur-général* Godard de Belbeuf and several other domiciles in the affluent parish of Saint-Godard. They crowd also turned on the chief tax-collecting office before destroying machines in a wide swath of territory stretching from Darnetal to Bondeville including the French version of Arkwright's water frame (built by Barneville).[81]

New incidents followed as part of a more generalized popular attack on all the symbols of the ancien régime. Beginning July 14, crowds attacked the residences or offices of authorities and of others (usually the rich) suspected of hoarding grain or bread. These crowds also destroyed any new labor-saving machinery that took the food out of workers' mouths. On July 19–20, machines were broken in Saint-Sever, in Oissel, and in Rouen (where English machinery purchased by the Bureau of Encouragement was broken into pieces and then burned). The municipality of Rouen attempted to stifle this round of machine-breaking with force. Volunteers headed by the Parisian actor Bordier and the local lawyer Jourdain were dispatched to disperse the crowd. They joined it instead. The next outbreak took place on the night of August 3–4 as the Great Fear cast its long shadow across the country. The intendancy and other tax assessors were overrun by a mob of 4,000. Later the crowd infiltrated a factory, removing a newly built English-model carding machine. It was brought to the Place Saint-Ouen and burned to the delight of a cheering crowd.[82] More machines were destroyed in Darnetal and Saint-Pierre de Franqueville in similar fashion.[83] A water frame operated by a spinner on the rue de l'Épée was dismantled and his shop looted on September 19. In mid-October, turmoil erupted again. Martial law was declared after another series of riots in Rouen and Sotteville led by artisans that began on September 17. Hundreds more spinning jennies were taken apart and the pieces consigned to the flames.[84]

In other parts of Normandy, machine-breaking took place in Louviers, where small masters and workers united to denounce and then destroy the machinery of innovating large-scale entrepreneurs (including Alexandre de Fontenay, whose mechanized production had been subsidized heavily by the central government). Machine-breaking also occurred in Argentan and in the *pays de caux*.[85] Violence spilled northward into Picardy. Machines were destroyed widely in and around the woollens center of Abbeville where stiff English competition after 1786 had agitated a formerly docile, rurally based manufacturing labor force.[86]

Popular unrest in northwestern France often featured intermittent machine-breaking. In July 1789, such activity was closely related to food shortages and to dissatisfaction with the political leadership of the city on Rouen. However, by October, according to a respected member of Rouen's legal community, artisanal mobilization stemmed almost exclusively from hatred of "the machines used in cotton-spinning that have deprived many workers of their jobs."[87]

In addition to its effect on employment, the laboring classes objected most strenuously to how the new machines transformed industrial practice. By lowering the cost of production, hand workers in competition with machine-based manufacture had to sweat even longer hours to make ends meet. In the *pays de caux*, complaints surfaced with increasing regularity beginning in the spring of 1788 as the workday reached 17–18 hours for the poorest families with all family members having to contribute to the spinning process. The aforementioned complaints in the cahiers de doléances reflected these concerns.

Yet machine-breaking was not solely the realm of those involved in the textile industries. Among the 182 (141 men and 41 women) arrested for machine-breaking during the summer of 1789, Jean-Pierre Allinne found that less than 30 percent (45 men, 8 women) worked in professions linked to textiles. In fact, the largest occupational group consisted of agricultural day laborers (28), and nearly 30 percent of the women arrested for machine-breaking were prostitutes. Nor could the presence of 16 soldiers (9 percent) among the machine-breakers have comforted either the authorities or the entrepreneurs. Such findings suggest that in 1789 the Normans' rage against the machine was an element of revolutionary agitation.

The authorities in Normandy—municipal, judicial, provincial, or royal—could slow but not halt unrest. Despite the death of five people on July 14, troops and the bourgeois militia could not maintain order. On July 16, property owners in Rouen, as in almost every other major city in France created a "national guard."[88] Rouen also declared its "autonomy" two days later. The city fathers hoped to isolate the city from the storm of events. As a result of the revolutionary crisis, all local authorities lost their willingness to confront the crowds unless threatened directly. Thirty were ultimately arrested and tried for their participation in the riots of July 14, but only because of the specter of a combination of attacks on the rich, on the authorities and on industrial machines, those necessary weapons in the fight for economic survival against British competition. Only this frightening concatenation of the targets of popular wrath from July 14 to July 20 and again in early August provoked a more spirited intervention by the royal garrison. A large number of textile entrepreneurs with new cotton-spinning machinery to protect whose establishments were located in Saint-Sever and Oissel took the drastic step of distributing arms to their workers.[89] The night of August 3, nearly 100 agitators were arrested preemptively in cafés or cabarets when news of unrest associated with the Great Fear reached the authorities. Bordier and Jourdain were arrested as instigators of the riots. They were hanged, with four others, on August 21. Textile workers and day laborers received lighter penalties. On the day of the hanging, the authorities positioned cannon to command the transit points into the city from its industrial suburbs. Events showed such attempts at intimidation to have limited effect until late October.

The rapid spread of machine-breaking in Normandy presented a far different initial pattern than in Britain where the emergence of organized opposition was more gradual. According to the report of the Bureau of Encouragement, "in a single day, the misguided people have destroyed the benefit of nearly 100,000 *livres* of expense and more than 15 months of work undertaken on their behalf."[90] But the more sporadic pattern later followed by the Luddites was glimpsed as more textile machines were destroyed in the suburbs of Rouen throughout July and into August, with flare-ups in September and October. When popular unrest finally ground to halt, more than 700 spinning jennies had been destroyed including nearly all of those that had been purchased in Britain or built on the English model.[91] Among those who lost their property were several industrial pioneers enticed from Britain to naturalize advanced textile machinery, such as George Garnett, whose workshop was sacked on July 14. Garnett and his workers fought off a smaller assault the day before, mortally wounding one attacker, but on July 14 the crowd was simply too large. The wooden pieces of his broken machines were burned publicly and the metal parts scattered.[92] Barneville's imitation of Arkwright's machine was also destroyed before it proved itself worthy of the enormous expense. The smoking debris of several years' government investment and entrepreneurial activity had a significant though subtle effect on the attitudes of economic decision makers in Normandy.

Northwestern France experienced intermittent popular discontent that expressed itself partially through machine-breaking in July–October 1789. In the rest of France, the pattern differed, depending on what region and what industries were involved—even more so than in Great Britain. In Saint-Étienne, violent worker response to innovation preceded the emergence of a revolutionary moment. Beginning in 1787, textile workers, miners, and artisans in the metallurgical industries repeatedly deterred the introduction of advanced textile machines and cast out foreign workers who brought new industrial techniques that might lead to short-term unemployment. Another incident took place on July 24, 1789, when a large group of miners and artisans in the stéphanois metallurgical industries, led by a former municipal official, marched on Roche-la-molière (a pit head situated atop the Rives-le-Gier coal basin). This riot aimed to prevent the company headed by the Marquis d'Osmond from opening a big new coal pit, which was to have steam engines and which was to employ some Germans. After the crowd demanded that all foreign workers be sent away, work stopped, thereby conflating the foreign with the innovative. The leader claimed that these measures had been ordered by the municipality of Saint-Étienne. When the directors demanded written confirmation, the crowd broke the windows of the buildings, freed the horses that were being used to power the water pumps until the steam engines arrived, chased away the company representatives, and sacked the installation. They then broke all the remaining machinery and set fire to the building.[93] This popular activism exacerbated the Great Fear in the Forez.

Subsistence concerns and fear of brigands were exacerbated on July 28 by rumors and a letter from the authorities of nearby Saint-Chamond describing the supposed depredations of an army of 4,000 brigands. These concerns inspired a municipal officer to distribute more than 2,600 weapons to a socially heterogeneous crowd. Augmented by the arrival of thousands of armed peasants from the surrounding areas, a militia of 12,000 germinated. A force of 2,500 was sent to succor Saint-Chamond, and the streets of Saint-Étienne were patrolled day and night. By the time the municipality of Saint-Étienne realized that there was no real outside threat, their fright concerning the potential activism of the popular classes had become all-consuming. An intense campaign to recover the distributed weapons successfully recouped nearly 1,800. However, the city fathers gathered in a "Permanent Committee for Public Security" swiftly decided on a different tactic to rein the people in. A national guard was created on August 8, and holders of weapons were obligated to serve alongside property owners.[94]

Attempts to straitjacket agitation by the laboring classes failed despite the easing of subsistence concerns. Ken Alder's succinct account of machine-breaking in September clearly depicts how an innovator could be derailed by popular defense of customary productive practices. Jacques Sauvade (1730–1806), a mechanic and entrepreneur from Ambert, sought to naturalize machines and processes for making metalwares he had seen in Germany in Saint-Étienne. After six years of expensive trial and error, he set up a large workshop to produce tableware, buckles, locks, and bolts. Stamping dies were used to cut through metal sheets produced by a water-powered rolling mill. Sauvade hoped to compete directly with the toy industry of Birmingham and the cutlery trade of Sheffield.

In a manner reminiscent of England, but not of Normandy, those directly affected by Sauvade's innovations took action. During the early evening of September 1, a group of artisans specializing in the making of forks gathered outside the workshop. Several municipal officers attempted to forestall popular violence. Sauvade recognized the threat to his investment of 5,000 *livres* and promised the crowd that he would "delay perfecting his establishment until the people believed it offered some hope of employing workers, and if it did not he would [from his innovations.]" He even dismantled two cylinders essential to rolling sheet metal and gave them to the mayor in hopes of safeguarding them. Appeased, the crowd dispersed. By the following morning, however, the crucial cylinders had disappeared, but that did not save Sauvade. A crowd of both men and women returned and dismantled the machines and waterworks, then burned the workshop. Perhaps by design of the authorities, eight companies of militia and one of dragoons arrived too late to stop the pillaging. That evening, some fork makers threatened to beat up and burn the home of one of Sauvade's mechanics if he helped to rebuild the hated machinery.[95]

The stockpile of arms in Saint-Étienne led to renewed popular "borrowing" of weapons in November. On November 10, arms workers accused the directors of the royal Arms Manufacture of sending weapons to émigrés who opposed the Revolution. When the leader of the accusers was arrested by the militia, a crowd materialized to demand his release. The militia commander was trampled in the ensuing clash. The following day, the Manufacture was despoiled of 5,612 muskets. The authorities fled the city. Upon their return, the weapons were recovered, but attitudes about the legitimacy of popular action in defense of custom had changed. Royal arms inspector Augustin de l'Espinasse reported: "The journeymen of various fabriques had risen against their masters. As a result everyone had seen the need to disarm the people."[96] The "threat from below" threatened to assume revolutionary proportions.

From the perspective of industrial development and technological modernization, the effects of this wave of machine-breaking were devastating for the region. The exploitation of the Rives-le-Gier coal basin remained crude, while the introduction of *métiers à la zurichoise* (Zürich-style ribbon-making machines) stalled until the Consulate. Sauvade's fate was most instructive. After two years of petitioning for recompense for his losses, he received only 1,500 *livres*. Sauvade claimed that some manufacturers had sanctioned the destruction of his machines to defend their own position. He also asserted that the authorities refused to find and punish those responsible for the machine-breaking because it would upset the uneasy social truce. Because he was an outsider, Sauvade's interests were irrelevant. He complained to the municipality that his sole transgression was that he was "searching for an ingenious means of combating the advantages of our rivals which have reduced the workshops to inaction. Although it is impolitic to do this, obstructing [people like him] damages the true interests of the state, French commerce and the city's prosperity. . . . The actions of this ungovernable group of unemployed workers would have been more justifiable if these machines, invented with such difficulty and perfected at such great cost, had been the true source of their unemployment. But you, Messieurs, you have knowledge to the contrary."[97]

Yet Sauvade continued to tinker. He developed a variation of the *zurichoise* ribbon-making machine that he patented in November 1791. But he soon recognized that the working classes would not permit its deployment; he relocated his metal-working operation to Mirecourt in the

Vosges, where Alsatian metal workers accepted mechanized production. This industry flourished well into the nineteenth century but died out in Saint-Étienne. The parallel to James Hargreaves (who, after his first workshop was wrecked, moved to an area noted for machine-breaking) illustrates how industrial conditions and the unwillingness or inability to discipline their labor forces impeded successful mechanization and retarded certain kinds of technological advance in some localities.

On the other hand, these events were a victory for the popular classes. Machine-breaking and the appropriation of weapons by the laboring classes of Saint-Étienne preserved the essential elements of customary productive practices. They successfully resisted mechanization or accepting "foreign" techniques even at the height of the Terror. A decade later, 18 metal workers contributed to the development of an inexpensive yet artisanal means of producing rosettes on knives similar to those made mechanically by Sauvade. The knives won a prize at the French industrial exposition of 1801.[98] In fact, despite repeated attempts by officials and entrepreneurs to develop the material and human resources of this potentially rich area, the industrial milieu of Saint-Étienne and its hinterland resisted most technological or productive innovation developed outside the region until the Restoration. No similarly favored English district could have done the same.

A final episode of machine-breaking in 1789 traumatized southern Champagne. Subsistence was a particular problem in and around the city of Troyes, sparking a relatively violent municipal revolution punctuated by a series of food riots. The Great Fear commenced a new more riotous phase of popular activity in southern Champagne.[99] A deepening political conflict within the urban elite sabotaged efforts to reassure the restive unemployed textile workers and poor of the city of Troyes. As in Rouen, local officials tried unsuccessfully to use the threat of violence to intimidate the hungry popular classes of Troyes into remaining quiescent.[100]

The embers burst into a major conflagration on September 9 with the public murder of Royal Mayor Claude Huez and the mutilation of his corpse. According to placards posted all over the city before the riot, the major charge against Huez, beyond those related to subsistence, was that he "had favored machines." Huez was killed trying to defend the actions of local entrepreneurs; Troyes' flour merchants stood accused of hoarding grain, adulterating its quality and failing to deliver sufficient supplies. The city's industrial entrepreneurs also ignited the ire of the mob for installing new cotton spinning machines despite the explicit opposition of Troyes' laboring classes as expressed in the *cahiers de doléances*.[101] Lynn Hunt closed her account of this event with the observation that "throughout the long night bands of desperate men roamed the streets of Troyes."[102] A little-known aspect of this incident was that, as these "desperate men" attacked and burned the homes of officials and notables, they also indulged their hatred of machines. The assaults penetrated the shop-front homes of several merchant-manufacturers who had workshops in their basements. The rioters targeted a number of prototype textile machines recently purchased from Paris and Rouen or imported directly from England.[103] All were destroyed. Popular scapegoating of machines and the officials responsible for overseeing subsistence issues reflected the many facets of *la menace d'en bas* to economic elites in 1789.

Although few machines were destroyed in Troyes, the effect of this incident loomed large for the city's industrial entrepreneurs. The day after the riot that killed Huez, the Provisional General Committee ruling the city banned mechanized spinning, hoping to prevent further unrest, even though the ban would throw 800 people out of work.[104] Over the next few months, plans by several leading textile firms to purchase Arkwright machines or to invest in other new technologies were quietly dropped.[105] When petitioning the National Assembly for financial support to revitalize the region's economy, a group of industrial entrepreneurs explained why they had not followed up earlier investments in new machinery: "These machines are often attacked during popular riots because those involved in hand-spinning fear that large machines will diminish their salaries, a fear which is frequently sustained by ignorance. . . ."[106] Recognizing the intransigence of Troyes' militant laboring classes, industrial entrepreneurs in Troyes generally decided not to invest in advanced machinery as they had been doing since the mid 1780s, spurred by heightened competition with England. Instead, with the support of municipal authorities and later of the département of the Aube, these entrepreneurs focused on maintaining total employment by shifting production to unmechanized sectors like linen and concentrating on the needs of the regional market.[107] Troyes' industrial entrepreneurs expanded the hand weaving of high-end cotton fabrics with thread made elsewhere instead of attempting to increase local spinning output. The Troyens increasingly emphasized quality not quantity, a major shift in their market orientation. These entrepreneurs hoped to avoid "any anxiety on the part of the indigent worker . . . because we want nothing more . . . than to ensure that they can earn their daily bread."[108] To reinforce the lesson, hundreds of female spinners in Troyes demonstrated to protest against the introduction of jennies in 1791, successfully preventing their installation.[109]

During the Consulate, officials seeking to explain the industrial problems of the early revolution blamed the unwillingness of Troyes' commercial firms to invest in mechanization on their fear that the machines would be destroyed by disgruntled textile workers and unemployed artisans.[110] As a result, despite a promising start in acquiring advanced machinery in the 1780s, Troyes stagnated technologically during the revolutionary decade. Investment in machinery in southern Champagne did not resume until the Consulate, and the flowering of local inventive abilities awaited the Restoration.[111] The contrast with the ability of English entrepreneurs to use this critical decade to forge even further ahead technologically could not be more stark.

Continued machine-breaking during the early years of the Revolution served to spread the attitudes exemplified by the Troyens. Although the mid-October riot in Sotteville, Normandy seems to have concluded direct action against machinery in 1789, sporadic flare-ups occurred over the next two years. The carders of Lille destroyed machines in 1790. The following year, jennies were attacked in Roanne and outside the critical experimental workshop housed in the Hôpital des Quinze-Vingts in Paris. At Vincennes, Honoré Blanc's pilot gun-making operation was heavily damaged in February 1791. Charles Ballot emphasized the repression of such events and the indemnification of the victims according to the decree of September 9, 1791. He concluded: "One cannot say therefore that the hostility of workers was a serious obstacle to the introduction of machines in France." Frank Manuel agreed with this assessment, as do more recent assessments of the effect of French labor relations on technological choice.[112]

Yet the evidence from Troyes indicated exactly the opposite. Nor were the Troyens alone in fearing to provoke labor militancy through mechanization. In 1792, local administrators in Amiens endorsed a suggestion made in Paris that they stop their pre-1789 use of "a portion of public funds to create workshops dependent on the use of new machinery" in favor of a strategy designed to permit a "progressive increase" in the number of workers employed through a "limitation of the number of machines . . . at work in the textile industry of the département of the Somme."[113] In the Year IV [1796], departmental administrators noted their inability to combat "the prejudice in public opinion against machines because they limit the amount of work available to the poor. . . . this prejudice against machinery has led the commercial classes . . . to abandon their interest in the cotton industry."[114] According to these administrators, and later according to the Somme's first prefect, fear of working-class reprisal played a major role in this shift in industrial entrepreneurs' attitudes toward mechanization.[115]

In the pivotal province of Normandy, the shift was more immediate and more drastic. In the textile town of Yvetot, a municipal commission charged with finding employment for the poor reported its findings on December 30, 1789. To avoid giving the unemployed a target, the commission recommended that the municipality support the production shift from cotton to linen because it required no new machinery.[116] Early in 1790, the intermediate commission of the province reported that, despite widespread interest in acquiring new machinery by both entrepreneurs and laborers before July 1789, since then they could not "propagate the use of new machines" despite offering "a gratuity proportionate to the talent of the worker." Furthermore, the industrial entrepreneurs' "abandonment of this type of [mechanized] work has discouraged the administrators."[117] Some innovators feared that popular opposition would delay the introduction of new machinery until schools could be established to educate the people about their "true interests."[118] Although there were sporadic attempts to import new English machines into upper Normandy during the revolutionary decade, local opposition to mechanization retarded the technological development of the most advanced industrial region in France.[119]

Even in Paris, a noted center of innovation and technological experimentation, artisans and textile workers mobilized to defend traditional methods and avoid mechanization.[120] The laboring classes recognized the critical role of the state. The ribbon makers persistently petitioned the legislature to "prohibit the introduction, construction and usage in every département of machines to make ribbons." If the state would not act, they threatened to break machines that rendered them technologically obsolete.[121] With further research into entrepreneurial activities during the revolutionary era, additional examples of hostility to mechanization, resistance to innovation and the defense of customary means of production from around the hexagon could be multiplied. Machine-breaking in 1789 contributed to the dramatic transformation of the *menace d'en bas* from rebelliousness into something new: modern revolutionary politics.

Labor Militancy and Luddism after 1791

Machine-breaking was only one aspect of laboring-class activity during the French Revolution, but it contributed to the emergence of a decisive break in the mentalité of innovating entrepreneurs after 1789. The extraordinary range of public activities undertaken by the laboring classes during the French Revolution is well known. But even a cursory glance will help to explain the shift in entrepreneurial attitudes to such issues as mechanization, labor discipline, the role of the state, technological innovation, and profit taking. This sketch also bridges the dramatic incidence of machine-breaking during the early Revolution and the rear-guard outbreaks that took place in the nineteenth century.

Spurred by English competition, mechanization in French industry bloomed in the 1780s. Not coincidently, workers' insubordination increased in the same period,[122] leading entrepreneurs in a variety of industries to focus their improving efforts on disciplining their labor forces as a means of expanding their profit margins and competing internationally.[123] The onset of Revolution amplified all these trends. In view of the current emphasis on improvements in productivity to explain total British economic growth during this era, labor relations deserve to be placed at the center of comparative examinations of the process of industrialization.[124]

State and entrepreneurial responses to working-class militancy must be situated in the context of an emerging revolutionary politics. Marx noted this phenomenon, but it has not been given much recent attention. The increased militancy of the working classes from 1791 is unquestioned. Most recent commentators have connected this militancy with the legislation that eliminated the rights of laborers to organize that was justified in the name of a laissez-faire conception of economics.[125] The popular classes, led by Parisian sans-culottes, achieved an ever-greater role in politics after the flight of Louis XVI to Varennes. They also developed or amplified organizational skills in clubs, in the sections, in the national guard, and in other public forums. Yet these same activists simultaneously lost their legal rights to band together to negotiate with their employers or to strike. Sonenscher asserted that "the Le Chapelier law is part of the political history of the French Revolution . . . more than to the history of the French trades."[126] This might be going too far, but the point that ostensibly laissez-faire labor regulation had a fundamentally political function deserves reiteration both for England and for France.[127]

The fundamental link between the activism of the popular classes and political radicalism after 1789 needs explication. For those in the know, the slave uprising in Saint-Domingue in August 1791 was a warning. The steadily increasing militancy of the sections led first to the fall of the monarchy on August 10, 1792, then to Louis XVI's execution in January 1793, and finally to the regime of the Maximums in 1793–94, as war engulfed Europe. If the image of thousands assaulting the Tuileries did not haunt the imagination, the 1792 September massacres in Paris did. The nationalistic fervor that emerged in response to foreign invasion frightened in its potency and potential. The Terror—with its revolutionary committees, its revolutionary armies, and its death penalty for hoarding— incited fear of popular retribution, which multiplied the effect of every one of the approximately 50,000 casualties. Although the Committee of Public Safety reined in the sectional movement in the spring and summer of 1794, a popular uprising on 9 Thermidor was still possible. On 1–4 Prairial, Year III (May 20–23, 1795), the faubourg Saint-Antoine revolted a final time.[128]

After 1795, popular involvement in revolutionary politics dwindled, but the nightmares of French elites did not fade.[129] The fact that these same elites had harnessed the power unleashed by revolutionary politics to fight off and control much of Europe forced them to recognize that the government could not put the genie back in the bottle. Comments like "All the rich are rascals, there are a million of them in Paris to punish" (shouted by a paper worker from the Pont-Neuf on March 22, 1795) exacerbated the situation.[130] The menace d'en bas assumed titanic proportions in the mentalité of French entrepreneurs as a result of the achievements of revolutionary politics. Post-1789 government policies to foster industrialization had to take this new situation—and the attendant new fears—into account.

Outside the political realm, continuing labor agitation during the revolutionary decade ensured that the economic implications of popular political militancy could not be disregarded. Michael Sibalis uncovered 85 labor disputes in Paris in the period 1789–1799. As would later be the case in Britain, the legal proscription against organization was rarely invoked.[131] In many urban areas, compagnonnages survived legal prohibition and even revived.[132] For example, just for the city of Bordeaux in 1805, evidence of either a group of compagnons who were part of a national network or who participated in a more informal yet still significant form of organization exists for thirteen separate trades.[133] Studies of Paris and investigations of various trades confirm such anecdotal evidence: labor organization, apprenticeship, and other labor rights and customs survived, albeit informally, the crackdown on popular political involvement after 1794.[134]

In view of the attention paid by historians to labor organization, the introduction of new technologies, and the strained economic circumstances, the absence of machine-breaking was as telling as its presence would have been. Machine-breaking appeared to have lapsed almost completely between 1791 and the Restoration (1814–1830) despite the persistence of organized groups of laborers, determined government efforts to create and adopt new technologies, and the boom-and-bust economic cycles of the revolutionary era. This hiatus can best be explained by the formulation of different, often more successful tactics by the laboring classes and the firm hand of the Napoleonic regime once mechanization resumed during the Consulate. For example, although workers attempted to destroy machines at Lille in 1805 and at Cholet in 1807, the government arrested the rioting workers before machinery could be wrecked and sent them before the courts for swift, heavy-handed punishment.[135]

In his examination of machine-breaking during the Restoration, Frank Manuel found that this tactic was largely confined to southern woollen towns that could not compete with more "modern" production techniques employed either in other parts of France or abroad. Organized by the masters, both large and small, the recrudescence of machine-breaking was an aspect of the competition between "progressive" and "backward" manufacturing centers rather than an example of direct action by the laboring classes.[136] Manuel's account of French "Luddism" was surprisingly elite-oriented. Dismissing the earlier outbreaks of machine-breaking as ineffectual, he focused on the overreactions of the authorities to rumors that machines were threatened with destruction. His version emphasized how small-scale masters in the southern woollens industry encouraged their workers to destroy machines as a means of preventing mechanization. These masters hoped to preserve their traditional (and profitable) means of production. They were unwilling and unable to invest in expensive new machinery. Only with advanced technologies subtracted from the productive equation could their increasingly outdated methods compete.

Machine-breaking revived in November 1816, when wrecking was threatened in Paris. As in the English West Country, however, it was the shearmen of the woollens industry who were the most capable of organized resistance to new machinery. In January 1818, workers in Clermont-L'Hérault, Saint-Pons, and Lodève protested against the use of carding and spinning machines. Placards with complaints and/or threats to the machines plastered many woollens districts. Minor riots erupted when machinery salesmen passed through some southern towns, but the main outbreaks took place in Vienne, Carcassonne, and Lodève between February 1819 and May 1821. These flare-ups concerned the introduction of a shearing machine with rotating blades that could do the work of twenty men. Developed during the Empire with an improved version patented in 1817, this machine was already in use in other parts of France. The shearing machine was the major target of machine-breaking in Restoration France, perhaps because its use would displace an entrenched and privileged group of laborers.[137]

Explicitly compared by contemporaries to Luddism in England, French events were striking for the lack of genuine proletarian involvement, minuscule property damage and the leniency of the authorities in dealing with rowdy workers. In Vienne, repeated meetings regarding the proposed introduction of the shearing machine led to state involvement in the affair: the prefect sought a non-violent solution at almost any cost. In Manuel's account, the only reason for direct worker action on February 29, 1819 was the ineptitude of the entrepreneurs who wanted to install the machine; it was compounded by the negligence of local authorities. Despite the best efforts of 8–10 workers who battered it with their 30-pound shears, the machine was only slightly damaged. However, the attack led to a confrontation with dragoons sent to protect the machine. With a hastily erected barricade in place and rocks flying, the workers rushed the troops who opened fire, wounding a few and dispersing the rest. The machine commenced operation a few days later, protected by 400 troops. Twenty-nine rioters were arrested, and eleven were tried. In part because of government recognition that a group of small masters had whipped up opposition to the machine for their own reasons, *none* of the rioters were convicted. In the repressive judicial environment of the Restoration, the verdict was a stunning absolution of the laborers.

On May 4, 1821, a shearing machine in Lodève was threatened by jeering crowds. Similar events had taken place in Limoux and Carcassonne, but this incident abruptly turned violent as the factory was broken into and the machine destroyed. The five rioters captured were all freed by the courts because the entrepreneur who owned the factory was suspected of having egged on the workers so that he would not have to pay the full cost of the machine. Another demonstration against the shearing machine took place in Castres on May 13, 1821. The situation was defused when the mayor and subprefect warned the small masters that they would be held responsible for the actions of their workers. Chronologically, this incident brings Manuel's account of French Luddism during the Restoration to a close.

Two broken machines (albeit expensive ones, at 20,000 francs each), no deaths, and no convictions hardly constitute a major occurrence of Luddism. The significance of these events is that they highlighted the difference between events in England in 1811–1817 and those in France in 1819–1821. At most, French Luddism during the Restoration was a rear-guard action in a troubled industry in a region sinking into industrial decline. In this era of growing worker organization and activity, threats against the machine were commonplace; however, the actual incidence of Luddism, even during violent confrontations between large groups of workers and the armed forces of the state, was minute. Shearing machines were introduced successfully, spurred by the challenge of foreign competitors who had adopted the new shearing machine wholeheartedly.[138] In almost every feature except the fact that machines were a target, the two movements were fundamentally dissimilar. Yet this last gasp of resistance to the machine suggests that French machine-breaking in 1789–1791 was analogous to English Luddism. Although the events discussed by Manuel were more contemporaneous, they do not compare in scope or importance to their British counterpart, much less to French events a generation earlier. Abbeville, Paris, Roanne, Saint-Étienne, Troyes, and numerous places in Normandy experienced machine-breaking in 1789–1791. The state had an important role in industry in these areas, all of which became large manufacturing areas and centers of technical innovation in the nineteenth century. The same cannot be said of the southern cities that resisted the introduction of shearing machines during the Restoration.[139] If the woollens industry of the West Country disappeared not long after the Luddites, Lancashire and the Midlands remained England's the industrial heartlands.

Timing remains the key to understanding the relevance of machine-breaking. Anglocentrism must not blind us to the importance of the wave of French machine-breaking in the years 1789–1791. Whether it was a larger or a more violent movement than English Luddism are secondary questions. The fact that machine-breaking was intertwined with growing labor militancy and the emergence of revolutionary politics gave a different twist to labor relations in France that proved extraordinarily significant to French industrial development. The importance of the gap of more than 20 years between the major incidences of machine-breaking in England and France cannot be underestimated. It is no coincidence that at the end of the Revolutionary and Napoleonic wars England was widely recognized, both by contemporaries and by current commentators, to be about a generation ahead of France in technological terms and in setting up the factory system. Machine-breaking in 1789 and the transmuted labor relations that emerged in the wake of the emergence of revolution-ary politics contributed powerfully to this lag and played a major role in derailing French attempts to industrialize on the English model.[140]

The rather pathetic recrudescence of machine-breaking during the Restoration only highlighted the different direction taken by French industrial and political culture in the intervening decades. The major uprisings of the silk workers of Lyon who protested their deepening hunger with direct action in October and November 1831 and again in April 1834 illustrate the point exactly. Despite some threats to do so, no machines were broken in 1831 or in 1834, but it was the organized potential of worker militancy that led to massive armed repression by the national guard in 1831 with at least 600 casualties and a death toll of 300 due to military action in 1834. Machines were left alone in France in 1848 during the far bloodier confrontations of the February Revolution and the June Days; French workers had learned to rebel against capitalism, the state, and the tyranny of the markets rather direct their ire at machines.[141]

To comprehend how powerful the social and political forces unleashed by the French Revolution were, it is worth reiterating that on the other side of the Channel the supposedly weak English state responded with massive repression to all laboring-class threats to entrepreneurial invest-ment and initiative. The deployment of 12,000 troops in Luddite areas even in wartime demonstrated the priorities of the British government more clearly than any pamphlet or statement made on the floor of Parliament. In France, what historians of science, technology, and indus-trialization have characterized as a hyper-involved, highly centralized, technocratically governed state reacted far less combatively. At its most confrontational, the French state responded to machine-breaking with a whimper rather than a whiff of grapeshot.

Is it any wonder that innovating French entrepreneurs, terrified by the risks posed by the outbreak of revolution, delayed mechanization until they believed that they had their labor force under sufficient control? Of course, the meaning of the French Revolution and the growing political significance of popular agitation more generally were not limited to France. An argument could be made that the same sort of reaction to the militancy of the working classes occurred in Britain. The essential difference, however, is one of degree. Without a capital-R Revolution, revolutionary politics, or a state-sponsored Terror, the "threat from below" was never as dire in Great Britain. With precautions that give the lie to any description of English industrial practice as laissez-faire, entrepreneurs on that side of the Channel could safely mechanize and/or experiment with innovative forms of industrial organization.[142] "Ned Ludd" did not have a guillotine at his disposal.

French hopes of matching the British at their own game were consumed by the flames of burning machines. The nineteenth-century British "machinery question" relevant to the laboring classes discussed by Berg troubled France a generation earlier.[143] In the 1790s, because the British working classes were rebellious but not revolutionary, England used its early advantages to forge a commanding lead in technology and productive practices. The next two chapters will be devoted to understanding the consequences of the French Revolution for technological change and industrial development. Many entrepreneurs preferred alternative forms of profit taking—smuggling, war profiteering, lowering wages, exploiting market niches, state subsidy—to mechanization and technological improvement. It is a fallacy to assert that France *could not* innovate in the same way as Great Britain or that such behavior can be characterized and then dismissed as "rent-seeking."[144] Rather, the threat of Revolution led French improvers to embrace different and longer-term means of achieving the same ends. These means were equally rational if not equally efficient. For France, the possibility of Revolution meant that the approach to industrialization blazed by England became the path not taken.

5

La patrie en danger: Industrial Policy in the Year II

Survival: Beset by civil war, a peasant revolt, great economic dislocation, and foreign invasion, the French Revolutionary government of 1793–94 focused on that most basic of goals in formulating and implementing industrial policy. The pressing wartime needs of subsistence and military production shunted aside long-standing efforts to improve productivity through the development of a more "liberal," more positive conception of unfettered market competition that imitated Great Britain. The objectives that underlay the catch phrase "laissez faire, laissez passer" were abandoned in favor of the bloodthirsty martial strains of the War Song of the Army of the Rhine.

The general outlines of how the revolutionaries recentralized French administration in order to seize the reins of an economy gone berserk are familiar. Politically innovative and administratively creative, the Revolutionaries harnessed runaway inflation, enforced nationwide food rationing, and vastly accelerated military production. They accomplished these herculean tasks through a systematic application of legal measures deliberately intended to frighten the general population into complying with the state's harsh demands. To protect the ideals of liberty, equality, and fraternity, the revolutionaries made "terror the order of the day" to coerce the French people to cede the central government more control over the day-to-day lives of the citizenry than any previous government ever had. This chapter investigates what use the revolutionaries made of their vastly increased power in the industrial realm.

For several decades, economic necessities and outcomes have not been the focus of historical investigations of the Reign of Terror. Beginning in the 1960s with the "revisionist" attack on the Marxist interpretation and continuing through the collapse of political Marxism in the late 1980s, the economy and its relationship to French politics were superseded by other issues.[1] The best-known recent economic reinterpretation of the Revolution, focusing heavily on the Year II, "does not depend upon original archival research."[2]

That is not to say that no work on French industrialization has been undertaken in half a century. Far from it. Local historians and historians of specific industries or industrial pioneers have made major contributions.[3] Those investigating the emergence of the market and especially its relationship to subsistence questions have transformed our understanding of the nature and means of commodity exchange in this highly politicized atmosphere.[4] In recent years, Jean-Pierre Hirsch, Philippe Minard, and William Reddy, among a host of others, have pushed for a reconceptualization of the way in which the French state and society understood a wide range of economic processes.[5] Sustained interest in the activities of the representatives on mission, as well as in Jacobinism, its means, and its methods have filled in portions of the puzzle.[6] The insights and concrete information provided by these studies suggest new approaches to industrial policy during the Year II and hint at new conclusions.

In one important sector, the French arms industry, old questions have been asked in new ways. Led by Ken Alder, Patrice Bret, and Charles Gillispie, historians of science and technology have finally gone beyond the foundational work of Camille Richard published in 1922. They both explored the longer-term economic, institutional and technological consequences of the decisions made under the Terror and situated the origins of the policies of the Year II in a broader framework.[7]

In this chapter, I will consider the making and implementation of industrial policy in the Year II in order to illustrate the trials and tribulations involved in the unprecedentedly thorough French mobilization for war. The raw, almost unbridled economic necessities that gave rise to the Reign of Terror allowed the revolutionaries to tame rampant inflation, feed the nation, and equip an army that dominated Europe for more than a generation. These overwhelming successes, achieved despite the breathtaking challenges of civil and foreign war, transformed the process of French industrialization by giving rise to a new industrial model to compete with that of Britain's in shaping the decisions of French policy makers. Despite the pervasiveness of the ideology of economic liberalism among policy makers and entrepreneurs, both before and after the Terror, the ever-present example of the accomplishments of the Year II ensured that not only the revolutionary French state but also its approach to industrial matters would survive. The state's forceful as to technological improvement in such important industries like steel making demonstrated conclusively that France's productive problems did not involve a lack of useful knowledge, but rather resulted from recalcitrant workers, insufficient economic incentives, and deficiencies in raw materials and transportation.

Maximums, Minimums, and Mobilization: War Production and Technological Choice

French policy makers in the Year II faced an unprecedented economic situation necessitating an innovative approach to economic regulation and industrial production. War broke out on April 25, 1792, when France, motivated almost solely by domestic political considerations, declared war on Habsburg Emperor Francis II. The French military was unwilling to fight, unprepared for combat, and more interested in politics than in invading Belgium as the Girondin/Brissotin ministry demanded. Humiliated in their initial encounters with the Austrians, inadequately supplied, badly organized, and poorly led, the French forces were on the defensive almost immediately. A Prussian army supported by émigrés violated French soil early in July. "*La patrie est en danger!*" declared the Legislative Assembly on July 11. Popular ferment and a subsistence crisis merged with patriotic fervor to give new impetus to the "patriot" movement in Paris. Supported by national guardsmen from all over the country who came to the capital to celebrate the Festival of the Federation and stayed to protect the capital, the "patriots" overthrew Louis XVI on August 10, 1792.

The fall of the monarchy enabled men espousing a more democratic, more populist Revolution to rise to power. Just as conscription and volunteer enlistment transformed the royal forces into a "people's army," the invasion of France transmuted traditional patriotism into modern nationalism, laying seeds for later French military successes. Although the Prussian advance was stopped at Valmy on September 20, the new legislators of the National Convention faced a difficult military situation when they became the governors of France a few days later.

Political considerations led the infamous *conventionnels* to first abolish the monarchy and then to declare France a republic. French armies, afire with patriotic martial fervor and taking advantage of their superior numbers and the Polish distractions of their chief adversaries, embarked on a Revolutionary crusade by going on the offensive along the Rhine, in Belgium, and around Nice. Despite these diversions, the embattled, bitterly divided legislature spent three of its first four months in office holding a trial for Louis XVI before executing him for treason on January 21, 1793. The death of the king transformed not only the political situation within France but also the military situation. The Dutch Republic, Great Britain, Portugal, Russia, Sardinia, Savoy, and Spain joined Austria and Prussia in the war against Revolutionary France in late 1792 or in 1793.

The crisis mounted with the resolution of the tense Polish situation. Prussian and Austrian troops returned to the war against France. Defeat at Neerwinden by Austria in March 1793 and the outbreak of a peasant revolt in the Vendée (touched off by a call to conscript 300,000 men into the army) precipitated a new domestic political crisis. The expulsion of the Girondins/Brissotins from the National Convention (May 31–June 2) led directly to Federalist revolts in the cities of Marseille, Bordeaux, Lyon, and Caen. By July, the Republic had lost all of its previous gains. Austrian, Spanish, Prussian, British, and Hanoverian troops advanced on France's frontiers.

A Committee of Public Safety, created on April 8, 1793 and consisting of twelve members of the National Convention, was progressively given vast powers to defeat the invaders, end the civil war, and put down internal rebellion.[8] Pressed by demonstrations by the sans-culottes[9] of Paris, the legislature, led by the Montagnard faction, established price controls on grain and bread on May 4. To enforce these "maximums," local authorities were empowered to search out supplies and to requisition food and transport. Once its membership stabilized in August, the Committee of Public Safety progressively dominated French policy making and administration. Over the next eleven months, the Committee of Public Safety, seconded by a host of other institutions, transformed how France was governed, created and equipped an enormous army, fed the country, crushed the Federalist revolts, and stabilized the anarchic economic situation. The successes and failures of French industry during the war crisis shaped the evolution of French industrial policy.

The economic history of the Revolution is linked inextricably to state imposition of wage and price controls (the Maximums). A law instituted on May 4, 1793 restricted grain sales to officially designated marketplaces. Prices were to be set by each district administration according to an average of recent market rates and then lowered month by month until the Maximums lapsed at harvest. Local authorities could requisition grain. Those who hoarded grain were liable to the death penalty. A thoroughgoing parochialism on the part of local administrators concerned mostly with feeding their charges combined with unwillingness on the part of producers not to make all the profit they could ensured disaster. Small-scale economic warfare broke out between municipalities, districts, and departmental administrations. Such competition was aggravated by the glaring differences between administrative boundaries and market supply areas.

With the cities, especially Paris, facing large-scale shortfalls, the popular movement pressed for drastic action. The *lévee en masse* was promulgated on August 23, and on September 5 the National Convention declared "Terror is the order of the day." By the end of August, the currency had fallen to 22 percent of its face value. A "Revolutionary army" of militant and hungry sans-culottes was formed on September 6. Legal pillaging ravaged surrounding départements as food was requisitioned by force so that Paris could eat. On September 11, 1793, the National Convention created universal price ceilings for grain and flour, set transport costs, and empowered local authorities to requisition food for the markets and to stock public granaries. On September 29 a Maximum was established for all "goods of first necessity." Noticeably absent from the schedules were transport costs and business profits. Anyone who withheld certain goods from sale was to be fined and placed on the list of official suspects. These goods included any quantity or quality of hundreds of raw materials or manufactured goods divided into four major categories: food; textiles; chemical products and hardware; and metals, ore, and fuels. Prices were fixed at the local prices of 1790 plus one-third.[10]

Wages too were regulated. Municipalities were to set wages rates no higher than 50 percent above the level of 1790, which in Saint-Étienne and certain other arms-making areas included entailed a little-known but significant reduction in wages.[11] In Paris, the wage maximums were much less strictly enforced until the popular movement was reined in, first in Germinal (March) and then again during the summer of the Year II (1794), when wages were again reduced.[12]

Chaos followed. Somewhere in France, during the week between promulgation and implementation, there was a run on nearly every item covered by a Maximum. The system favored producers at the expense of small retailers. By undervaluing transport costs, sellers were encouraged to bring goods to the nearest market not the one that needed it most.[13] Within the month, reasonable levels of profit had to be established: 10 percent for the retailer and 5 percent for the wholesaler. A sliding scale of transport costs based on means and distance also was factored into the final cost. A Subsistence Commission that reported to the Committee of Public Safety, usually in the person of Robert Lindet, was founded on 1 Brumaire (October 22) to oversee the formulation and execution of national price controls. The Surveillance Committees were empowered to conduct domiciliary visits to search for hoarded or "surplus" items. After the publication of the price schedules on 2 Ventôse, Year II (February 20, 1794), at the direction of the Committee, agents of the central state standardized maximum prices by fiat and redistributed resources via requisitions. Although the loopholes remained substantial and a black market flourished, until their abrogation on 4 Nivôse, Year III (December 24, 1794), the Maximums assisted in provisioning the cities, especially Paris. Thanks to the Maximums, during the Year II, the French central government was able to boost industrial production dramatically as part of a general attempt to optimize the economic war effort. Taken together, these activities represented a drastically different approach to industrial policy which shifted the course of French industrialization.

The Maximums did not control all commercial activities. Mercantilist conceptions of foreign trade were subordinated to the war effort. The best example was the Navigation Act, written by the Committee of Public Safety's usual spokesman, Bertrand Barère; it was modeled on Oliver Cromwell of England's well-known measure of 1651. Enacted on September 21, 1793, the law restricted overseas commerce to ships flying the French flag. For all intents and purposes the measure was suspended on November 7. Although the British naval blockade limited French trade, particularly in grain and industrial raw materials, which were declared war contraband, it did not choke off all trade, particularly during the blockade's first years. Neutrals were welcomed in French ports. They were even permitted to sell goods at prices above the General Maximum and allowed to carry away luxury goods like high-quality textiles, especially lace and silk, along with fine wine and alcohol, all left deliberately unregulated by the Maximums. The property of émigrés, that of the clergy, and that of the former royal family was sold to neutrals to pay for war materiel. After 21 Ventôse (March 11, 1794), anything not subject to the Maximum was liable for export. Agents were sent to Hamburg and especially to Switzerland and Genoa to acquire needed arms and horses and funnel them to France through these entrepôts. Overseen by the Subsistence Commission but largely left in the hands of a subordinate Trade Commission of experts, French exports increased substantially and steadily from the spring of the Year II (1794).

Not even the archrival Britain was exempted from the expansion of French foreign trade. Although French anti-British rhetoric increased exponentially on 18 Vendémiaire, Year II (October 9, 1793), when the National Convention decreed that all British citizens were to be interned, British goods confiscated, and the English language obliterated from public use, the reality was that the trade restrictions between the two nations were effectively rescinded on 30 Prairial (June 18, 1794). The Committee of Public Safety and its subordinates made efficient use of external means during the Year II, but the preservation of the Revolution depended primarily on domestic resources.[14]

The first requirement of the Revolutionary government was manpower for the army, for defense work, to staff the military supply network, and for the harvest. In the course of a prolonged debate sparked by petitions from the sans-culottes through the medium of the sections (wards) of the city of Paris on how to organize national defense, Barère reminded the National Convention on August 23 that, with only 9,000 muskets in the Paris arsenal and almost no timely means of procuring more, the country was in need of more than just soldiers: "All France must be stirred up against the tyrants [attacking France]; but only some citizens must leave [for the frontiers]. Thus all are requisitioned, but not all shall march. Some will make weapons, others will make use of them. It isn't enough to have men, men will never be lacking for the defense of the republic. Weapons! Weapons and supplies! The resolution of these needs must be the constant object of our actions. . . ."[15] The legislature's response to Barère's rhetoric was to adopt the Committee's call for a *levée en masse*. All men were drafted into national service, and unmarried men and childless widowers between the ages of 18 and 25 were liable for military service. The rest were to manufacture arms and supplies, to work in transport, or to help provide food. This was total war. Women were to make clothes and work in hospitals; children were to roll bandages; older men were to exhort them to greater efforts. All horses and public buildings were also requisitioned, and the Committee was given nearly unlimited powers to win the war.

An initial 30 million *livres* (soon raised to 100 million) was voted to be spent on arms production. With the main arsenals either menaced by the enemy (Charleville and Klingenthal), already in enemy hands (Maubeuge), or controlled by the Federalists (Saint-Étienne), Paris was to become the center of arms production for the republic, manufacturing a projected 1,000 muskets a day.[16] This was a purely political decision intended to provide 6,000 jobs to sans-culottes, thereby rewarding the popular movement in the capital. The Committee envisioned turning the artifacts of vanished royalty into symbolic sites of patriotic production. The palaces of the Luxembourg and the Tuileries along with the Place de la Révolution where Louis XVI had been guillotined were to be festooned (collectively) with 250 forges. Sixteen public arms workshops would be established. Ten large forges situated on barges moored in the Seine would use water-powered lathes and millstones to grind and bore artillery pieces. In terms of economic efficiency and technological capacity, however, initiating large-scale arms manufacturing in Paris was a grave mistake.[17]

Despite the hopes and plans of the Committee of Public Safety and other policy makers, Paris was poorly situated to become the "arsenal of democracy." As the central government had learned from its attempts to manufacture 40,000 pikes and javelins in the capital, launching this sort and scale of production from scratch was an organizational nightmare. The basic raw materials, particularly iron and coal, had to be brought long distances, which distorted long-standing commercial patterns already in flux because the transport system was in disarray from civil war and was focused on getting food to the armies and to Paris. When the distant mines and the overstrained transport system could not deliver enough raw materials, especially iron, homes of émigrés, former churches, and former royal buildings were stripped of any "unnecessary iron," which was turned over to arms manufactories along with the coal stored in the caves under the Abbey of Saint-Germain.[18] Skilled labor was also seriously lacking. Although Paris had a number of master armorers and/or gunsmiths, these men were accustomed to making high-quality pieces. They were unwilling to switch to high-volume production.

The Committee delineated the scope of the problem. On 10 Brumaire, Year II (October 31, 1793), they declared that the goal of 1,000 muskets a day required 250 gunsmiths, and despaired that there were only 22 in Paris. Workers fleeing from Maubeuge made up a useful cadre, as did a group of gunsmiths from Liège who offered their services to the Republic. Some went to Charleville and others came to Paris. This infusion of trained armorers numbered 1,200. Pike production was stopped and the workers reassigned. To encourage the gunsmiths in public workshops to train more workers, a decree awarded a bounty of 100 *livres* to master gunsmiths for each of their students who finished training. In December, master armorers were also obliged to train students, but their less relevant experience and more traditional training methods based on the "outdated" *corporate* model were considered less worthy of recompense: they received only 50 livres per student. Each student underwent a two-day trial, and a student's capacity had to be certified by an expert before the bounty would be paid.[19]

At first, the relationship between the Revolutionary government and the arms workers of Paris was based on cooperation. To implement the decrees of August 23, even before the creation of the General Maximum, the Committee of Public Safety initiated an extraordinary means of setting wages that foreshadowed the wartime arbitration boards of the twentieth century.[20] Each section choose deputies who then met with representatives of the workers from Maubeuge, the city, the départe-ment, and the central government's Arms Administration to determine a just daily wage for every task involved in arms manufacture. The first two of these assemblies accomplished nothing because of a lack of detailed knowledge of the work process and the tasks involved in arms making. Recognizing the isolation of the worker delegates among their fellows, the Committee of Public Safety asked each of the 48 sections to choose six delegates, who met collectively to select 24 emissaries to negotiate on their behalf. The Committee sent the chemist Jean-Henri Hassenfratz and a *conventionnel* (the former gunsmith Noël Pointe) to provide expertise and lend authority.

At the meetings, held on 24–27 Brumaire (November 14–17), mili-tary necessities were once again subordinated to political realities. Administrators and experts preferred piece rates. They also suggested creating short de facto apprenticeships to explain the intricacies of mus-ket making to metal workers (even experienced ones). Due to pressure from the sections, these recommendations were rejected in favor of high daily wages. The workers rejected apprenticeships unless there was some sort of additional financial support: heads of families needed their full salary in a time of scarcity. It was also decided that the Republic would furnish tools, raw materials, and a place to work. Political expedience even protected inefficiency: workers who made mistakes in the manufac-turing process were not charged for the supplemental raw materials needed to fix them.

The cause of efficiency made some gains. All work was inspected before being accepted. As a means of quality control, workers had to mark each gunlock or other component they made. The standard of pro-duction was set high: a worker was expected to make five gun barrels every two days, for which he was paid 40 *livres*—nearly twice the pre-war wage. In the name of uniformity, higher wages were denied for those who worked at home. The outrageous salary demands of the workers from Maubeuge were rejected; they were so exorbitant that "Citizen Olivier" was censured for pressing his demands too hard. The agreed-upon schedule of wages exceeded that dictated by the Maximum, but the unique skills of the arms workers and Paris' devotion to the war effort made that the exception acceptable to the population of the capital. Although piece rates did not become as generalized as the directors of the Small Arms Administration hoped, the issue would be revisited in three months. At least temporarily, such measures enabled the Arms Manufacture of Paris to ratchet up production.[21]

Eventually the Committee, led in this domain as in most other facets of military supply by former military engineer Claude-Antoine Prieur-Duvernois (known as Prieur de la Côte-d'Or), turned to even more drastic, less accommodating measures that foreshadowed the strictures of a "command economy." The Committee ordered each district to send three laborers experienced in working metal "either with a file or at the forge" (with a strong preference for the former) to Paris within five days of receipt of the orders. The workers were to come on foot, "like all good defenders of the fatherland," but would receive food and lodging at public expense along the way. This was not to be a tour de France sponsored by the Revolutionary state: local administrators were to put a reporting date on their letters of introduction based on their estimate of how long it would take to walk to the capital. Clearly, the Committee of Public Safety "was taking every possible measure to multiply the manufacture of arms." Ten months after the capital was ordered to become a center of arms production, it was observed that "Paris . . . needs a large number of [skilled] workers." The members of the Committee were not only willing but forced to raid the provinces to get them.[22]

This demand for skilled labor and the multitude of small workshops executing subcontracts permitted workers to switch jobs constantly in search of higher wages or better conditions or to find a new position after being let go—usually for incapacity to execute this type of metalwork. On 23 Frimaire, Year II (December 13, 1793), the Committee enacted stricter controls over the arms workers, forbidding them to change jobs without the express consent of a supervisor.[23] Pirating workers from another shop or putting qualified workers to a non-arms-related task was made punishable by two years in irons.[24]

Chaos reigned on the shop floor, exacerbated by fear that spies and/or foreign intriguers would "excite disorder, hold up work, cause workers to waste time, give rise to [worker] agitation and inflame passions by making adroit use of individual interests." The Committee's response was to assert ever-stricter control over the arms workers. Wages were standardized by task throughout Paris, no matter the experience level of the worker. The mix between piece rates and daily wage rates for each workshop was to be adjusted every 10 days by an overseer. Those paid daily had to remain in that workshop for 30 days and adhere to the posted hours of the establishment (6 A.M. to 8 P.M., with hour breaks at 10 A.M. and 3 P.M.) instead of coming and going as they pleased. Although this measure would diminish the political involvement of these men, the Committee was serious: delinquent workers were to be sent before a police court, a frightening prospect when "Terror is the order of the day." All coalitions or assemblies of workers were forbidden, nor was there to be any contact between laborers in different workshops except with the express permission of all the heads of all the shops involved.

Civisme ("good citizenship") was made a requirement for workers, but in this case civisme was defined as "integrity and the tendency to be orderly and tranquil." Those who could not consistently display these virtues, so far removed from those needed for Revolutionary political action, would be fired even if they were experienced and productive gunsmiths. Complaints about conditions or management were acceptable and genuine grievances would be heard and rectified, said the Committee, but only if the objection went through proper channels. Action in the streets to put pressure on arms administrators was made punishable by law, either by entrepreneurs or by workers. "The directors or managers . . . of each workshop . . . were also made responsible for all occurrences in the shop and will be punished if they do not anticipate and prevent them."[25] From the perspective of the ancien régime, these measures were a Foucauldian pipedream, but for the laboring classes, ensuring the survival of the Revolution, with its participatory principles and material advantages, made such draconian discipline temporarily acceptable.[26]

The Committee of Public Safety, like other regimes both before and after, could made decrees, but were they followed? Outside rebellious areas, this question seems to be generally ignored by the historical literature which assumes that the combination of the Terror and the war crisis worked a kind of alchemical magic that got the French to do what the government told them to do. Compliance explains how France survived the onslaught of the seemingly unbeatable First Coalition. Making 1,000 muskets a day required enormous quantities of raw materials, the skilled labor to work them, a reliable transport system, and an effective administration to coordinate production and to ensure that what was produced was usable on the battlefield. Improving the reliability of transport, securing the necessary flow of raw materials, and finding additional skilled laborers were almost intractable problems[27] that the Committee could not attack directly; they were delegated to subsidiary commissions. Most tasks were delegated to deputies of the National Convention, including members of the Committee who were sent out to the provinces as representatives. In Paris, the Committee itself exercised oversight.

For many months, Paris produced neither enough nor very good quality weapons. In early November, Carnot reported that to date only six muskets had been fabricated completely in Paris and that most of the 200 muskets coming out of Paris daily had, in fact, only been repaired.[28] The Committee placed some of the blame on administrators who played favorites with workers, did not enforce work standards, refused to expand forges or workshops, and skimped on the materials for needed improvements to factories, forges, and workshops. Several entrepreneurs given the task of constructing either buildings or machines for arms production had to be jailed for negligence. In the middle of Nivôse (January 1794), another group of subcontractors were imprisoned temporarily for not delivering the required number of gun parts. On 7 Thermidor (July 25), after 3,000 gun barrels were rejected by quality control, the Council for the Administration of Arms wrote: "We cannot hide that most of the [production] agreements contracted . . . [by previous administrations] are illusory, except for objects that are not very difficult to make." These problems were serious enough that an additional deputy—François Legendre de la Nièvre—was detailed to oversee construction.[29]

Labor unruliness, particularly among skilled workers, plagued all the manufactures of Paris. The workers at the gunpowder factories at Grenelle and Saint-Germain-des-Prés were so fractious that they had to be overseen by a deputy of the National Convention.[30] On 22 Frimaire (December 12), the Committee issued a sort of factory code that defined a 14-hour working day with only every tenth day off, established fines for late workers and again forbade any form of coalition. Output was to be inspected, and managers were made responsible for the behavior of their men.[31] In arms making, these problems were even more intractable because of the chronic shortage of skilled labor.

The Périer brothers (Jacques-Constantin and Claude-Augustin) made bronze cannon and some naval artillery at their famous metalworking establishment at Chaillot. They also had a plant on the Isle de Louviers in the Seine where a steam engine was used to grind gun barrels. The arms workers insisted that although they had been drafted by the government to make weapons, their "patriotism" and their "skill" merited wages far beyond those set by the Maximum. The Périers granted the wage increases, but the workers continued their disorderly ways, taking little care in their work and leaving the worksite at will. In Floréal (May 1794), the Committee wanted to restore worker discipline and improve productivity, not only at the Périers' establishment, but more generally. The earlier regulations were taken a giant step further. With the express approval of the Committee, the Périers decreed that the door to every workshop would close at 6 A.M. and five minutes after each of the two scheduled meal breaks. No one would be allowed to enter once the door was shut, and therefore a worker would not be paid if he was not in his place. Laborers were not permitted to leave the worksite between 6 A.M. and 8 P.M. Subcontractors who suspended work for more than 24 hours or who did not fulfill their allotment in a timely fashion would not be paid for any work at all, even work already completed. The Périers claimed and the Committee agreed that these measures were necessary "to re-establish the order indispensable to the maintenance of workshops executing such important work and to recall the workers to their duty."[32]

Arms workers in Paris did not accept either the Périers' or the Committee's vision of their "duty." The workshop on the rue Saint-Dominique was beset by an "insurrectionary movement against the overseers." The Committee responded by inviting the workers to take up their tools "with courage and quickness." If they did not, the leaders were to be arrested by the revolutionary committee of the section and declared traitors to the Revolution. The skilled armorers from Maubeuge and Liège led an insurrection over wages in Frimaire, Year II (December 1793) at the Capucin workshop. Even after six worker-elected foremen were arrested, unrest continued. On Christmas eve (14 Frimaire), the workers refused to stay until 8 P.M. Although the director of the Marché-aux-Poissons gunlock workshop was jailed because the workers mutinied, more stringent oversight could not prevent another mutiny from breaking out in the shop in the Maison d'Aine on 16 Frimaire (December 26). In a manner reminiscent of the activism of 1789–1791, five other workshops convened general assemblies to express their discontent with salaries, the long and increasing hours, and the administration's attempts to control work practices and the job site. Rumors spread that a general strike would be called in Paris' Arms Manufacture.[33] Alder connected this agitation to the Enragés/Hébertist movement, but, even though they did not resort to machine-breaking, arms workers were clearly unwilling to accept the discipline of the state, no matter how pressing the war crisis. Nor did the arms workers learn any lessons from the destruction of the Enragés/Hébertists. At the height of the Great Terror, a number of experienced but young gunsmiths who had been held back from the military draft were incarcerated for "insubordination." They asked for higher wages. Only after a month of detention were the leaders of the "cabal" centered on the rue des Carmes near the Place Maubert judged to have learned their lesson.[34]

After 9 Thermidor, agitation in the arms workshops was just as common and was clamped down just as strictly as it had been before. Carnot and Prieur remained the driving forces, providing continuity in the political direction of military production and supply. When they were replaced, it was by members of the same technocratic elite: Louis-Bernard Guyton(-Morveau), who was Prieur's uncle, and the chemist Antoine Fourcroy, the director of the Commission of Powder and Saltpeter. Although some of the most draconian measures enforcing workplace discipline were tempered and a number of armorers released from prison, the new Maximum issued on 22 Thermidor (August 9) that increased the wages for many trades by about half specifically excluded the already well-paid arms workers. In Frimaire, Year III (December 1794), when production was scaled back, the protests of arms workers grew even more vocal. In addition to protesting the early start to the workday, worker-delegates from various workshops complained that high production costs stemmed from inept managers who did not understand the difference between major and minor flaws and attempted to impose outdated and unfair ancien régime work standards. Without the pressure of the popular movement and the exigency of the war crisis, the legislature refused to give in to the arms workers' demands. Hoping to save their jobs, 400 arms workers demonstrated on 20 Frimaire, Year III (December 10, 1794). A larger demonstration, involving perhaps 6,000 workers, occurred two days later.[35]

Despite labor unrest and structural impediments, Paris emerged as an impressive arms manufactory. According to the oft-cited culminating report delivered by Guyton-Morveau on 14 Pluviôse, Year III (February 3, 1795), Paris had produced 145,630 pistols or muskets in the preceding 13 months at a cost of 1.8 million *livres*. This was more than the other seven major arms manufacturing centers put together where an additional 2 million *livres* had been spent. At its peak, Paris fabricated 600 gun barrels and 190 gunlocks daily. Artillery production, both for the navy and for the army, had also blossomed. Paris' four foundries had cast 1,500 bronze cannon in the same period. At least 5,400 workers labored in at least 39 different public workshops, which did not include the plethora of subcontractors' private workshops and individual homes where accessories were made.[36]

Although these notable successes had helped the Republic not only to survive but to return to the offensive, Guyton-Morveau recognized that production in Paris remained problematic. Not only had the goal of 1,000 muskets a day never been achieved; chronic shortfalls in certain vital components (especially gunlocks) prevented rapid expansion. Thanks to the advance of French forces, the gunsmiths from Maubeuge were headed home, depriving Paris of its best, most efficient workers. The experienced metal workers of Paris, "those skilled with file or hammer," also wanted to return to their professions. Hundreds had already done so. The Committee was also concerned about ongoing agitation in the arms workshops. Unstated, but not unknown, was the fact that the average cost of making a musket was 60 *livres* in Paris and only 43 *livres* in Charleville. A bayonet made for 15 *livres* in Paris could be made for 4 *livres* in the provinces. Alder computed that the cost of each gunlock made in Paris was ten times that achieved under the late ancien régime and that the gunlocks were made at a mere one-fifth the speed while also being of poorer quality.[37] Since most of the Parisian arms workers were paid a generous daily wage rather than by the piece, the Committee believed that with the Revolution no longer in danger of being destroyed, arms production could be reduced dramatically and could be shifted to sites closer to the sources of the necessary raw materials. "The moment has come when old and new manufactures in the départements will receive the fruit of the apprenticeships undertaken in the Manufacture of Paris," stated Guyton-Morveau. Eight hundred trained workers would be transferred to the provinces. From 5,400 workers employed in state workshops in Frimaire, Year III (December 1794), the Committee planned to maintain only one state-run workshop for each component of a musket and/or a pistol and to employ 1,146 workers, each of whom would be paid according to what he produced. This was not a "dismantling" of Paris' Revolutionary arms-manufacture (as Alder termed it) so much as a privatization. Twenty-four hundred state arms workers were to work for the former masters of Paris' *corporation des arquebusiers* and an additional 1,200 were transferred to private subcontractors of the state-run Arms Manufacture. Although concern over labor militancy in Paris certainly played a role in these decisions, the Convention was also returning to the market-oriented economic principles that so dominated policy making during the early years of the Revolution.[38] The measure reorganizing arms production in Paris concluded: "The manufacture of muskets will not be slowed down or diminished, but there will be fewer workers employed at it. . . . This will result in incalculable savings."[39] Nor did this attitude end there. When the district of Autun petitioned the Committee to take over the arms manufacture there, the response of 8 Pluviôse, Year II (January 27, 1794), crafted by Prieur and Carnot, stated: "The interests of the Republic are in opposition [to taking over the manufacture]. The interest of the establishment. . . . requires a measure contrary to what you propose. . . . The Committee has only one principle in this regard: no workshop ought to be run for the profit of the Republic: all of them ought to be run under contract [by a private entrepreneur]."[40]

Although Paris-based production made an intentionally impressive display, it was not the only arms-production center of the new French republic. If much of the arms production in Paris was crisis-oriented, in the provinces more attention was paid to improving technological capacity and training and to sustaining production. In part, this emphasis was the result of a decision taken in August 1792 to allow private entrepreneurs to run most of the provincial, nationally organized arms manufactures.[41] The impetus for a more far-sighted approach generally came from the men on the scene, the representatives-on-mission (see below), but the Committee of Public Safety also played an important role in fostering innovation.

In *Engineering the Revolution* (1997), Alder followed up an observation made by Charles Ballot in 1922 on the importance of the activities of Honoré Blanc. Alder focused on Blanc's promotion of an innovative system of interchangeable parts in the manufacture of gunlocks for muskets in the years immediately preceding the outbreak of the Revolution. When the long-standing state patronage from the Ministry of War dried up with the change in regimes, Blanc attempted to convince a new set of military policy makers of the possibility and advisability of interchangeability through a demonstration held in November 1790. Blanc's achievement of limited interchangeability was endorsed by the Academy of Science in October 1791. The Legislative Assembly, however, chose not to fund Blanc's proposals. For reasons that are not apparent, Alder emphasized that there was no economic reasoning behind this rejection, yet he himself admitted that Blanc's gunlocks cost 30 percent more than those made in Saint-Étienne and twice as much as those made at Charleville or Maubeuge.

Instead of considering cost effectiveness or economic rationality in his discussion of why Blanc's methods were not given further state support, Alder highlighted the potentially dangerous "*social* effect of substituting cheap labor for skilled artisans in the highly charged climate of Revolutionary France." But even after Blanc shifted his efforts from getting the state to adopt his productive system to setting up a private arms manufacture, and after he received official authorization to do so (in April 1792), he did nothing. After nearly two years of dithering, Blanc purchased a former convent in Roanne. But he did not leave Paris until March 1794, when Prieur explicitly ordered him to leave the capital in the hope that he would finally use his talents to benefit the Republic. From Roanne, Blanc complained repeatedly about recalcitrant and untrained workers, insufficient tools, and persistent shortages of food and raw materials. By Frimaire, Year III (December 1794), 150,000 *livres* specifically authorized by the Committee of Public Safety had been spent on this manufactory, yet Blanc's gunlocks had not progressed past the testing phase. Ultimately, Blanc did begin production, but not until 1796. The plant ran until 1807, when Blanc died. Contemporaries were interested in protecting the revolution with well-made weapons and in demonstrating that the added costs of investing in interchangeable parts were worth it. From both perspectives, Blanc's performance must be counted a failure. It is little wonder that no other contemporary arms maker adopted his industrial methods or his approach to manufacturing.[42]

The Committee's greatest influence on arms making was in matters of supply, broadly conceived. In the provinces, finding qualified personnel to administer a shop or to labor in arms production was an overwhelming task. The Republic was desperately short of large-caliber naval cannon. Metal was found by taking the radical step of ordering several ports to melt down their clocks, but the Committee of Public Safety could not find anyone who knew the craft of forging such large weapons to run the proposed arsenals in either Bordeaux or Toulouse.[43]

The Maximums played a major role in stabilizing the French economy so that the newly marshaled battalions of the Republic could defeat both rebels and invaders. From a logistical standpoint, however, it was a nightmare of localized contradictions and loopholes. With prices set locally at 10 percent above 1790 prices, district administrations could create enormous imbalances of supply and demand. Industrial areas, which tended not to grow enough food to feed the native population much less an influx of arms workers, were hardest hit by market distortions. To cope with these problems, the Committee ordered the military supply network to send the 350 arms workers at Klingenthal full rations of bread and of fresh and salted meat at the levels set by the Maximum: they could not acquire the food for themselves at any price. Shortages also forced Thiers, Indret, and Moison to use field rations for arms workers. André Jeanbon of the Committee of Public Safety gave naval rations to workers at the Arsenal of Toulon. Even in Paris, some foundry workers, notably at the Arsenal and at the Barrier d'Enfer, received military rations.

Because of the poor quality of the wrought iron and the pig iron being delivered to Le Creusot for weapons production, the Committee explicitly ordered a large group of iron masters to furnish their *best* product at the prices required by the Maximum. National agents were commanded to furnish the foundry with food and raw materials of the necessary quality or face all the harsh penalties stipulated by the Terror. Uncertainty over the sources of faulty arms production in Paris led the Committee to ask the workers of Paris to inform them directly of the specifics regarding "great number of vices stemming from the speed of creation [of the manufacture there], the evil-minded [workers] and the unfaithful administrators . . . who hinder the fabrication of arms."[44]

Shortfalls in quality were a major problem under the Maximum. During the spring and summer of the Year II (1794), local administrators from all over France reported to the Committee that "cloth production had experienced a noticeable deterioration" in quality. Led here by Robert Lindet, the Committee feared that "poor quality renders the Maximum illusory." Premium prices were being paid for inferior grades of cloth that wore out rapidly and had to be replaced. The Committee restored a regulatory form used under the ancien régime, the *marque d'étoffes*. Each bolt of cloth had to display a thread count, and the manufacturer had to affix a mark indicating where the fabric was made and by whom. Merchants were forbidden to remove the mark. If the trademark was fraudulent, the manufacturer had to pay a fine of 100–500 *livres*. Under the watchful eye of the Committee of Public Safety, and enforced by local administrators backed by the violence apparatus of the Terror, neither workers nor manufacturers were permitted to slack off on quality to take advantage of profitable opportunities created by the Maximum. Fear of the guillotine and the realities of the war crisis protected the vulnerable machinery.[45]

The Committee recognized that the strictures of the Maximum could limit war production. District administrators did not necessarily understand or care about the cost of supplying large foundries or making quality weapons. The entrepreneur of Le Creusot, Ramus, and his compatriot at Indret, another considerable foundry, informed the Committee that the steam engines used by these plants required coal to be brought long distances and that large naval cannon required double smelting adding to fuel costs. Ramus could produce cannon at the price he had negotiated with the Ministry of Defense in 1792, but not at the Maximum set by the district administration of Autun. The Committee allowed the Arms Commission to disregard the limitations of the Maximum and return these foundries' payment rates to their pre-Maximum levels. Almost every ironmonger for dozens of square kilometers was also ordered to send not just his best but his entire output to Le Creusot. Where patriotism did not suffice, the incentive provided by regular rations was amplified by the threat of denunciation to a Revolutionary Tribunal.[46]

Dispensations were not accorded solely to those involved in arms production. At the height of the Great Terror, supply problems similar to those of the naval cannon foundries prompted Carnot, Prieur, and Lindet to suspend the Maximum for the metal workers in the toy industry of the Commune des armes (the temporary name of Saint-Étienne). The Committee also accorded prices 10 percent above the Maximum to the makers, merchants, and sellers of twisted yarn from Lille. That dispensation was then applied throughout the Republic. Such flexibility on the part of French administrators from the top down facilitated the operation of this nascent command economy, enabling France to defeat its multiple enemies.[47]

This flexibility could not mask major French industrial deficiencies, notably in the production of steel.[48] This deficiency involved both quantity and quality. The Committee orchestrated a multi-faceted campaign to improve and increase French steel making. In September 1793, they recognized that "in ordinary times, France could not furnish all the steel needed for the fabrication of arms. It would be impossible that it furnish that much now or to increase output significantly." The first step was to improve quality. To that end, the Minister of War was ordered to commission three scientific and/or industrial technocrats to write a how-to manual. Mathematician and former Naval Minister Gaspard Monge, chemist and former dyeing director of the Gobelins Claude Berthollet, and mathematician Alexandre Vandermonde had all been members of the ancien régime Academy of Science; now they applied their scientific skills to solving the Republic's industrial problems.[49] Monge received a special dispensation from other pressing duties which signaled how important this project was to the Committee. Fifteen thousand copies of the resulting 34-page pamphlet depicting how to make both "natural" and cementation steel were distributed in 1794–95.[50]

Carnot and Prieur recognized the limitations of the utility of learned accounts in producing something as complicated as steel. They knew that craft practice was often rather different than what theory or so-called best practice might suggest. This perspective had been confirmed by the Arms Jury which discovered that productive practice in Saint-Étienne differed significantly from the methods used either in Paris or in Charleville. Thanks to a decree of 9 Pluviôse (January 28, 1794), Hassenfratz and a group of engineering students went to the various workshops and drew detailed engravings of tools and machines related to arms manufacture. Hassenfratz and Vandermonde were ordered to report on "advantageous improvements that should be introduced into the manufacture of arms, either in using machines or in changes hand-processes"; this report was to be used to organize an Atelier de perfectionnement (improvement workshop) in Paris, which operated from 1794 to 1796. It was soon put under the supervision of the directors of what became the Conservatoire des arts et métiers.[51]

The findings of these savants substantiated the need to inspect provincial manufacturing processes. Carnot and Prieur were particularly concerned with finding out how well the directions given by Monge, Berthollet and Vandermonde worked. In Ventôse, Year II (March 1794), a Mining Engineer named Anfry was sent to the départements of Allier, Nièvre, and Cher, where "natural" steel was made. Anfry was "to go himself to each steel foundry and have a sample made using normal techniques and then he was to have them make a sample according to the perfected practices [recommended in the *Avis*] and send these two samples to the Committee." He was to note how much each process cost and what difficulties the workers had with the perfected processes. An engineer named Duhamel was sent on a similar mission to the Isère.[52]

Carnot and Guyton considered the possibility that certain craft methods might be superior to the "best practice" established in the guidelines sent out to the provinces. A father-and-son duo of file makers in the arsenal at Tours named Dinaron had made steel using the new methods, but believed that they could make better and cheaper steel using a process they had developed themselves. They built a furnace at their own expense and sent a sample of their steel and some files and a sword of their manufacture. When tested, the steel was found to be of good quality although the tools and weapons had small but "easily correctable" flaws. The Committee sent these illiterate workers 300 *livres* to continue their experiments and to teach their methods to others.[53]

This direct approach to improving the state of technical knowledge on steel making was complemented by creating new centers of steel production. The poor quality of the output of the existing steel foundries of Souppes (Seine-et-Marne) and especially Amboise (Indre-et-Loire) despite large outlays by the central state magnified the need to increase capacity.[54] The Committee acknowledged that even the most patriotic entrepreneurs and workers could not simply be ordered to begin making steel, even if there were directions to follow. Led by Carnot, Prieur, and later Guyton, the Committee recognized that steel making was as much craft as scientific process and required experience with the industrial use of heat energy that was rather uncommon in France.[55] As a result, in places where high-quality iron was mined, the Committee initiated a special program for entrepreneurs who wanted to set up a steel foundry. The Committee paid for entrepreneurs and foremen to visit places such as Rives where they could watch the steel suitable for military needs produced.[56] For those arms workers who could not leave the job site or who lived far from other centers of steel production, the Committee ordered a Commissioner to demonstrate how to make natural steel in the southern départements, notably the Arriège, Aude, and Haute-Garonne. An experienced worker accompanied the Commissioner to illustrate craft practices or other information that was difficult to convey in written form. The Committee established two national steel foundries: one at Angers and one at Drigny, near Sedan, the latter directed by Jean-Baptiste Clouet, a former abbé and former member of the Academy of Science who was also a former professor at the Royal Mining School. A school of steel making staffed by German workers was also founded at Miremont in the Dordogne.[57]

Once the immediate military crisis passed, experienced steel makers particularly those who had worked for the state were encouraged to establish their own foundries as a way of building up capacity and ensuring that France would never again be tributary to Britain and the German lands. Esser and Vicary, steel workers and file makers at the arsenal in Allais, were loaned 40,000 *livres* to strike out on their own. In Toulouse, military production had so monopolized supplies of cementation steel that it was not regulated by the Maximum.[58] With local supply increasing beyond military needs, Lafforgne the younger and Baqueris had to ask for a price to be set so they could legally sell their steel to instrument makers and tool makers. The Committee gave 10,000 *livres* to Berard with no strings attached. This proven producer was to expand his steel manufacture located at Chantemerle in the Haute-Alpes. Constant Lenormand established a foundry at Gravilleau to make cementation and cast steel. Jean LeComte and Company were ceded space rent-free for 27 years in the former royal military college of Brienne (once attended by Napoleon Bonaparte) in the Aube to construct a foundry to make cementation steel and to manufacture files, other tools, and nails. LeComte and Company were also given all the tools used in this former arms manufacture, and the Arms Commission was directed to ensure that this fledgling firm could get enough quality iron to perfect its techniques and establish its manufacture on a firm basis. When the Taxation Commission protested and delayed implementing these generous terms, the Committee of Public Safety insisted on immediate implementation on the grounds that improving and increasing the production of cementation steel was "an absolute necessity." Sponsored and supported directly by the central government in Paris, steel making overseen by military engineers commenced at a host of new sites: at Bizy and at Pont-Saint-Ours, both near Nevers in the Nièvre (using the expertise of three English prisoners of war), in the Côtes-du-Nord, Côte-d'Or, Doubs, Eure, Gard, Indre-et-Loire, Haute-Marne, Haut-Rhin, Lot-et-Garonne, Manche, Mont-Terrible, Saône-et-Loire, Seine-Inférieure (using German prisoners of war), Var, and Vosges. In the Ardennes and in several places in the Doubs, state aid also supported the concentration of several forges or blast furnaces under the direction of owners who had proved to be quality producers. Existing facilities (most notable among them Le Creusot) received large subsidies, and provincial officials were forced to neglect other duties and distort rational and normal commerce in raw materials in order to keep coke production high and cannon production at four per day.[59]

Once Revolutionary forces regained the offensive, the resources of the conquered territories were diverted to French military production. Agents of the Subsistence Commission were told to purchase as much "natural" steel in the Palatinate (the Rhineland) as they could. Conscious that the fortunes of war could, once again, shift suddenly, these agents were told to send the steel to the Strasbourg Arsenal quickly; it could be distributed from there to arms makers, who could use it to make cavalry sabers and bayonets.

Extensive use was made of Belgian resources. Arms-making equipment, raw iron, potash, and bells were requisitioned in French-controlled areas. Belgian entrepreneurs were encouraged to speculate in these products at prices set by the Maximum. Steel production resumed quickly around Namur, and coal was soon mined around Mons and Charleroi. Cotton-spinning machines either made in England or built on the English model were refurbished and run by "extraction agents" of the Commerce Commission for the profit of the Republic. By Thermidor, Year II (July 1794), Belgium was filling a significant and increasing portion of France's needs for these critical industrial materials. Belgium's contribution to French military supplies necessitated that a branch of the Agency of Arms and Powders be established there. After the battle of Fleurus, the gunsmiths from Liège agitated to return home. Only seven weeks after the fall of Robespierre, the Committee give the Liègois contracts to make muskets. Belgium also manufactured gunpowder for the Republic.[60]

The Committee and its subsidiary administrations expected the conquered territories to develop their strengths in the production of needed tools and of semi-finished and manufactured goods to benefit the Republic. It is often overlooked that Switzerland was important to this process. In the canton of Zürich, a tinkering metal worker named Goldschmid had invented a method of milling gunpowder that had earned him compensation from the Arms Jury. Goldschmid hoped to recruit proven metal workers in Geneva, where he had lived for 15 years, and to found a file manufacture on French soil. He asked for no compensation for himself and claimed that "these workers had a genuine desire to live in the land of liberty and equality and to get them to move, the government had only to assure them of their support and the intention to repay relocation costs and for them to make the needed tools." The Committee decided to send these Swiss workers to Annecy (Mont-Blanc), where the iron was of high quality and where they could make use of fast-running mountain streams to power their machines. They were domiciled in the homes of émigré priests. Experienced workers received generous daily wages plus the promise of substantial further payments based on output. Supervised by the National Convention's Committee on Agriculture and the Arts and supported by the French agent in Geneva, Goldschmid wooed the needed workers. Those workers who made the short journey to Annecy remarked that they needed steel to be able to make this sort of file. Goldschmid promptly found two Swiss foundry workers who knew how to make cementation steel and arranged to bring them and all their equipment to Annecy. The Committee diverted local iron ore from exclusive use for the navy. Initial testing of the steel made by the Swiss in Annecy found it "of the best quality." This promising start was not followed up during the economic confusion of the Directory (1795–1799), and the venture ultimately failed.[61]

The conquered territories were expected to transfer not only workers but also technological expertise to France. In Frimaire, Year III (November 1794), the Committee of Public Safety directed the Committee of Agriculture and Arts to create a needle-making workshop in Paris "which would serve as a school for the entire Republic." Since this was a specialty of the newly reconquered region around Aix-la-Chapelle (Aachen), they sent an agent there to find experienced workers. This task proved to be enormously difficult because local manufacturers did not want to lose the lucrative French market and would not sell them tools, especially in exchange for depreciated French currency. Unlike the Swiss referred to above, these workers were reluctant to come to Paris. One reason was that they did not speak French. They also worried about the near-worthlessness of the *assignat* and about the regularity of supply and the price of the high-quality steel they needed. After weeks of traversing countless workshops, the Committee's agent convinced "one of the most able workers who was as well acquainted with making the needed tools" and his two children to work at the proposed school. This worker believed that 50–100 others would follow because of the stagnation of the industry around Aix-la-Chapelle caused by lack of steel. Sent across the Rhine to look for fresh supplies, the agent was given 8,000 francs to purchase the steel needed for instruction. The workshop lasted three years, after which the workers returned to Aix-la-Chapelle, now incorporated into France. Despite the Committee's attempts to spread these techniques, this region remained the premier producer of needles in Continental Europe. Thanks to territorial expansion, Revolutionary France benefited from this area's unique expertise.[62] Because of the scope of the efforts made there and their vast potential resources, the left bank of the Rhine, Belgium, and Switzerland must be included in any attempt to survey industrial development in the Year II.

The measures depicted here continued long-standing French government objectives. The Committee of Public Safety used the opportunities provided by the war to spread technical knowledge, to increase the number of places where steel was made, and to ensure that the raw materials were available to those who knew how to make steel. Yet it is well known that French steel production could not compete technologically with its British counterpart and that France imported steel for generations.

Incremental innovation improved French steel making. Several individuals (including Nicolas Pradier, an arms inspector at the Paris Arsenal) developed refinements of the process of making "natural" steel, using charcoal in the "German" manner in Catalonian-style open-hearth forges. Instituted widely during the Year II thanks to the teaching of several agents of the Committee of Public Safety, although with uneven results, these methods yielded enough steel of sufficient quality to equip the Revolution's armies. A fundamental understanding of the process and of the practices used to make steel proved much more difficult to forge even as informed French observers from the Bureau of Consultation of Arts and Crafts noted the utility of methods like Pradier's for making weapons and tools.[63]

Claude Régnier, a master armorer originally from the Côte-d'Or and one of most successful troubleshooters for the Committee in industrial matters identified the limitations of this type of steel making. He asserted that most French metal workers believed that domestic craft practices were as efficient or effective as those of Germany or England. After conducting a national survey to determine if French steel could be substituted for German in arms making, Régnier noted recent improvements to craft practice, but his praise was circumscribed by this observation: ". . . to continue steady and prompt improvement [of techniques], I propose that continual trials, always time-consuming, bothersome and expensive, be avoided." Instead, he suggested that experienced men from Germany be sent to *show* inefficient producers how to make both natural and cementation steel, because the instruction manual written by Monge, Berthollet, and Vandermonde needed some demonstration to be of practical use. Even if this was done, the French would not be able to produce as efficiently as the Germans, "because these newly established foundries have not yet developed the rapid methods of foreign manufactures." If German workers could teach native workers, as they did at the best French foundries, Régnier hoped that the Revolutionary Republic would soon be able to compete with German producers.[64]

These reports, particularly Régnier's, highlighted some factors limiting French steel making.[65] Workers clung desperately to traditional methods in many important metalworking districts. Heightened by Revolutionary nationalism, more efficient methods taught by German-speakers clashed with customary conceptions of proper work practices, particularly in the isolated metalworking areas of southeastern and central France. Even those establishments that made quality cementation steel with advanced methods suffered from irregularities in the supply of industrial raw materials and foodstuffs stemming from the overstrained and underdeveloped transportation network, erratic payments for their output (often in depreciated or even worthless currency), and an almost complete lack of coke to smelt the iron. With charcoal the only available fuel, technological progress proved chimerical. The almost complete failure of the French state under the Directory to pay its bills to steel makers or to many arms-making establishments discouraged innovation and forced a number of inventive and patriotic producers into bankruptcy.[66] It is worth noting that France did produce the overwhelming bulk of the steel it needed, even during the darkest moments of Revolutionary military fortunes during the Year II. Theoretical understanding of steel making expanded almost exponentially, held back only by lack of craft experience (and sufficient supplies to experiment) with using coal and/or coke as fuel. France soon conquered efficient steel-making districts in Belgium and the Rhineland. The Republic and later the Empire could supply itself without technological improvement. In sum, systematic state encouragement during an era of Revolutionary political ferment could not overcome long-standing French impediments to developing internationally competitive steel production. As with many other facets of technological change, basic knowledge in the hands of those who needed it was not the most profound barrier to be surmounted. In steel making and in other industrial domains, the Committee of Public Safety had seen to that.[67] Understanding how to do something was not the most daunting stumbling block facing French producers: the acquisition of useful knowledge was far outweighed by intractable workers, a lack of economic incentives, insufficient supplies of essential raw materials, and a deficient transportation network.

The Missions of Noël Pointe, Worker-Deputy: Incentives, Supply, and Innovation under the Terror

The Committee of Public Safety masterminded economic planning and arms production during the Year II, but its effectiveness came from the energy and creativity of the representatives on mission. Without the contributions of the representatives on mission—those members of the National Convention delegated to enforce the will of the central government—French war mobilization would have been far less effectual. The role of the representatives on mission (including the notorious Joseph Fouché, Jean-Baptiste Carrier, and Claude Javogues) in the conduct of the provincial Terror has long been the focus of scholarly attention.[68] However, the activities of the representatives on mission in forming and implementing French industrial policy has been ignored. By making use of local knowledge, political savvy, and personal charisma, the deputies solved some knotty production problems by requisitioning supplies, tools, and workers, by demanding transport, and by commanding workers to follow orders or adopt different techniques. Perhaps the most effective industrial troubleshooter of the Years II and III was Noël Pointe.

Pointe (1755–1825) was among the rarest of political animals in late-eighteenth-century France: a genuine member of the working classes who rose to the highest levels of influence outside of Paris. Largely self-educated, he labored as a blacksmith in the arms industry of Saint-Étienne. Pointe, a well-known journeymen, was elected a deputy to the city assembly that wrote the *cahier* in 1789. He soon became the syndic of the journeymen's *corporation*. After the corporation was abolished, he became influential in the city's four Jacobin clubs. Only when the fall of the monarchy eliminated income restrictions on officeholding was he eligible to hold a government post, but it was an impressive beginning: he was elected a deputy for the département of the Rhône-et-Loire to the National Convention in September 1792. Pointe became a member of the Jacobin Club of Paris, sat with the Mountain, and voted for the execution of the king without appeal. During the Federalist revolts, while on a mission to the arms industry that began in late June 1793, he consistently sought to undermine the rebels; he was imprisoned for his trouble. After the fall of Lyon (October 9), he returned to Paris and was almost immediately sent back to Nièvre, Cher, and Allier (October 15, 1793–January 30, 1794). From February until September of 1794, Pointe was on a mission to the Nièvre, Allier, Saône-et-Loire, and Yonne, with only two interludes in Paris. In May, he defended himself against denunciation as a moderate because he had freed 98 prisoners in Nevers. He returned to the capital for the two weeks before 9 Thermidor. Finally, beginning on February 2, 1795, his fourth mission took him to the Ain, Saône-et-Loire, Doubs, and Jura until June. For 14 months, Pointe was responsible for overseeing all facets of arms production, including increasing supplies of essential raw materials and improving the technical skills and knowledge base of the arms workers. His successes and failures while on mission demonstrated the strengths and weaknesses of French technological and industrial policies in 1793–1795.[69]

Better than almost any other Jacobin representative on mission, Pointe understood both the way the productive deck was stacked in favor of those in charge of an enterprise and how much innovation depended on the creativity of the labor force. Upon arriving at the arms manufacture of Moulins in Allier on 27 Brumaire, Year II (November 17, 1793), he first assembled the workers and listened to their complaints seriously, writing many of them down. Only then did he approach the local authorities. They criticized the entrepreneur Brillantais, who then offered to give up the post. Pointe accepted.

Pointe ignored the Maximum on salaries. On 7 Frimaire (November 27), he wrote to the Committee of Public Safety that the complaints of the workers at the naval foundry of Guerigny were legitimate. The district administration had set their wages at 36 *livres* a month (25 *sols*, 8 *deniers* a day). "The Convention did not intend that those who labor for the Republic lack for bread," he noted. "In consequence, I have authorized the naval supply administration to pay them 50 *livres* a month." This, he claimed, was "justice." The same motives led him to allow an even more drastic revision to the Maximum on salaries on 9 Nivôse (December 29). The high cost of food and fuel led the new directors at Guerigny to petition to increase wages again. Most workers were to get 63 *livres*. Those who stoked the fires were to get 50 *livres*, and the *frappeurs* (beaters) were to get 40. The increase was made retroactive to 1 Brumaire.[70]

Even after the Committee of Public Safety reined in the Parisian sans-culottes, Pointe and other representatives on mission continued to increase arms workers' pay. But now, increases were in the form of indemnities, bonuses, and other non-salaried payments, because "the law opposes any increase since they are already paid the *maximum* rate." All who labored at the various workshops being set up in Nevers received two days' extra wages on 29 Messidor (July 17, 1794) as a bonus when the furnace was lit ahead of schedule allowing production to begin early. The experienced carpenters sent from Paris to build the foundry and the metal shops were given an extra 200 *livres* for working so rapidly. On 10 Fructidor (August 27), Pointe awarded an additional two days' wages to the laborers of Cosne in the Nièvre, "to give the workers all the encouragement necessary" to speed the manufacture of anchors and other iron fixtures for the navy. When the entrepreneurs of the manufacture at Guerigny again sought to increase their workers' pay even after the revision of the Maximum in Germinal, they justified it by saying that the Maximum had been calculated for six workdays in four seven-day weeks rather than nine in three ten-day weeks. A further increase was sought for the workers who had to perform tasks in addition to their nominal ones because the workshops did not yet have a full complement of laborers. Experienced iron workers overseeing other laborers could earn up to 100 *livres* a month, and even the lowly *frappeurs* earned a relatively princely 48 *livres*.[71]

These actions were part of a general strategy of giving financial incentives to productive workers in essential areas to bolster morale during times of dearth and defeat. Pointe instituted payment by the piece whenever possible, particularly in newly formed arms manufactures, even though this form of payment was unpopular with the workers who were the basis of his local support. "To accelerate the manufacture of files for gun barrels and improve their quality as a means of ensuring the safety of defenders of the fatherland," he wrote, "encouragement bonuses to the workers in this sector will be the best means of arousing strongly the patriotism of those honored to serve the Republic in this manner." Accordingly, on 11 Nivôse (December 31, 1793), he instituted a program of bonuses in eleven different foundries located in the départements of Nièvre and Cher to be accorded to the team of workers at the tilt hammers where the files were ground and hardened "that had worked the best and most reliably" during the month. The third prize was 20 *livres*. The second prize was 30 *livres*. The first prize, 40 *livres*, would go to those who had made the most files with the fewest rejects. The prize was to be divided equally by the entire team and was to be distributed by Pointe himself on the overseers' recommendations. In Pointe's absence, the departmental directories would award the prizes, which were to be paid from the extraordinary funds voted by the National Convention for arms production.[72]

If the opportunity to earn more money was the carrot, Pointe did not hesitate to resort to the disciplinary stick when the workers' behavior did not meet his expectations of how a patriotic sans-culotte should act. On 22 Germinal (March 11, 1794), he forbade 100 workers at Querigny to go on a hunt slated to last for several days. On the surface, this seemed obvious, but given the problems of supplying food to the region's arms manufactures, this measure was not necessarily simply a matter of "pleasure or personal interest." The fact that the overseer approved the request suggested a subsistence-related interpretation or that this administrator was bullied by his workers. Rather than leave the matter in the hands of this ineffectual manager, Pointe required the national agent to enforce his orders.[73]

As a general rule, Pointe made entrepreneurs responsible for their workers' output as a means of enforcing standards and maintaining quality production, particularly for workers in training. Pointe and most other representatives on mission also held entrepreneurs accountable for finding food for their workers and raw materials to keep their establishments running. The founding statutes for the arms manufacture at Moulins mandated that the entrepreneur had to ensure the stock of raw materials. If production could not be continued because of shortfalls, the entrepreneur still had to pay his workers. He also owed the Republic an indemnity for slowing down essential military production. When Ruffray, the entrepreneur at Charbonnière in the Nièvre, shut down his foundry even though he had received considerable supplies of scarce raw materials, including coal and high-quality iron ore, Pointe responded angrily, clapping this entrepreneur, ironically, in irons, and proclaiming that the only possible cause for this "inactivity" was "a blameworthy malevolence on the part of Ruffray." Eight days later, his temper having cooled, Pointe released Ruffray "considering that the slackening [of production] of his furnaces could be regarded as a lack of experience rather than as wickedness." Ruffray was warned that it was up to him to prove his patriotism by fulfilling defense contracts on time and under budget.[74]

Continued frustration with the entrepreneurs running the various arms manufactures under his supervision led Pointe to appoint Robert the Elder "inspector-general of all the foundries and boring mills existing in or to be created in the département of the Nièvre" on 11 Germinal (March 31). Robert asked for the assistance of Rousset in Paris because he knew this mechanic could construct all the requisite machines and buildings needed for arms production. Pointe enthusiastically supported this request, recommending that the Committee of Public Safety provide funds for Rousset to leave immediately for Nevers. Over the next six months, Pointe entrusted Robert with the authority to requisition laborers and to direct the efforts of apprentice engineers sent from Paris, and even with the training of his own son, the patriotically named Libre Pointe. When Pointe returned to Paris, Robert took over direction of military production in the region.[75]

Pointe complained that the reason why he needed Robert and Rousset was that the apprentice military engineers already sent to him "have, for the most part, little use. These men do not know how to work with either wood or iron and cannot in a mere 15 days wandering through the

workshops of Paris become experienced. . . . I cannot count on them."
"How," he continued, "can one believe that a professor of the German
language, a lace maker, a shoe maker, [and] a printer, etc., etc. can
acquire the experience necessary to oversee manufacturing in so short a
period? . . . I need workers more than managers, but they do not understand how to [do this kind of] work." Pointe did not think these men
were worth their 9 *livres* a day. It would save the Republic money if they
returned to Paris. He sent Teste, the most promising of the apprentice
military engineers, to Montcenis "to acquire the understanding and
experience necessary to become the director of this type of manufacture."[76]

Pointe came to trust some of the military administrator-engineers on
technological issues, but he did not think that innovation was a panacea.
He recognized that new techniques and/or processes often met resistance
or incomprehension from generals and admirals on one side and artisans
and manufacturers on the other.[77] Upon his arrival in the Nièvre, Pointe
rejected the preferred Parisian technological model of using a steam
pump to bore artillery barrels in favor of the traditional water-powered
method used locally for the past 30 years: "Drilling machines powered
by means of steam engines are susceptible to disruption and are quite
expensive. Accidents are frequent and very quickly a cylinder needs
repair, then a balance, etc., etc. The drilling machines at Nevers do not
have to worry about these troubling and inevitable inconveniences."
(The river did not freeze over at Nevers, so water power was available
year-round.) But when Gazeran, a commissaire sent by the Committee of
Public Safety to increase iron production and maximize existing supplies,
wanted to construct a steam engine at Moulins, Pointe considered it, particularly because of the strong support of local authorities. According to
Gazeran, it would take two months to build the engine and rig it to the
drilling machines, at a cost of 15,000 *livres*. Before slowing down production and expending that much money, Pointe took the unusual step
of asking for the approval of the Committee of Public Safety. For Pointe,
tried and true technologies were more appropriate for fabricating the
arms needed during the Revolutionary crisis.[78]

The termination of the Maximum allowed Pointe to become more supportive of entrepreneurs and/or administrators interested in technological change. Colombier, a military engineer serving as the inspector of the

arsenal and foundry of Autun, wanted to build a new boring machine of his own design, but the entrepreneur in charge, Olinet, did not have the funds. Pointe approved Colombier's request to present his plans and supporting letters (including his own and the entrepreneur's) to the Republic's technological decision makers in Paris. The district allocated generous funds for the trip. With the arms budget slashed, Colombier's machine was not built for several years, but not because of Pointe's opposition.[79]

Pointe was more interested in training than in technological improvement. To build up the work force at the new manufacture being created at Moulins, he ordered masters to take on 50 apprentices—"more than had been trained up to that point." The local authorities were charged with requisitioning 30 young men between the ages of 16 and 18 suitable for learning metalwork. Twenty more youths whose fathers worked with steel or iron were also to be selected. Each would be paid 3 livres a day, except for the married men, who would receive 4 livres and 10 sous. Pointe warned: "The workers of Moulins are not those of Paris; the intelligence is not the same, it will take a long time to train arms workers." Seasoning workers was even part of his day-to-day life; the Committee of Public Safety ordered him to take "two young sans-culottes, the children of workers or skilled craftsmen everywhere he went to give them indispensable knowledge of natural history, physics, mechanics and chemistry while illustrating how to conceive and execute the different processes to follow to make gun barrels of cast iron." Such training could only have an effect over the long term, they did not contribute much to Pointe's missions.[80]

The only way that Pointe could get skilled workers to construct the buildings and the machines needed to get arms production underway in the Nièvre was through requisition. Explaining to the Committee of Public Safety why he had commandeered every experienced woodworker or metalworker he encountered, Pointe stated that unilateral and "irrational" decisions made in Paris concerning what was needed outside the capital did not correspond to real problems in the provinces. Even before this, Pointe had resorted to hijacking "all the workers who passed through the city [of Nevers] . . . who can be useful in constructing an establishment where artillery cannons can be made." The city watch was charged with finding workers and bringing them to the director of the arsenal. By requisitioning these workers for only a month, he hoped to avoid significant distortion of the labor market or stalling of essential production.[81]

Pointe seconded Ferry's idea to use prisoners of war for public works in the départements of Nièvre and Loiret, particularly for improvements to canals and deepening the river passages linking the Loire River and Paris. Ensuring that this crucial communications link was not severed by the severe drought plaguing France in the summer of the Year II (1794) was one of Pointe's most difficult tasks. When the boats could not get over the rapids, scarce horses and even scarcer wagons had to be found to transport goods.

A great deal of Pointe's correspondence concerned this sort of requisition and problems of supply in general. Despite the regulations requiring entrepreneurs to provide such things, often they were unable to find them because payments at the Maximum were an insufficient incentive to guarantee supplies. Pointe intervened regularly to ensure the provision of essential raw materials, food, and skilled labor. He also had to rebuff the attempts of other representatives on mission and various agencies in Paris to requisition the workers and resources employed on the projects under his purview. Censuses of what supplies existed locally were conducted to justify refusals to comply with orders, but Pointe held firm, earning the approval of local administrators, entrepreneurs, and workers concerned by such welfare issues.[82]

Pointe's actions were not always so well received. He was censured by the Committee of Public Safety for intervening in local politics, purging departmental administrations, and dispatching 29 former nobles and clergy to be judged by the Revolutionary Tribunal. Pointe justified himself forcefully; he argued that it was necessary to ensure the political dominance of those most dedicated to a definition of the public good that prioritized arms production. Yet these concerns did not explain why he protected a rich and underperforming arms manufacturer in Moulins during the crackdown in Lyon or his increasingly frequent commands at the height of the Great Terror to release prisoners or suspects who had nothing to do with the war economy. What garnered the greatest degree of disapproval from Parisian authorities was Pointe's intervention in religious matters, especially his dogged pursuit of refractory clergy. That Pointe became embroiled in these matters was far from unusual for a representative on mission. That he devoted so much of his attention to arms production was what made him an effective and long-lasting industrial troubleshooter.[83]

On his fourth mission, as the Revolutionary legal and institutional framework for war production unwound ignominiously, Pointe's priorities shifted. Ensuring the availability and continuity of supply for the arms manufactures was his most urgent task. Even Le Creusot was producing only "half of what it ought, and of very poor quality" because of the penury of iron, coal, and foodstuffs. No longer did Pointe blame the entrepreneur for these deficiencies; instead he praised the "intelligence" of Ramus for finding useful ways to keep some workers occupied despite the shortages. But in a request to the district of Autun for food, he worried that the English would remain the "masters of the seas" if the iron workers supplying Le Creusot could not be nourished. Pointe suggested that grain requisitioned for the army be diverted to the workers. He also gave 30,000 *livres* of state money to Ramus to buy food for his workers.[84]

Pointe did not laud many entrepreneurs' honesty or patriotism. He described his mid-Germinal (April 1795) inspection of the small-arms workshop of Deroche at Autun in the harshest possible terms. Most of the 30 workers were 15–18 years old and "did not really know how to hold a tool." Considerable force was needed to prime and load muskets with defective gunlocks or crooked barrels made of poorly tempered iron that he found in stock. He claimed that he "had not seen a single piece without faults." Pointe concluded: "If the Republic is waiting for arms from this supposed arsenal to end the war, we will have to fight a long time yet and perhaps forever." Deroche had also taken advantage of various local administrations. He had convinced those in charge of setting prices for the national lands to drastically undervalue the buildings he wanted for his establishment by as much as 50 percent, thereby depriving the Republic of desperately needed income.

Despite this scathing report, Pointe did not take direct action. A year earlier, Pointe would have sent Deroche before the Revolutionary Tribunal or, at the very least, imprisoned him. Now, because of the White Terror that put the Revolutionaries on the defensive, Pointe did not single out Deroche's faults so much as place them in a broader (potentially explanatory) context. In particular, he noted that "new arms manufactures" had often acquired national lands at half their true value. In a different letter on a related topic, Pointe explained that the chief reason that entrepreneurs believed that their methods were legitimate was the long time lag and frequent default on the payments due them. For Deroche, Pointe's tepid, timid recommendation was that "his acquisition could be annulled along with the agreement to purchase his arms." To protect himself from reprisal or charges of terrorism, Pointe closed as follows: "I will leave you [the National Convention] to decide, I have only given you my opinion." Hamstringing a troubleshooter drastically limited his effectiveness.[85]

Power replaced skilled labor as the most urgent lacking productive factor. With the rivers at low ebb from the heat of the previous summer or frozen over, the demand for coal was urgent in many arms workshops, but nowhere was it more pressing than at Le Creusot. A major reason for the shortfall was that the upkeep of nearby mines had been sorely neglected to maintain production. The low wages and high demands for output established by the Maximum did not encourage timbering or maintenance. Because of military requisitions, mines and manufactories were also chronically short of horses to transport coal. Teams that required two or even three horses to convey heavy loads up and down steep grades often had only one, drastically slowing the transport of needed coal, particularly in winter.

The solutions devised by Pointe on 6 Ventôse, Year III (February 25, 1795) relied on requisition, borrowing from Peter to pay Paul, and an enormous amount of wishful thinking. Two of Le Creusot's prized steam engines were shifted from industrial use to drawing water and raising loads of coal at a local coal mine where new seams of easily exploitable coal were to be quarried as rapidly as possible. All of the plant's remaining horses were diverted to carrying coal to keep the furnaces going. Unspecified and/or imaginary other horses were to be found to haul supplies and supplement the teams carrying coal. Pointe ordered that if any coal mined in the region was suitable for transformation into coke, it was to be reserved for that purpose. Local entrepreneurs ranging from iron masters to military suppliers were ordered to provide large quantities of raw materials. The districts of Châlons-sur-Saône and Autun were to advance 100,000 *livres* to coal-mine operators who had stopped production because they could not pay their workers. Pointe also requisitioned a large number of workers who had been recalled to active duty after the termination of the Maximum because other representatives had not understood how crucial the jobs of wood cutter and charcoal burner were to arms production. Continuing production shortfalls at Le Creusot suggested that Pointe's problem solving provided little relief in the rapidly deteriorating economic situation of the Year III.[86]

Pointe believed staunchly in the use of the Terror to awaken the power and potential of popular patriotism. But without the Terror, Pointe was disappointed with how the people were using their restored freedoms. With the Maximum dismantled, farmers gambled heavily that the price of grain and other foodstuffs would rise even higher. They also insisted on payment other than in depreciated *assignats*. In many towns and cities of the region, there was very little food available even at six times the former price ceilings. According to Pointe, such speculation threatened famine. He feared that prolonged shortages would lead to the revival of institutions such as the Revolutionary armies. When urban dwellers began to discuss arming themselves and searching for food in the surrounding countryside, Pointe confined himself to inflammatory rhetoric. He was not always able to restrain himself.[87]

When he learned in late Floréal, Year III (mid-May 1795) that thieves came every night to Le Creusot, Pointe went into a rage. Although the thieves left the tools, the copper, and the iron, they destroyed wagons, cut cables and ties, and stole the chains from machines. They had also broken the door to the powder magazine and stolen funds from the strong box. He exhorted the workers to take action and stop the thefts which he described as "truly counterrevolutionary" because "they doubly compromise the interests of the Republic by stealing objects which interrupt the machines and stop production." To repress these "brigands," Pointe mobilized the National Guard. They were to maintain a small picket and conduct an armed search for the missing objects throughout the commune and in surrounding villages. Pointe proclaimed that the gravity of the crime mandated that the guilty be sent immediately before the Revolutionary Tribunal.[88] This was among Pointe's final acts as a representative on mission. Perhaps it was coincidental that only a few days later, he, along with three other deputies, was denounced by the département of the Nièvre. Former terrorists, including Fouché and Jean-Lambert Tallien, successfully defended his actions before the National Convention. But when the legislature finally held new elections, Pointe failed to win a seat to either of the successor legislative bodies. Under Napoleon, he became a tax collector, but lost his job as a result of the First Restoration. He supported the Hundred Days, which led him to go into hiding after Waterloo. Given amnesty by Louis XVIII, Pointe like many of his surviving colleagues from the Convention, lived an impoverished but quiet existence until his death in 1825.[89]

The French Republic, created to defend the ideals of liberty, equality, and fraternity, weathered the multi-faceted crisis of the Year II. Through a combination of extreme measures, patriotic fervor, and luck, the Revolutionaries kept the economy running and produced enough arms to equip the large army created by the *levée en masse*. Only the Reign of Terror could have enabled most Revolutionary legislation to be enacted, much less successfully implemented. Because of its accomplishments, the paradox of a regime based on such lofty ideals making "terror the order of the day" in order to survive has haunted all subsequent French regimes.

From the related standpoints of technological advance and industrial development, the Year II marked an under-appreciated crossroads for French policy makers. Because of the successful mobilization of French scientific expertise, entrepreneurialism, and administrative genius by the Committee of Public Safety and its collaborators, many impressive industrial establishments were created, stocked, and staffed. In several critical fields, both cutting-edge scientific knowledge and craft-based practical knowledge were spread much more widely than ever before. Tens of thousands of workers in several essential industries were trained in advanced techniques. Institutional mechanisms and financial incentives were devised to encourage collaboration among the labor force, entrepreneurs, and state administrators. The French state proved that, even during a Revolutionary crisis of epoch-making proportions, it was capable of directing a staggeringly widespread and impressively successful industrial effort.

Noël Pointe's four missions as an industrial troubleshooter on behalf of the Committee of Public Safety and National Convention illustrated both the problems faced by provincial arms producers and the dependence of this industrial effort on the legal, political, and administrative apparatus of the Terror. Pointe recognized that new industrial circumstances applied once the Maximum was dismantled and with legalized government terror replaced by an informal White Terror that struck back at former terrorists. Despite his recognition of the dangers of practicing a heavy hand in his stewardship of arms production, he had few effective alternatives.

The industrial edifice created during the Revolutionary crisis of 1793–1795 collapsed during the succeeding regime, the Directory. The collapse of this edifice was largely political in nature; the disintegration of arms-making institutions should not be allowed to disguise the lessons that the Republic's triumphant performance in the Year II conveyed to policy makers in France and abroad: that a strong state governing a rich nation undergoing a Revolutionary crisis *could* jump start industrial production and compete technologically at the highest international level. The English model was no longer the only possible pathway to industrial development. To contemporaries, both approaches seemed to feature activist states that controlled working-class militancy while fostering technological advance and improvements in industrial processes. The difference between the two models concerned the state's methods of intervention and the government's relationship to the market.

Choices about industrialization were shaped politically and militarily. French bureaucrats, entrepreneurs, and scientists confronted a very different problem than the one portrayed by the literature. The "threat from below" ruled out the English model in France. The experiences of the Year II demonstrated that a command economy enforced harshly from above could achieve impressive results, at least in the short term. For French economic elites, neither means of achieving productive success was worth it: the Reign of Terror and the popular revolution of 1789 conditioned their thinking on industrial matters. The next three chapters depict the next two generations' efforts to steer between Scylla and Charybdis. The distinctive approaches to individual initiative, state oversight and scientific expertise that came to characterize nineteenth-century French industrialization were created during this stormy crossing.

6

From Allard to Chaptal: The Search for an Institutional Formula for French Industrialization, 1791–1804

France was at a turning point in 1789. Historians on both sides of the Atlantic have examined the nature and the ramifications of that transformation from various ideological and institutional perspectives. This chapter examines the central government's unceasing search for an industrial policy suitable to Revolutionary France. That policy had to foster technological advance and economic growth while maintaining social peace. The upheavals of the period 1789–1795 complicated this search by ruling out the British model of industrialization and giving rise to a different approach tainted by the Terror. Nearly a decade of experimentation culminated in the creation of a new institutional structure under the Consulate that would enable French industry to compete internationally.

Most interpretations of the significance of 1789 for industrialization focus on the month of August. The abrogation of privilege on August 4 and 5 and the promulgation of the Declaration of the Rights of Man and Citizen on August 26 made possible radical changes in the nature of labor relations, protection for inventors, and thorough regulation of markets. The revolutionary occurrences of August 1789 have structured explanations of the French state's industrial policies in the early years of the Revolution and beyond. Jean-Pierre Hirsch and Philippe Minard articulate a view of French mentalité claiming that fundamental change was possible in 1789–1790 for both ideological and institutional reasons.[1] They state that in the early years of the French Revolution "laissez-faire" actually meant "Laissez-nous faire" ("Let us do it"). French economic actors, from capitalists to artisans, did not reject state intervention in the economy per se; instead, they wanted to help to make the rules and participate in their implementation. Regulation should emerge from an internal negotiation among the key players within an industrial sector or market and an external negotiation between sectors or states rather than being imposed from above. At the same time, said Hirsch and Minard, in the aftermath of the Eden Treaty, these same entrepreneurs and laborers increasingly demanded the state's "protection," which really meant "protectionism" (i.e., defense against foreign competition). In light of their unquestioned mastery of the sources, both at the local and national levels, this argument both provokes and convinces.

Focusing on market regulation from Colbert into the nineteenth century, Hirsch and Minard believed that French consumers wanted to ensure the availability of adequate supplies of accessibly priced industrial goods. How to accomplish that demanding (pun intended) goal was one of the twin objectives of the project of French industrial regulation. (The other was self-regulation by producers.) The rights enshrined in the Declaration of the Rights of Man and Citizen seemed to conflict with the lingering privileges of those ancien régime institutions that regulated the urban marketplace for labor and goods: *corporations* of masters and *compagnonnages* of journeymen. For many, the state remained the essential guarantor of just or fair economic relations throughout the period; those espousing a more "liberal" or supposedly British perspective, however, preferred that individuals perform this task through the dynamic mechanism of the marketplace. In 1791, the French state, by implementing laissez-faire policies that suppressed the *corporations* (the Allard Law in March), forbade coalitions (the Le Chapelier Law in June), and eliminated the inspectors of manufacturing (in September), among other measures, eliminated the intermediary institutions that might have functioned as a substitute in overseeing the market. The main point underlined by Hirsch and Minard was that, by implementing a liberal doctrine, French policy makers knowingly and against the inclinations of most, made the state more powerful and more necessary to properly regulated markets. In sectors ranging from metallurgy to textiles to leather, various levels of the French state sought to arrange access to raw materials and admittance to foreign markets, spread advanced technical and craft knowledge, and regulated shop-floor work practices. Thus, the French state may have officially championed laissez-faire, but, paradoxically, legislative actions designed to create a liberal market environment instead guaranteed the policy makers of the central government a preponderant role in industrial development exercised through their domination of a growing assortment of powerful institutions ranging from the Consultative Chambers of Manufacturing to the Society for the Encouragement of the National Industry where the pivotal technological and economic decisions were made. The most significant institutions that emerged between 1795 and 1804, however, were generally intended to manage if not necessarily master the market.

Hirsch and Minard built their interpretive edifice on the well-known insights of Steven L. Kaplan, Michael Sonenscher, and William H. Sewell Jr. Kaplan concluded that, by outlawing all forms of coalition and eliminating all the intermediary institutions that regulated the labor market in 1791, the French state had created a new "liberal" environment for "policing" labor relations in which laissez-faire was the governing ideology if not always the dominant political or institutional practice.[2] Sonenscher investigated many of the same issues through the prism of natural-law mentalité, which implied the existence of natural rights that stood for individual rights. He also judged that 1791 was the turning point as natural law was increasingly codified into positive law and political institutions.[3] Sewell, on the other hand, anticipating the "linguistic turn" so prevalent among contemporary historians, focused on discourse in the form of an enduring *corporate* idiom and on institutional continuity among journeymen. He downplayed the influence of the laws and practices developed during the early years of the Revolution in favor of emphasizing the ideological and institutional links between the Revolutionary decade and the Revolutions of 1830 and 1848.[4] The goal of all three of these outstanding historians was to provide a more autonomous, less reactive history of working-class organizations and the development of working-class consciousness.

All these accounts identified the events of 1789–1791 as pivotal for France's future industrial development. Each of these historians explored pieces of an intricate puzzle created amidst the stunning political and administrative decentralization enacted by the National Assembly in 1789–90, when the French central state voluntarily surrendered its control over politics and the provinces.[5] Hirsch and Minard contended that the most important feature of liberalism was how it affected market regulation. Kaplan concentrated on the position of the individual in society. Sonenscher saw the implementation of natural law as the most important shift; Sewell argued for a fundamental continuity linking the old regime to 1848. What is most striking about these interpretations of this transitional period was the central place of ideology, whether liberalism, Physiocracy, natural law, corporatism, or socialism. Even when examining the market—whether as a mechanism for the distribution of goods and services or as a means of setting prices or even as an ideal—the essential notion of competition, and particularly the state's role in how to manage that competition, was largely absent.[6] In these accounts, analyses of legal transformation and political upheavals were not related directly to questions of technological innovation, international competitiveness, and especially production.

This chapter seeks to complement these influential explanations of French industrialization by locating them within larger frameworks of social, economic, and technological change. Ideology was at the heart of institutional foundation and change, as these authors made abundantly clear. However, the attempts to develop a coherent French industrial policy after 1791 were also based on shop-floor realities, on the *patrie*'s international competitiveness, and on the availability of profitable entrepreneurial opportunities. Material issues affected the formation of industrial policy, particularly with regard to the role of the state in managing the market, mechanization, and labor relations as much as ideology. The revolutionary decade altered French industrial possibilities: the approach to industrial policy that emerged under the Consulate was a compromise among competing impulses that cannot be fully explained by ideological or legalistic imperatives.

Technological Policy during the Revolutionary Decade

If the Revolution constituted a new beginning in French industrial policy, surely one of its most important elements was the new patent system enacted on January 7, 1791.[7] In the waning years of the ancien régime, French administrators exhibited a deep understanding of the importance of British patent regulations in fostering technological experimentation.

Yet the Constituent Assembly deliberately rejected a British regulatory model in favor of a more "open" but more centralized system resting on a different ideological foundation. French patent regulation, its roots in British practice, and the importance of patent law have been investigated extensively in recent years in response to H. J. Habukkuk's challenge to historians concerned with the process of invention.[8]

Christine MacLeod and Liliane Hilaire-Pérez have led the flood of inquiries into the history of eighteenth-century patents.[9] Their complementary findings suggest that the British system of regulating patents provided no better an incentive to invention than the French approach, either before or after 1791.[10] Their work also reminded historians that the number of patents cannot be used as a reliable guide to either the quantity or the quality of inventive activity.

The new patent regime embodied the same laissez-faire conception of economic development that underpinned the Allard and Le Chapelier laws. The Patent Law guaranteed the right of ownership to exploiters of their inventions for either 5, 10, or 15 years. A master list with detailed descriptions of each invention was to be available for public consultation in the capital of each département to prevent inadvertent infringement on an inventor's rights. The French patent list was fundamentally a registry; under the ancien régime, there was no technical examination to limit the excessive influence wielded by "experts."[11] Approximately 750 patents were registered over the next 20 years.[12]

On September 12, 1791, the Legislative Assembly established an impressive fund to compensate inventors who released their discoveries to the public domain. This fund was to provide material support for invention and for industrial development in general. The Legislative Assembly continued a royal policy goal going back decades, if not centuries: to liberate and support the French inventive genius.[13] The real break with the ancien régime was not in the policy but in the implementation.

The new legislation seemingly democratized the process to protect inventors. Patent protection was now a right available to all who met the requirements rather than a privilege granted by the royal government. However, the supervisory provisions concentrated decision-making authority over innovation even more than before 1791.[14] No longer could parlements, provincial estates, or municipalities grant privileges of varying geographic extent. Nor did the royal government have recourse to experts on a case-by-case basis. In theory, high officials, favorites, and members of the royal family could no longer utilize either undue influence or the powers of patronage to sway the process.

From October 16, 1791 until 1797, a thirty-member Bureau of Consultation, with half its membership coming from the Academy of Science and the rest from other learned societies, investigated *all* claims to government support and made recommendations to the legislature. Led by an executive group composed of six secretaries and a presiding secretary, the initial membership included some of the great names of French, and indeed European, science and technology. The chemists Berthollet, Lavoisier, and Hassenfratz, the mathematicians Jean-Louis Lagrange, Pierre-Simon de Laplace, Jean-Baptiste Leroy, and the inventors Nicolas Leblanc and Jacques-Constantin Périer were joined by such knowledgeable and experienced administrators as Nicolas Desmarest and Alexandre-Théophile Vandermonde. They were complemented by contingents of engineers (both civil and military), medical doctors, apothecaries, and agronomists.[15] The centralization of the decision-making process combined with vastly overstrained financial resources to limit dramatically not only the support given to inventors during the revolutionary decade but also the number of individuals protected.[16] Despite these constraints, the Bureau dispersed 1,157,100 *livres* to nearly 300 petitioners.[17] The Bureau of Consultation and its successor, which operated under the aegis of the National Institute, thoroughly insinuated the revolutionary "technocracy" into industrial affairs. Under their management, the "laissez-faire" economic policies implemented in 1791, abrogated in 1793–94, and returned to favor after Thermidor prevented industrial chaos despite the recrudescence of inflation while producing some practical benefits.[18]

The members of the Bureau generally preferred to let the market dictate not only the pace and direction of technological innovation but also the course of industrial development.[19] In response to a request for financial support from Widmann in October 1798 to re-establish his bankrupt woollen manufacture in Colmar (Bas-Rhin) and/or receive a government job in some technologically related field, the Bureau issued a kind of policy statement: "If the manufacture is in such a state that it cannot survive without the intervention of the government, then from this fact alone, funds cannot be risked on an enterprise which has no promise of success. Experience over the last ten years has shown that enterprises constructed on the debris of failed businesses have only an ephemeral existence either because of the inadequate finances of entrepreneurs or because they are incapable of running that type of enterprise. The Bureau also observes that encouragements should go only to those who demonstrate the spirit of discovery or of perfection in the application of their art."[20] Other organs of government did not always follow this policy. Under the Directory, the most important exception to laissez-faire policies was the Ministry of War. Defense contracts kept the French woollens industry afloat during the Revolutionary decade, but orders were placed or withdrawn for political reasons rather than favoring either the most advanced or the most efficient producers.[21]

One of the first tasks facing the new regulatory regime was to determine precisely what rights had been granted to inventors under the ancien régime. Although the multiplicity of privileges granted by parlements, estates, and municipalities had disappeared with their authority, compiling and disseminating a definitive list of the sometimes overlapping prerogatives conferred in the past 20 years proved surprisingly difficult. With record keeping complicated by the diversity of agencies with competence in matters of privilege, desperate requests for more information were received from befuddled administrators until the Consulate. This information was a safeguard against compensating a process or device twice. This task was closely related to the task of gathering the various mechanical models developed over the years in hopes that inventors would be inspired or might confirm the originality of their approach. Begun in 1782, the process accelerated when the second patent law (March 29, 1791) mandated the creation of a centralized repository of all models and records pertaining to invention. Only when the National Convention appropriated the necessary funds (October 13, 1794) did the Conservatoire des Arts et Métiers take shape. The Conservatoire symbolized how "liberalization" ultimately led to centralization during the age of revolution. Sometimes viewed as the culmination of Jacobin views on science and technology, the Conservatoire actually emanated from a longer-term process in which new institutions were developed to meet the uncertain social, technological, and industrial situation prevailing during the Revolutionary decade and beyond.

After England's declaration of war in 1793, French administrators identified an opportunity for domestic industry to monopolize the vast domestic market. These decision makers recognized that, with judicious encouragement to advanced producers and continued legal prohibitions, the profits accruing from domination of the French market could fund major technological breakthroughs by domestic industries.[22] The annexation of Belgium and the Rhineland promoted that view. Although the prohibition of English goods was renewed on 10 Brumaire, Year V (October 31, 1796), it quickly became apparent to officials overseeing the industrial economy that this measure was both thoroughly ineffective and counterproductive. France benefited from the importation of English machines, tools, and other goods such as books. These officials also acknowledged, albeit unhappily, that the smuggling of English goods, particularly cotton textiles and metal goods, was probably greater under the Directory than it had been before 1786. Such profitable alternative possibilities to production diverted entrepreneurial initiative from innovation and mechanization.

The Bureau of Consultation coordinated, supported, and reported on the success of sub rosa wartime contact with Great Britain, thereby challenging the reduction of French state action to either mercantilism or liberalism. Books on technology and science from Britain and elsewhere were acquired, translated, and disseminated. Despite the legal ban on the export of British machines, they continued to be acquired by French spies and entrepreneurs, who included such important Continental innovators as François-Bernard Boyer-Fonfrede and Liévin Bauwens.[23] British artisans, mechanics, and tinkerers and natives of the Netherlands, Switzerland, and the United States also trickled into France, joining the new-made citizens in the German and Belgian territories in search of opportunity. The Bureau lauded and occasionally compensated these immigrants, but such compensation was balanced by controls ensuring that aid recipients fulfilled their obligations.[24]

Although the excesses of revolutionary wartime rhetoric must be taken into account, the office-holding savants who evaluated French inventive activity during the Revolutionary decade argued that France possessed all the requisite theories to build and operate even the most advanced of English industrial machinery, as the successes of Bauwens and other entrepreneurs illustrated: "The Bureau thinks that it is less an understanding of the machines appropriate for spinning cotton that France lacks than the means of multiplying them [the machines] and propagating their use. The English don't have any inventions in this area that France doesn't have the theory and the art of operating it, but France doesn't have the multitude of establishments that the islanders have. France hasn't spread, as the English have, in the countryside, all the different types of jennies that replace in country households the bad habits [of poor quality production] developed by our female spinners. That, perhaps, can only be the result of a general education and, as a result, it will take time [to eliminate these habits] and it is the government that must encourage it."[25] The Bureau blamed the temporary distractions of the revolution for fits and starts visible in the mechanization of French industry and the lack of technological progress that actually found its way onto the shop floor.

In their evaluation of the relative technological positions of French and English industry, delegates of the National Institute (which replaced the Bureau of Consultation as originally constituted) focused on the carding machine, the key artifact in mechanizing textile production. They launched an extensive and intensive investigation into the most advanced carding machines extant in France or being constructed nearby (for example, in the machine-building center of Verviers in Belgium). Their report on the existing options juxtaposed cultural and material explanations for British advantage. The major French handicaps were identified as less devotion to producing a quality product on the part of some machine builders and fewer opportunities for entrepreneurial profits. Building cards domestically was a major difficulty they said—but not because France did not have the necessary raw materials, but because "there is a prejudice that French iron is not as good as German. It is this prejudice that must be demolished—and demolished immediately— among our Enlightened artists. It can be done by demonstrating the excellent cards that have been made for several years with French wire." They continued: "Perfect wires for cards depend first, on the quality of the iron, second, on the quality of the draw-plate and, finally, to the attention paid to the reheating process. The less that the wire is reheated, the more elasticity the iron retains. Every time even a slightly intelligent French card maker seeks to imitate English cards, for example, Faget, Leclerc and Company [of Toulouse], they put the wires through the draw-plate several times without reheating before they cut them and shape them into hooks. The general inferiority of French cards is attributable to a unwillingness to employ this simple and well-known process. As this practice is applied by higher-quality card makers, there is no doubt that eventually they will force all carders to imitate them or they will have that branch of industry taken from them."[26]

The views of revolutionary technological evaluators strongly resembled those of their ancien régime forebears. They were convinced that France could compete with England in every field of scientific and technological endeavor, either currently or in the very near future. The impressive accomplishments of the "secret" scientific testing center and training school at Meudon provided convincing evidence of that competitiveness to French administrators.[27] However, just as in the 1780s, that conviction did not lead them to believe that competitiveness could be achieved without the thorough involvement of the state.

Competitiveness acquired a different meaning during the crisis of the Year II. War prioritized short-term military needs over longer-term technological and industrial development. Even so, the revolutionary state drew on the Enlightened scientific establishment and directed these men to improve the technological and scientific knowledge basis of French industry. The efforts of these individuals are well known[28]; here I emphasize the consequences both intended and accidental outside of the military-industrial complex (discussed in chapter 5) for French technological development. The Committee of Public Safety and its subordinate organizations may have initiated the mass mobilization of France's scientific "best and brightest," but it was in technological diffusion that the Revolutionaries pioneered a different approach.[29]

The related but independent technological decisions of the Bureau de Consultation and the Committee of Public Safety in 1793–94 suggested that, while some important innovators (including Claude Chappe with his semaphore telegraph and the balloonists gathered at Meudon) were thinking in new ways, the scientific and technological elite were focused on elaborating inventions and ideas developed under the monarchy. For them, "Revolutionary science" was an extension of "Enlightenment science." The list of entrepreneurs or tinkerers who received support for non-arms-related projects during the early years of the war reads like a "who's who" of the late ancien régime. From Nicolas Barneville (who received a building and 200,000 *livres* to set up a muslin manufacture) to George Garnett and John Macloud, a surprising number of British expatriates who had begun enterprises in the 1780s continued to receive support from the Revolutionary state. Macloud bolstered his claim by inventing a loom for muslins. Modernization of French production practices through the introduction of new technology adopted or adapted from the British remained the chief goal.[30] Reading the files and the reports on inventions in various stages of development, one is struck by how many decisions were based on the author's ancien régime reputation, credentials, or accomplishments.[31]

This continuity can be glimpsed in Vandermonde's report on Lyon's industrial situation. As a member of the Academy of Science, Vandermonde had been a pillar of Bourbon technological decision making. Co-opted by the Revolutionary state and put to work managing arms production in Paris, Vandermonde was drafted by Carnot and Prieur for other duties, such as writing reports on technological subjects for the Committee of Public Safety. Vandermonde emphasized that the Revolution—particularly its wars, the Federalist revolt, and the continuing political violence—had done horrible damage to Lyon's infrastructure and commercial prospects. A significant number of highly trained silk workers had already fled to Spain, Prussia, and Austria in response to judicious recruitment. But the chief causes of Lyon's long-term problems were greedy workers who stole raw materials, failure to observe quality regulations, and, most important, "the insufficiency of government protection both domestically and abroad." Vandermonde criticized Adam Smith and James Stewart for not recognizing the need for state industrial regulation to ensure quality production. Piedmont was his counterexample of beneficial state management of the silk industry through the enforcement of regulations. He also defended the role of *corporations* in maintaining standards and training workers: he believed that formal seven-year apprenticeships were the most effective means of teaching young men a craft. Vandermonde suggested that technical schools be revived to teach design and other subjects, including foreign languages, and that women be taught arithmetic and writing. He claimed that the English economist Josiah Child inspired his proposals. This outlook on state involvement could have been expressed before the Revolution.[32]

With Vandermonde and other technocrats in charge of formulating and implementing policy, despite the liberal views of many decision makers at the highest levels, France could not transcend certain biases in its development that had little to do with economic path dependence and sunk capital. Most French scientists and technologists trained under the ancien régime favored state control, high standards of quality in production, and the fostering of change through formal education. In fields contested either by foreign competition or shop-floor turmoil, private-sector competitiveness was not the stepping stone to either technological change or international industrial success. The manufacturing achievements of the Year II cast long shadows over the policy makers of subsequent French regimes.

The experiences of the Maximum and the problems associated with the "command economy" of the Year II shaped French technological policies. When it came to protecting infant industries, fostering export markets, supporting entrepreneurs who suffered war losses, or maintaining the right to work in the face of skilled laborers seeking to limit access, even the most "liberal" French technological decision makers could not envision the French industrial economy without the involvement of the state. Officials espousing a "Jacobin" perspective asserted that the most vital activities of the French revolutionary state were educational and vocational, but must be overseen by a centralized bureaucracy. The absolute right of inventors to the fruits of their discovery would guarantee the market's ability to determine which individuals would succeed. If such attitudes about science and society, centralization, or the market mechanism were reported on the other side of the Channel, they would be seen as a perfect embodiment of recent formulations of proper laissez-faire state action in this era. That such actions occurred in France again reveal the practical limits of contemporary conceptions of laissez-faire and demonstrate how far along the same path these rival countries traveled.[33]

Under the Directory, the Bureau of Consultation confronted the industrial heritage of previous regimes. This legacy included the luxury enterprises of the ancien régime taken over by the state and the industrial ventures undermined or destroyed by the Maximum. For example, the former royal tapestry manufacture of Aubusson, located in the mountains of the Creuse, had collapsed early in the Revolution. A grant of 80,000 *livres* from the National Convention in the fall of 1792 to start a woollen manufacture there was intended to revive industry in the impoverished region. It failed, leaving 1,500–1,800 workers unemployed. One of the former directors suggested it be revived with the Bureau's financial support. Before ruling on the request, the Bureau consulted a well-connected former inspector of manufacturing who resided in Aubusson. Pierre Laboreys de Châteaufavier rejected the proposal to revive the woollen manufacture but suggested that the production of tapestries resume.[34]

While rejecting a request for a fresh infusion of capital for the porcelain manufacture at Sèvres, the Bureau articulated their views on such enterprises. They rejected the notion that the production of luxury goods had a place in revolutionary France, even to showcase national technical skills or aesthetic prowess. The only valid criteria to judge an enterprise were utility and profitability. Before its demise, Aubusson had depended on two schools of design to train professional designers and an inspector, the annual salaries for whom ran to nearly 8,200 francs. Even if these support mechanisms were restored, production technology had evolved to the point where "engravings, painted linen and other forms of decoration have passed tapestries by." The Revolution did not cause the failure of the tapestry manufacture, stated the Bureau's report; rather, "it had an infallible link to the unique attraction for the consumer of newly created fashionable products." Instead, the Bureau believed that, as a general rule, government should facilitate industrial development by means of "favorable laws, by subsidies, [and] by drawbacks on the customs," not direct financial support. The state should accept petitions only from entrepreneurs or local organs of government that would make use of locally available raw materials, employ those who otherwise would be needy, or stimulate other productive sectors. Enterprises that focused on the needs of the domestic market were the most useful. Of the other old-regime national manufactures,[35] only the Gobelins had survived. ("The inimitable superiority of their work puts them in a separate class, but the elevated price of their products makes them inaccessible to the masses. The Gobelins will remain an honorable testimony to the industrial honor of a powerful nation, but it will never furnish a lucrative branch of commerce.") But the Bureau repeated its view that the government should funnel its energy, its capital, and its technological support to more labor-intensive sectors with greater potential for developing a mass market and in which France was technologically competitive. This view of the national luxury manufactures underscores the practical limitations of a "liberal" approach to market relations and the role of the individual in the age of revolution.[36]

The Bureau criticized the National Convention's habit of creating large-scale enterprises and then showering them with too much capital. This had a negative effect, "because it is a secret of the human heart that men will dissipate what they have acquired without effort. . . . [and] if they acquire the impudent hope of always getting what they want without work, they will abuse the resources that have been so prodigiously supplied." This group of technological decision makers clearly preferred smaller-scale production aimed at a mass market: the best means of increasing industrial output was to envision and execute products with vast potential sales and then allow entrepreneurs to take advantage of France's low wages and artisanal creativity to fulfill demand. Meanwhile, they minimized direct financial support to inventors except in areas where France was not internationally competitive.[37]

If the Bureau began its tenure by recommending the use of government money for projects such as building textile machines, experiments in steel making, and general tinkering by those with academic recognition from the ancien régime, that attitude did not survive the fall of the Committee of Public Safety and the concomitant decline in the influence of savants.[38] Perhaps because of their desire to funnel what limited funds could be spared to establish the Conservatoire on a firm footing, the Bureau and its successors increasingly rejected requests for state investment in mechanization or technological advance. Instead, they funded publicity. Large subsidies were given to various journals, including the *Mém- oires du Bureau de Consultation des Arts ou Journal des inventions, découvertes et perfectionnements dans les sciences, arts et métiers* (of which 98 issues appeared in 1793–1795) and the influential *Journal des inventions et découvertes* (September 1794–December 1795), published by the Lycée des Arts.[39]

In Floréal, Year VI (May 1798), the Bureau stated as a maxim that the government "could not give financial encouragements to an already existing manufacture without damaging all the other establishments of the same genre and fostering comparisons to the privileged manufactures of the old regime." Yet acquiring "new" knowledge was part of the scientific justification for the French expedition to Egypt in 1798, led by that noted member of the Institute, General Napoleon Bonaparte.[40] As would become the norm once Bonaparte assumed direction of the government, what funds were provided directly to innovators during this era tended to come directly from the Minister of the Interior, unmediated by the views of "experts." On numerous occasions, the Bureau only learned that such funds had been dispersed when an inventor deposed a copy of the machine, description of the process or product at the Conservatoire where some members worked. In scientific, technological, and industrial matters, liberal economic principles were clearly at odds with French state practice.

The chief exception to this rule was tools—especially those essential to perfecting the manufacture of other metal goods, such as files. The Bureau inspected the file-making workshop run by Morbach, who had worked in the large steel and file-making operation at Souppes (Seine-et-Marne) as an inspector employed by the Minister of the Interior. They watched him make a set of files to ensure that the pieces they had seen were not masterworks but rather the product of everyday practices; then they proposed that Morbach receive 900 francs "because the government should encourage those artists whose knowledge allows them to teach their art to students." He was also to depose a set of files at the Conservatoire.

Such views of the proper role of the state in the industrial economy marked a clear departure from the early years of the Revolution. They reflected a more English, though hardly classically "laissez-faire," approach to development. The Bureau accepted many of the "liberal" concepts regarding the essential regulatory power of the market identified by Hirsch, Minard, Sonenscher, and Kaplan, but always with a twist. The case of Vidal, a manufacturer of blue dye in Brumaire, Year VII (October 1798), was a case in point. Vidal asked for protection from the central state after the municipality of Calais closed his enterprise because some neighbors complained about the chemical residues. The Bureau leaped into the breach. Citing the principle that "the free exercise of profession is a common right," the Bureau recognized that local authorities unjustly restricted this right as a matter of course when "particular animosities are given the pretext of the common good." Complaining that "the existing legislation is incomplete on such points," they stated that, "the law consecrates the liberty to exercise a trade, but the laws necessary to guarantee it have not been enacted." Their response, however, was moderate; the Bureau wanted only to restrict the arbitrary administrative power of local governments. Here "liberalism" was subordinate to "Jacobinism." The Bureau recognized that only the state's police power could ensure "the cleanliness of the streets, the salubrity of the air, the peacefulness and safety of the citizens" as "essential rights of the citizen that must be respected."

A restricted state role in economic matters in favor of market regulation was reiterated constantly in the waning years of the Directory. When the noted chemist and manufacturer Jean-Antoine Chaptal, a member of the National Institute, reported (also in October 1798) that, because there was no longer any inspection of verdigris, manufacturers were skimping on quality to such a degree that consumers were turning to Spain and Italy rather than "buying French," the Bureau remained true to the principle of liberty. Their reply to Chaptal's recommendation that government-sponsored quality inspection be reinstated was that it was forbidden by the constitution. They added that the creation of inspectors or regulation of any kind slowed or even prevented improvements induced by market conditions in chemical manufacturing. These views were expanded into a general policy statement later in Thermidor, Year VII (August 1799). A formal policy position was necessary because "the principles that the trades and that commerce must be completely free . . . lend themselves to so many diverse interpretations that no advantages are drawn from having these principles recognized." The Bureau said that this "liberty" had two distinct characteristics: "1. That of freeing the trades and commerce from every sort of constraint or hindrance that could prevent their exercise or their progress. 2. That of assuring artists and those exercising the trades or commerce the right to their work." What led the Bureau to insist on this interpretation of economic rights was a request from the numerous small-scale metal workers of Thiers to regulate the industry by restoring a version of the *corporations*. Such a law, said the Bureau, would "surpass the natural limits on liberty" by menacing the right to work and impinging on the property of individuals.

The metal workers of Thiers wanted to force each enterprise to conform to regulations that would regulate the quality of each item. The article would then be stamped with a mark specific to Thiers to protect the manufacture from "the audacious counterfeits that since the Year III (1795) have stolen the trademarks of the best-known enterprises and have sold by means of this fraud, works so detestable that they have discredited the metalwork of Thiers, hindering sales and ruining its reputation." Disdaining the case for the re-establishment of the *corporations* or of the royal *Lettres-patentes* that had governed the trade, the Bureau's only interest in the matter was that the counterfeits of the goods of Thiers represented a "theft" of the absolute right of an individual to the fruits of their labor. Likening the case to patent infringement, these officials asked the legislature to guarantee those manufacturers who chose to follow regulations and mark their goods with a distinctive sign the exclusive right to that sign. By having metal workers, not just in Thiers, but anywhere in the country register their marks and enforcing them in the same way as patents, the Bureau sought to generalize the approach to property embodied in the patent laws to material artifacts. Budget constraints meant that enforcement would have to be administrative and judicial rather than by inspection, leaving manufacturers to defend their trademark.

This general attitude about the market and the role of the state led the Bureau to take a narrow and paradoxical view of their duties as the overseers of French technological policy. Despite the various regime changes and political ideologies that held sway at different points during the Revolutionary decade, as a group, the Bureau increasingly understood their role to be that of defenders of liberty or facilitators of technological improvement rather than the directors of a centralized development policy.[41] In their conception of economics, the market—strangely divorced from the impetus to make the most profit possible—was more successful than the government in advancing French technological competitiveness. The state should focus on allowing the market to work more efficiently to encourage technological change. Publicity generated administratively by the Directory depicted a society in which the pace of invention seemed to be accelerating, at least in the eyes of contemporaries.[42] While collecting books and other material for distribution to disseminate technical knowledge more widely, the Bureau decided to "prepare an industrial and manufacturing geography of the Republic" to better inform the public as to the existing state of affairs. The next section illustrates how ideas and project were taken over and expanded by state administrators to fashion a wholly new institution.

The First Industrial Exposition: The "Visible Hand" and the Dilemmas of Competition

The technological technocrats gathered in the Bureau of Consultation and its successor agencies believed that the market should direct French entrepreneurs to the most productive sectors of the economy. Other French bureaucrats, however, had a different approach to industrial development. While sharing many of the same "liberal" attitudes about the power and function of the market, Minister of the Interior Nicolas-Louis François de Neufchâteau enacted a more statist approach to fostering technological innovation and accelerating industrial expansion. This former colonial military officer, Physiocratic fellow-traveler, amateur agronomist, and mediocre playwright wholeheartedly embraced the task of assembling an industrial geography of France. This information would allow entrepreneurs to evaluate the uniqueness of their product, identify unfilled market niches and earn a sufficient profit for their investments of time, money and expertise. François de Neufchâteau and other bureaucrats believed that entrepreneurial profits were essential to France's ability to protect the domestic market, a necessary first step in the Republic's economic rivalry with Britain. He also wanted to publicize the measures undertaken to address France's economic shortcomings.[43]

As Minister of the Interior, François de Neufchâteau viewed the French economy through the prism of a revolutionary politician. Transparency among realities, intentions and actions was a dominant political theme of the first half of the revolutionary decade which this influential bureaucrat linked tangibly to accountability and especially to the reproducibility of applied science to inspire innovation in the means and methods of production.[44] Bureaucrats and theorists deemed transparency essential to the free play of the market, but believed that the state must nurture it. Revolutionary festivals had been among the most important means of fostering transparency and mobilizing the public during the first half of the revolutionary decade.[45] It is little wonder, therefore, that François de Neufchâteau breathed new life into the whole system of festivals[46] and created an institution designed to publicize the industrial situation, spread technical knowledge, and provide new opportunities to entrepreneurs.

The impetus behind François de Neufchâteau's creation was his fundamental dissatisfaction with the economic situation and his recognition that market forces alone could not revive a French economy beset by civil unrest, war, and revolution. The depth of the economic crisis put this Physiocrat, along with many other French officials espousing "liberal" views, in the paradoxical position of having to find ways to use state resources to improve the economy. Constrained by revolutionary ideals, domestic politics, and military needs, "liberals" searched for ways to facilitate the workings of the market. Although these "liberals" did not usually remain in office long enough for their vision to mature, their ideas and their conception of the necessity of increasing competition through state action was shared by their successors. Between 1795 and 1804, innovating officials used institutions such as the industrial exposition to manage if not master the market.[47]

The first European exposition of industrial products was held at the end of the revolutionary Year VI (September 19–October 3, 1798). François de Neufchâteau envisioned both political and economic benefits. Politically, the exposition was a response to the Directory's need to mark the annual festival for the founding of the Republic with something fresh—"an unknown means of amusement"—to capture the attention of jaded Parisians. Having organized the festival unveiling General

Napoleon Bonaparte's booty from Italy some months before, François de Neufchâteau recognized that an ordinary festival without some extraordinary spectacle would be ignored.[48] Although the five Directors suggested an exhibition of new paintings with revolutionary and republican themes, François de Neufchâteau preferred to celebrate the achievements of the mechanical and industrial arts.[49] He concluded that a series of public expositions could improve the French economy through the spread or introduction of new technologies and by strengthening the market for industrial goods.

For François de Neufchâteau and a significant number of other revolutionary bureaucrats with liberal leanings, building up French industry was vital to preserving the Republic against both foreign and domestic threats.[50] He asserted that only economic strength could "consolidate [the] victory [of French arms] and assure peace."[51] Such strength could be developed only by recognizing the economic component to France's military rivalry with England; therefore, "glory belongs to the ingenious inventor just as it does to the intrepid warrior. . . . The people must persuade themselves that our manufactures are the arsenal from which will come weapons fatal to the power of Great Britain."[52] These views affirmed the axiom that through technological innovation, French industry could beat the English at their own game. However, the military reference suggested another aspect of economic competition. Under the Directory, an ever-increasing number of French officials recognized that the state would have to issue the marching orders—literally—because a market dislocated by the revolutionary crisis and wartime demands could not be entrusted to re-energize the French economy or encourage technological innovation sufficiently to compete with the arch-enemy, England.

Refocusing French energy toward technological solutions and improving the market orientation of industrial production were François de Neufchâteau's chief economic priorities. In his welcoming speech to the entrepreneurs exhibiting their wares at the exposition, François de Neufchâteau emphasized that such festivals were "a tool for public instruction." He told the 97 exhibitors: "You can fulfill a sacred duty by teaching the citizenry that national prosperity is inseparable from the health of the [mechanical] arts and manufacturing. . . ." Because his speech would be published and distributed nationally, François de Neufchâteau clearly had a broader audience in mind than the paltry group of exhibitors assembled before him. Recent scientific advances by members of the National Institute, said François de Neufchâteau, demonstrated that the Republic's abolition of privilege and restrictive regulations allowed "technology to open a vast new domain for the expansion of the human spirit."[53]

A circular from the Minister of the Interior to the départements dated August 26, 1798 announced the exposition.[54] All agents of the central government, the locally elected departmental administrations and the Consultative Bureaus of Commerce were charged with encouraging local innovators and entrepreneurs to participate. Any manufacturer or artisan could enroll, but the only place to do so was the Bureau of Arts and Manufactures of the Ministry of the Interior at quayside in Paris. Then a jury appointed by the Ministry inspected the goods and decided if and where the wares would be displayed.[55] A separate jury, drawn from the National Institute, the Council of Mines and the Agricultural Society of the Département of the Seine, chose the 20 prize winners.[56] François de Neufchâteau directed the jury to designate products "which will strike the most deadly blow to English industry." They "will receive the gold medal for the perfection of their manufacture and the extent of their commerce."[57] A sample of all the medal winners' goods was to be displayed permanently in the Conservatoire to instruct those who could not or did not attend the exposition.

The main exhibition was held on the Champ de Mars in southwestern Paris. François de Neufchâteau spent an impressive 60,000 *livres* to have 68 arches constructed in the form of a parallelogram around a mock Roman temple dedicated to the achievements of French industry. Designed by François Chalgrin, the future architect of the Arc de Triomphe, these papier-mâché arches were almost finished when a storm destroyed them, forcing a two-day delay of the opening ceremonies.[58] In accordance with the Directors' wishes, an exhibition of painting, sculpture, architecture, music, and dramatic poetry ran alongside the industrial exposition; it was little noticed by contemporaries.

François de Neufchâteau contrived special efforts to publicize the virtues of those French wares in competition with goods customarily purchased from England. The program made the government's point with authority that "buying French" was not only patriotic, but possible and even fashionable: "Everyone admitted to the festival, even government officials or to participate in the games, is warned that they will not be able to enter if they wear foreign-made clothes. Citizens of both sexes *must wear clothes of French manufacture.*"[59] In an explicitly nationalistic appeal, a military parade opened the festival and army bands played patriotic airs throughout the day.[60] The first day also featured athletic games and horse races with nationalist themes. To enter the games, it was necessary to be a citizen, a member of the national guard and to have voted in the last election. Horse owners even had to prove that their animals were French-born! Because of the war, the laissez-faire attitudes toward competition that dominated the early Revolutionary years were subordinated to a nationalistic, mercantilistic approach to industrial policy reminiscent of Colbert.

The products presented at the first exhibition ran the gamut from steel razors to fancy porcelain and from tapestries depicting revolutionary military victories to plain white cotton shirts, but the jury favored machine-spun textiles, improved metalwork, and new technology. Of the 97 exhibitors, 35 displayed some sort of textile, overwhelmingly machine-spun cottons, and 24 exhibited some sort of precision metalwork, of the sort essential to building the machines vital to industrialization. The chief metal employed was iron, but several exhibitors made their own steel and two used copper. A further 18 exhibitors demonstrated new inventions, including a leather-cutting device, a better safety lock, artificial limbs, and a cotton-carding machine patterned on the most recent English model.[61]

The notable demonstrators included Nicolas Clouet of Paris, who displayed "iron refined into steel and then made into razors." Raoul, France's leading file maker, had already received 3,000 *livres* from the Bureau of Consultation in 1793. He made files with a machine of his own invention that stamped out the shape to then be hand-finished. He also made his own cementation steel with domestic iron. Experts from the Bureau and numerous entrepreneurs confirmed that his files were "equal to those of England."[62] The cotton industry so critical to international industrial competitiveness was prominent both in numbers and in quality. Jacques Payn of Troyes exhibited "well-made cotton stockings of a silken texture dyed a beautiful white." Denis Jullien from the Seine-et-Oise presented an "assortment of machine-spun fine cloths made with cotton from Cayenne." Boyer-Fonfrede et fils, a large manufacturing firm from Toulouse, took advantage of their Paris agent's stocks to demonstrate the tangible results of their illicit acquisition of English machinery.

Perhaps the most notable of the newly invented products were Nicolas Conté's graphite pencils of different colors, which, the program boasted, were a "triumph over English industry."[63] In 1795, as Minister of War, Lazare Carnot directed Conté (a Norman of peasant stock who had served in France's military balloon corps) to replace English plombagine in the making of pencils. In only eight days, Conté devised a new substance composed of different mixes of potter's clay (argile) and mashed graphite (according to the hardness of the desired pencil), which was then heated in a ceramic crucible and molded into cylinders. Afterward, the cylinder was put into a grooved semi-circular cedar tube and sealed with glue. Conté also tinkered successfully with textile machinery. Later he went to Egypt with Bonaparte. He was one of the first members of the Legion of Honor.[64]

For officials and entrepreneurs, the product mix of the exposition departed from the high-priced artisanal luxury goods that were the pride of French industry before 1789.[65] For example, there were no silks on display and only a smattering of luxury items. Instead, the exhibitors' wares demonstrated a confirmed practicality by emphasizing everyday needs.[66] The means of appealing to a potential mass market also broke with the ancien régime. French industry had focused on expensive luxury goods sold mostly to French consumers and on low-priced merchandise of very low quality produced for export to the French or Spanish colonies and the Levant. In 1798, the jury rewarded products that offered reasonable quality at reasonable prices that could compete with prohibited but widely available English imports.[67] The scientists and manufacturers of the jury also preferred ordinary machine-made products to the artisanal masterpieces that typically had won government prizes before the Revolution. Increasingly, the intrinsic worth of a product was conceived in terms of the relationship to its price, with all that shift implied for calculating how the ratio between the price and quality might appeal to an intended market.[68] Chaptal, who wrote the jury's report, asserted that the timing for the exposition was fortuitous. He declared: "The progress of French industry needs the continuation of this new institution which the government has founded. They [the jury] announce that the moment has come when France can escape its servitude to its neighbors' industry."[69]

The twelve gold-medal winners included four textile manufacturers; three inventors of cheap, high-quality steel products, such as razors, pans, and smokestacks; two porcelain makers; and three makers of consumer products, among them a new illustrated edition of Virgil.[70] Paradoxically, the prize winners in this exposition, which emphasized everyday merchandise, were given luxury goods from the national manufactures: hunting pistols from the arms factory at Versailles, porcelain from Sèvres, watches from Besançon. Although the organizers intended these prizes to display the quality and range of products made by these establishments and to save money,[71] these prizes also revealed the lingering ambivalence about emphasizing the potential quantity of sales, not the quality of manufacture.

The exposition's emphasis on everyday goods led to friction between artisans and machine workers over who deserved to be recognized by the government, but the spectacle made a strong impression on the viewing public. A Danish observer, the scientist Thomas Bugge, wrote: "The whole exhibition . . . is well worth viewing; besides the idea is new. You see there many proofs of the industry and ingenuity of the nation. Whoever views them with an impartial eye must, however, acknowledge that they fall far short of that perfection of which they are capable; but when coupled with [the war's] circumstances . . . it is a matter of surprise that anything worthy of public attention could be offered. Let France once enjoy the blessings of peace . . . then manufacturers, handicrafts, commerce and the fine arts, will daily gather strength, shoot forth and expand into luxuriance."[72] Here Bugge identified the first exposition's chief problem. Despite François de Neufchâteau's dedication to spreading information on technical advances and helping to support entrepreneurs who adopted these advances through incentives, the Directory's pressing need to build support for the war effort led him to hold the exposition with scant preparation. As a result, representatives of only 16 of 119 départements were present. Because the exposition opened only a month after the first announcement went out, many entrepreneurs did not even hear of the exposition until a few days before it began. Such hasty preparation explained the absence of many important French products made outside the Paris basin, such as silks.

François de Neufchâteau recognized the shortcomings of this initial effort. Even as he opened the first exposition, he was planning the second.[73] He was encouraged by the fact that the Exposition of the Year VI was at least a partial economic success for the exhibitors thanks to increased sales, free advertising in Paris, a licensing agreement with a manufacturer from a different region or a government subsidy to continue research or experimentation. Contemporary accounts, such as the exhibitors' petition to remain open longer to maximize profits, emphasized these benefits.[74]

Assisting the initial set of exhibitors was, however, only a secondary goal of the government; its primary goal was to spread useful knowledge, which required much broader participation in and attendance by artisans, manufacturers, and merchants along with mechanics, inventors, and scientists. Jean-Pierre Hirsch and Christopher Johnson, among others, have noted that, thanks to the Terror, many economic actors were alienated from the revolutionary state and were unwilling to risk their assets and reputations to support government initiatives in an uncertain economic environment.[75] The political reasoning that led many to refuse to attend any republican festival meant that broadening and deepening participation in the expositions required either financial incentives or coercion. The market alone would not suffice.[76]

François de Neufchâteau fell from power on 8 Messidor, Year VII (June 26, 1799) because of factional infighting at the highest levels of government. He was replaced by Nicolas-Marie Quinette, who later emerged as an important supporter of industrial expositions. But as Minister of the Interior, Quinette stated that, although the exposition was "an institution very advantageous to the arts," he was "obliged to postpone the exposition until the year VIII" because of the financial demands of the war effort.[77] Instead, during the five complementary days of the Year VII (September 18–22, 1799), all of Paris' libraries and museums and the Gobelins and Sèvres factories were opened to the public. In the courtyard of the Conservatoire, models of the machines invented or refined by medal winners from the Exposition of the Year VI were displayed.[78] Quinette's well-meaning attempt to keep the public's attention focused on industrial progress was no substitute for the publicity and commercial opportunities generated by an industrial exposition.

This pathetic follow-up effort demonstrates that, despite the efforts of François de Neufchâteau and others, the Directorial regime was too poor, too unpopular, too involved in the war against the Second Coalition, and suffered from too much ministerial instability to follow a consistent industrial policy capable of modernizing the French economy enough to compete successfully with Great Britain. As in the Year II, French policy makers, in direct opposition to the technologists gathered in the Bureau of Consultation, believed that the state must intervene much more deeply in technological and industrial matters than their liberal attitudes toward individualism and the role of the market would recommend. Only through state intervention, these policy makers believed, could France achieve the same success in its economic rivalry with Great Britain that French arms achieved regularly on the battlefield. Liberal conceptions of the state's proper role in the industrial economy were increasingly sublimated in favor of a centralizing, statist view of international competition. This shift had important consequences for French industrial performance in the first decades of the nineteenth century.

The Work of Jean-Antoine Chaptal: Forging a New Industrial Order

Jean-Antoine Chaptal (1756–1832) was Napoleon Bonaparte's most influential Minister of the Interior, serving from November 1800 to July 1804.[79] From this powerful post, Chaptal initiated a harmonious and creative approach to improving French industrial competitiveness that melded Physiocratic notions concerning resources and liberal attitudes about the rights of the individual with an activist vision about the necessity of state action. Thanks to improved military and economic circumstances, Chaptal and his collaborators could develop institutions designed to develop and direct a coherent industrial policy in a manner that would have been inconceivable during the revolutionary decade. The military successes and territorial expansion of Bonapartist France provided both resources and political stability, but they also bedeviled those who thought that the best means of defeating a "nation of shopkeepers" was by manipulating the warp and weft of a loom, not the point of a bayonet.

In the first decade of the nineteenth century, Chaptal inaugurated an industrial policy that had more in common with the mode of state activism characteristic of the ancien régime than with the liberal intentions of the revolutionary decade. His ministry sounded the final death knell for an "idealist" interpretation of French liberalism. The revolutionary decade magnified "the threat from below," successively bankrupted many investors in new technologies or production processes, and initiated a seemingly never-ending conflict with England. Chaptal's reaction to these developments was to continue to escalate the role of the state in industrial matters.

Chaptal's distinguished career, both before and after his turn as Minister of the Interior, demonstrated a consistent approach to encouraging technological change. He recognized that dynamic entrepreneurialism was essential to fostering economic growth and that effective state involvement was necessary to industrial development.[80] From a solid bourgeois family with close ties to the world of international commerce, Chaptal was educated within the vibrant medical community of the southern city of Montpellier. There he was influenced by vitalism, freemasonry, and Physiocracy. He completed his education by studying medicine and chemistry in Paris. Institutionally supported by Montpellier's Royal Academy of Science and the Provincial Estates of both Languedoc and Gévaudan, Chaptal emerged as a leading scientific figure. Named a professor in 1780, he reported, on behalf of official bodies, on a number of chemical reactions with industrial applications. He published, in several different forms, the outline of his chemistry course. In 1789, he also published a chemical textbook, a natural extension of his earlier sociability and publications. For the rest of his life he produced treatises on the practical application of chemistry and on educational reform. He also disclosed the fruits of practical researches into bottlenecks in French technology. Believing firmly in the link between the search for scientific knowledge and its industrial application, Chaptal founded a highly profitable enterprise at La Paille to manufacture various salts and acids desperately needed by Languedoc's textile industries.

Although ennobled on the eve of the Revolution, Chaptal hoped for political reform. He joined the initial Jacobin Club of Montpellier. Arrested as a moderate and a Federalist sympathizer during the Terror, he was released (thanks to the intervention of Carnot and Berthollet) and put to work by the Committee of Public Safety. For most of 1794, he oversaw French gunpowder production. With the easing of the war crisis, Chaptal returned to teaching medicine in Montpellier and rapidly rebuilt his fortune by selling chemical products at grossly inflated prices. Elected to the National Institute in 1798, he soon replaced Berthollet as the chemistry professor of the École polytechnique. With the accession of Bonaparte, Chaptal became a councilor of state (replacing Laplace) thanks to the patronage of a fellow Montepellerin, Second Consul Jacques-Régis de Cambacérès. Chaptal was charged with improving the educational system. While serving on the Council of State, he impressed Napoleon, who in November 1800 named him the provisional Minister of the Interior, replacing Lucien Bonaparte.

In 1804, Chaptal left the Minister of the Interior and entered the Senate, where, for the next decade, he usually served as an officer. Named hereditary Count of Chanteloup in 1808, he expanded his chemical manufacturing activities to include three large-scale workshops in the vicinity of Paris and purchased a vast estate where he became deeply concerned—both scientifically and commercially—with improving the process of distilling sugar from grapes and the naturalization of both the sugar beet and merino sheep. During his stint as Minister of the Interior he had discovered the process that bears his name. Still used today, chaptalization consists of adding sugar to wine while it ages to promote fermentation. He was appointed to the Council of Commerce and Manufacturing created in 1810. During the Hundred Days, he served as Minister of Agriculture, Commerce and Industry. In 1819, Louis XVIII named Chaptal a peer of the realm. The same year, Chaptal published one of the first accounts of the nascent industrial revolution in France. Throughout the Restoration (1814–1830), Chaptal actively promoted improvements in scientific and technical education while employing his precepts to increase his fortune.[81]

Such a career illustrates an impressive degree of consistency—a trait that also characterized Chaptal's stint as Minister of the Interior, no matter how contradictory some of his ideas and actions appear to be to those who emphasize the liberalism of the early years of the Revolution or who condemn Bonaparte as an unreconstructed authoritarian. Before the Revolution, Chaptal had been strongly influenced by the Physiocrats, and he had read Adam Smith's *The Wealth of Nations* with approval not long after its publication in 1776. With the advantage of hindsight, he saw himself as a lifelong adherent of liberal economic theories, particularly, "the most sacred and most inviolable of individual rights"— i.e., the right of individuals to make their own economic decisions, most essentially in the disposal of their labor. Chaptal criticized earlier administrators from Colbert to Trudaine, stating that their excessive regulation of production was the cause of France's retardation relative to Great Britain.[82]

Yet this self-proclaimed disciple of Adam Smith played a major role in founding most of the state institutions that firmly guided the French economy throughout the nineteenth century. To understand this seeming paradox, we must begin with Chaptal's emphasis on the necessary bond between theory and practice in science, in education, and in matters of production. To apply theory successfully, and especially to achieve the proper balance between theory and practice, the market alone would not suffice; the government had to take a hand.[83] The seemingly limited state goal of achieving this balance had vast consequences for nineteenth-century France.

For Chaptal, the government's role in industry had three basic components. First, state involvement must mend damage dating from before 1789. By depriving those involved in the production and distribution of goods of their just "*considération*" in society, the Bourbon regime harmed French public spirit while alienating those involved in manufacturing and commerce from the government. Only active state sponsorship of the social value of commerce and industry could repair centuries of contempt. Second, in good Smithian fashion, he asserted that with regard to industry "the actions of government ought to be limited to facilitating supplies, guaranteeing property, opening markets to manufactured goods and to leaving industry to enjoy a most profound liberty.

One can rely on the producer to pay attention to all the rest." Yet after 1789 no government could be quite so "hands-off." To maintain social peace, the state must guarantee all French citizens gainful employment and apply the Revolutionary ideal of equality under the law to economic practice. Chaptal's justification for this third precept was—again— French "public spirit." He contrasted the French emphasis on equality guaranteed by the state with Britain, where "private interest directs all actions" and "whose selfishness . . . offers us a terrible example of what I claim." To Chaptal, the state must mediate the myriad private interests for the public good.[84] In practice, liberal theories required arbitration to function in a manner appropriate to the post-1789 French polity. Nor could contemporaries forget that market functionality was only a theoretical axiom. France was at war—on the battlefield, in the laboratory, and on the shop floor. Ultimately, every high-ranking servant of the Bonapartist state was dedicated to victory, not competition. Thus, the context for liberal practices required revision of laissez-faire doctrines.[85]

Revision did not, however, mean abandonment. The institutions developed by Chaptal or with his support had a laissez-faire core. Sometimes this liberal marrow was sucked away, but the later alteration of an institution must not blind us to the original intent of the founders. Recognizing the liberal basis of the actions of Chaptal and his collaborators, along with the practical limitations of liberalism, both in broad economic terms and in scientific and technological terms, is an essential step toward resolving the seeming paradox.

Educational reform was fundamental to Chaptal's approach to improving France's competitive position in technological matters while enabling the market to function more efficiently.[86] He benefited from his teaching experiences at elite educational institutions such as the Polytechnique and Montpellier's Medical School and from his survey of the problems facing French education undertaken for the Council of State. Chaptal perceived two educational deficiencies vulnerable to systematic state intervention. First, he wanted French workers to acquire practical experience in the use of advanced machinery and the latest production techniques. Second, he hoped to increase the quantity and quality of interaction among scientists, tinkerers, artisans, and entrepreneurs to permit and facilitate the development of practical solutions to shop-floor bottlenecks. Both goals reflected a desire to imitate what Chaptal perceived as the wellsprings of English technological prowess.[87] Nor did Chaptal ignore the uncertainties and harsh realities of the French political situation: education must "recreate the unity of the French, replace the passions with reason and reinforce the authority of the Republic."[88] For Chaptal, the secondary-school curriculum should center on cultivating citizens more than elevating individual capacities.[89]

Improving the mechanical knowledge of artisans was a difficult and necessary task if France was to construct more efficient machines or even build enough machines of known types. To address these related challenges, Chaptal rejected proposals to revive formal apprenticeships, in part on the authority of Adam Smith.[90] Instead, workers should learn the fundamentals of design, mechanics, and mathematics in primary schools to be established in every large municipality. Having mastered the basic knowledge necessary to "exercise a mechanical profession," workers would then be directed toward one of an elaborate system of trade schools according to the aptitudes and abilities identified at the primary level. These new schools were to be tied to production. Many were associated with one of the national manufactures, such as the Gobelins for chemistry or Sèvres for the uses of coal and furnaces. Under the watchful eyes of not only their teachers but also experienced entrepreneurs and foremen, these trade schools would demonstrate the application of scientific theory to industrial problems through heavy doses of hands-on experience working with and on the best available machines.[91] The educational goals of this system anticipated the Mechanics Institutes that appeared in Britain during the second decade of the nineteenth century.[92]

Spreading best-practice technique and providing access to advanced machinery were national priorities complicated by fragmented markets and disjointed transportation networks in a nation as large—and expanding as rapidly—as Napoleonic France. In December 1801, following the example of Parisian arms-making apprenticeships in the Year II, Chaptal proposed to bring experienced weavers (nominated by local authorities) from all over France to the renowned workshops of the Bauwens brothers at Passy outside Paris to learn the use of an improved flying shuttle. After finishing their apprenticeships, each weaver was to be given two of the new machines with the proviso that they instruct others at home.[93]

If such a grandiose scheme proved financially unfeasible, developing centralized depots for the teaching of best-practice technique was more reasonable. On May 24, 1802 (Prairial, Year X), the Musée des arts et métiers opened. Three large workshops were soon formed to provide hands-on instruction on the vast collection of machines. One workshop specialized in wood working, another in iron making and steel making, and a third in making precision instruments. The government intended to open branches of the Musée and its attendant workshops in Toulouse, Brussels, and Lyon, all of which would have a broad selection of up-to-date industrial machinery. Here Chaptal illustrated his commitment to spreading knowledge widely in a geographic sense, but also his recognition that there were important regional and local technological traditions that influenced the adoption patterns of new processes and machines. In June 1804, Chaptal revived the Écoles gratuit de filature (free spinning schools). In these schools—common under the ancien régime—foremen and experienced spinners (usually women) taught groups of 25 students (a mix of men and women) how to use and maintain new machines copied or acquired from Great Britain. When some of his organizational ideas for primary schools were rejected by the Legislature and the Council of State in February 1801, Chaptal ordered the Prefect of the Seine to pay a yearly salary of 1,200 francs to every primary-school master in the département. The same directive ordered those teaching girls to instruct them in spinning, knitting, and sewing. Finally, thanks to pressure from Chaptal, all students at the École des mines, the École des ponts et chaussées, and other specialized technical schools began to spend part of the year in the field applying their ideas and skills.[94]

Improved education in mechanics and mathematics and hands-on experience in the practical use of machinery were not to be restricted to the trade schools. In the course of their instruction, all boys ought to be exposed to machines actually used in production, said Chaptal. In his plan to improve the Directory's educational system, Chaptal sought to put these artifacts in every classroom, because "machines, once absorbed by eye and touch, give the advantage . . . to fix or form one uniform language for the arts [permitting more efficient] . . . transmission of discoveries in mechanics."[95] For most of the nineteenth century, secondary schools were the preserve of a broad socio-economic elite. To supplement these schools and their mostly classical curricula, Chaptal proposed to form specialized schools: an École spéciale d'agriculture et d'économie rurale, an École spéciale des arts mécaniques et chimiques, and more focused mining schools at Pézay in Savoy for lead and Gailautern along the Rhine for iron and coal. All featured practical training designed to foster technical knowledge among foremen and engineers and likely or potential entrepreneurs. Through the systematic application of chemistry and the mechanical arts, Chaptal expected that the productive forces of the nation would be unleashed to benefit French prosperity.[96]

Chaptal was not the sole architect of the new state institutions uniting educational and industrial goals. In 1780, the Duke de La Rochefoucauld-Liancourt founded an experimental school at his chateau that mixed primary education with hands-on mechanical experience. During the Revolution it was turned into a military school and transferred to Compiègne. Upon his return from exile in Great Britain and the United States, La Rochefoucauld-Liancourt sought to transform the establishment into a technical training school for skilled workers, foremen, engineers, and scientists. At Chaptal's urging, Bonaparte visited the school in 1799 and decreed that the current hodgepodge of subjects should give way to one that would "train petty officers for industry." Based on the recommendations of Monge, Berthollet, and Laplace, the establishment at Compiègne was reborn as the École des Arts et Métiers (School of Arts and Trades) in 1803. It opened with 503 students and a half-theoretical and half-practical curriculum. The top five graduates matriculated to the Conservatoire for more intensive training to earn the degree of ingénieur des arts et manufactures. In 1806 the school was moved to Châlons-sur-Marne and thoroughly militarized. La Rochefoucauld-Liancourt became inspector-general and instituted a more coherent curriculum intended to reap the industrial benefits of mixing theory and practice. A new branch opened at Beaupréau (later moved to Angers) in 1811, and a third was projected. Such a trajectory illustrated the collaborative nature of the institutional edifice created during the Consulate.[97]

Chaptal's other major institutional focus aimed to provide expert advice to the politically minded, technologically ignorant, and sometimes heavy-handed state administrators at every level of government who, increasingly, had to implement the government's economic dirigisme. He hoped to create a unified technological environment and a single productive climate just as the Revolution had made major strides toward creating a national political unity in France. For models of how to accomplish these tasks, Chaptal looked back to the ancien régime for state institutions and across the Channel for private organizations with an authoritative government presence.

For Chaptal, the interaction of scientists, entrepreneurs, and bureaucrats thought to be so common in England was the key both to solving production problems and to spreading useful knowledge. The Conseils d'agriculture, des arts et commerce founded in each département in June 1801 were so active that a Conseil supérieur du commerce with significant provincial representation was set up in December 1802 to coordinate the flood of suggested actions and improvements. In addition, Chaptal sponsored the revival of 23 municipal Chambers of Commerce (December 1802) in the largest cities—except Paris which was exempted for political reasons—and the formation of more than 150 Chambres consultatives de Manufactures, fabriques, arts et métiers (April 1803) in the small and medium-size urban areas. These institutions ensured that technical knowledge and market opportunities were known to those who might put it to use. They also pointed out bottlenecks and spread improved manufacturing methods.[98] Chaptal played an important part in the formation of the Society for the Encouragement of National Industry (November 1801). Though this was ostensibly a private organization, most of the initial 300 members were state officials recruited by Chaptal.[99]

In addition to bringing together "officials, scientists, merchants, manufacturers, artisans and inventors," the Society sought to "excite emulation, spread knowledge and support talent." At the first meeting held in Paris in January 1802, Chaptal was elected president. He was reelected unanimously each year until his death. The 1804 statutes identified six primary means of accomplishing the Society's lofty goals:

1. Distributing models, designs, or descriptions of new inventions.
2. Creating written instructions for useful and little-known processes.
3. Experiments and demonstrations to judge new methods which will be announced and published.

4. Reimbursements, advances and encouragements to meritorious artists.
5. Publication of a *Bulletin* distributed only to members containing the deliberations of the Society, its correspondence and announcing discoveries made in France or abroad.
6. The distribution of prizes.

These activities were supported by an annual subscription of 36 francs and supplemented by funds from the Ministry of the Interior and by generous personal donations from Chaptal and others. The Society contributed to a number of French technological advances and improved techniques with signal successes in the perfection of the Jacquard loom for silks in 1808 and the naturalization of the sugar beet. Later recipients of funds from the Society included Pasteur, Beau de Rochas, and the Lumière brothers.[100]

Politically, these institutions attempted to counter the widespread resistance among entrepreneurs and workers to government intervention favoring either the unfettered rule of market forces or technical innovation. According to Chaptal, Bonaparte was infuriated by such defiance.[101] But, thanks to Chaptal's influence, the lack of success in overcoming resistance by fiat from the central government led, at least initially, not to harsher or more dictatorial methods, but to renewed efforts to co-opt local people and local methods to accomplish state goals.[102] For Chaptal, these institutions emulated the English model of technological development anchored by the London Society for the Encouragement of Arts, Manufactures and Commerce (1754) and the radiating network of provincial scientific societies. In the absence of a parallel degree of private initiative, and in keeping with the long-standing tradition of French state activism in industrial matters, Chaptal and his bureaucratic collaborators substituted themselves, much as their predecessors under the ancien régime had when forming the Bureaus for the Encouragement of Industry in Rouen and Amiens in 1787. The parallel tracks of public and private institutions created during the Consulate demonstrate the difficulties Chaptal experienced in creating an institutional environment that would improve technological competence, motivate entrepreneurs to innovate, and support France's competition with England. Nowhere were these difficulties more apparent than in the legislation regarding the world of work enacted under Napoleon. As was mentioned above, Chaptal adopted from Smith a dedication to the essential right of individuals to make their own economic decisions, particularly with regard to the disposal of their labor. His conception of labor relations was fundamentally individualistic—on both sides of the bond. The infamous law of 22 Germinal, Year XI (April 12, 1803) outlawed all forms of coalition, including patrons' uniting against workers. Although the penalties prescribed for employers was much less severe than those for workers, this vision of the proper economic relationship between master and man owed more to the ancien régime than to the Revolution.

The popular restiveness associated with the Revolutionary legacy was, however, readily apparent in Napoleonic labor relations. Under the Directory, workers increasingly left employers who imposed or enforced the high degree of industrial discipline deemed necessary to maximizing the efficiency of production. A second difficulty was that many entrepreneurs, pressed to find competent and disciplined laborers, simply stole them from others with promises of higher wages and better conditions. Vociferous complaints from entrepreneurs all over France prompted the government to intervene in the name of international competitiveness.[103]

The ancien régime provided a model. On 9 Frimaire, Year XII (December 1, 1803), the *billets de congé* (discharge papers) were dusted off and revived as the *livret* (report book) issued by the municipality. By forcing all laborers to present their *livrets* before beginning employment, this mechanism would assure the "honesty" of workers and minimize their nomadism while also making it more difficult to pirate workers from another entrepreneur. A worker's entry date was inscribed in the livret by the employer, who then took change of the document. If the worker wished to leave, the employer had to certify that all commitments had been fulfilled before filling in the date and returning the *livret* (which also served as identification papers). The worker then had to inform the mayor or the deputy-mayor where he was going.[104]

The attempt at "social control" embodied by the *livret* and the responses of the laboring classes will be discussed in the next two chapters. Although the disciplinary apparatus deployed by the Napoleonic state was extensive, it was not completely one-sided. The law instituting the livret also delegated the power to adjudicate many kinds of conflicts between employers and laborers to the mayor, making justice more accessible to workers and less expensive. To permit conciliation and to minimize the employers' advantage, lawyers were forbidden to appear before this body. A further step in this direction was the creation on March 18, 1806 of a *conseil de prud'hommes* (arbitration board) composed of employers and skilled workers to arbitrate conflicts in the Lyon silk industry. Imitated informally throughout France, *conseils* were established in other industrial centers on June 11, 1809 to adjudicate disputes up to 100 francs.

Chaptal was no longer in office when the arbitration boards were created, but he was determined that workers, particularly skilled workers, not be permitted to dictate terms to their employers. To that end, on 29 Brumaire, Year XI (November 20, 1802) he created worker placement bureaus in major industrial centers to replace this important function of organized labor. For Chaptal and his collaborators, the market could not be trusted to allocate France's precious stock of potentially violent skilled workers with maximum efficiency and minimum social unrest.[105]

Markets were the key to improving France's industrial competitiveness. Chaptal believed that monopolizing the domestic market would provide the profits French producers needed to compete internationally, just as it did in Great Britain. In 1802–03, he intervened repeatedly to raise tariffs. But the domestic market alone would not suffice; Chaptal had an imperial vision of French potential that required France to recover its position as a great trading nation. In July 1803, to facilitate and encourage international commerce, he gave significant tax concessions to seven commercial entrepôts in major transit centers. In addition, he revived the great month-long commercial fairs of Saint-Denis and Beaucaire.[106]

Such concerns about trade had been shared by many of Chaptal's predecessors as Minister of the Interior. Chaptal fulfilled or revived many of their programs, including the industrial expositions. A circular sent to every département on 21 Brumaire, Year IX (November 13, 1800) announced: "The government's devotion to this noble cause stems from liberal principles, which have been ignored or neglected [during the revolution]. Now, this task has become the most important responsibility of the [new] government."[107] Chaptal, with Bonaparte's support, intended that industrial expositions occur regularly to display the results of the nationwide effort to improve French economic competitiveness. The expositions would display French successes, inform people of new technologies, and allow other industrial and scientific institutions to publicize their endeavors.

Expositions were also perceived as an important way to generate orders for entrepreneurs who developed or adopted advanced technology, thus providing an important profit incentive for innovation. Chaptal directed the prefects to appeal to the profit-seeking instincts of provincial merchants, manufacturers and artisans: "Explain to the manufacturers that their products are not destined to be part of a sterile display. Experience has shown that the first impulse at such a spectacle is curiosity, but the second impulse . . . is an irresistible attraction to buy the objects which arouse the curiosity. The reputation and the fortune of many manufacturers dates from . . . the Exposition of the Year VI."[108] The market would rule, but in a manner that required the systematic state intervention.

The industrial Exposition of the Year IX (1801) ran from September 17 to September 23. In the central courtyard of the Louvre, 243 exhibitors displayed more than 400 different products under 104 arches.[109] The exhibitors came from 33 French, three Belgian, and two Swiss départements.[110] As at the first exposition, the juries rewarded entrepreneurs displaying textiles or new technology with prime locations near the entrances. Exactly one-third of the exhibitors were textile producers, and textiles ultimately won seven of the twelve gold medals.[111] Most of the textiles, including flannels and calicos, were machine-made and were considered to be particularly notable because they were types or styles that had been exclusive English preserves. A further 34 participants (14 percent) displayed new machines, ranging from steam engines to an improved device to imprint designs on cotton cloth. Many machines were subsequently awarded brevets of invention. Chaptal informed the jury that a product's quality was at least as important as the price. This policy allowed the return of luxury goods and artisanal crafts, the traditional strength of French industry. The consumer products displayed in 1801, ranging from fine pottery to jewelry and from tapestries to high-quality furniture, signaled a broader conception of France's industrial promise.

The exposition held the following year had 540 exhibitors from 73 départements. Chaptal instructed the jury for the Exposition of the Year X to reward the application of new technology to the making of commercially viable everyday items.[112] More than 150 medals were awarded in explicit recognition of how well manufacturers had responded to the commercial challenge. Recognition by the jury also stimulated sales. The exhibitors included 139 textile producers (26 percent) and 39 individuals (7 percent) who displayed new machines or industrial processes.[113]

Exhibitors profited in several ways from their participation in these expositions. If they had sufficient stock on hand, they could sell their wares directly to the public. Nicolas Conté sold tens of thousands of francs' worth of his new graphite pencils at the exposition of the Year X.[114] Such sales seems to have been a major attraction for many provincial exhibitors. Others, mainly Parisians, garnered new orders from foreign and provincial wholesalers. For example, the medal-winning Paris-based firm of Richard and Lenoir whose twills were "judged infinitely superior to those of England" took orders worth more than 400,000 francs.[115] Lyon's silk embroiderers benefited impressively from their appearance at the exposition.[116] Public exposure at an exposition could also lead to government support, which might include a free brevet of invention, a financial subvention, rent-free use of state property, or even a no-interest loan. Mere participation could attract the patronage of government officials who could provide technical, commercial or advertising support. These powerful incentives superseded those offered by the market alone and enticed all kinds of entrepreneurs to participate in an industrial exposition.

This résumé of Chaptal's industrial activities is far from exhaustive. The *Bulletin de la Société d'Encouragement pour l'industrie nationale* and the *Annales des Arts et manufactures ou Mémoires technologiques sur les Découvertes modernes concernant les Arts, les Manufactures, l'Agriculture et le Commerce* (both of which received considerable subsidies from the Ministry and were distributed by it) provide further glimpses. These journals described prize contests for specific inventions and announced encouragements given to inventors or entrepreneurs who adopted advanced machinery. Chaptal targeted certain industries, such as woollens, and subsidized the construction and use of English-model machines.[117]

No matter what the state mandated, there were clear constraints on what could be accomplished on the shop floor. The *tondeurs* (cloth shearers) of the woollens industry of Sedan illustrated what government intervention could and could not accomplish. Enjoying a stranglehold even tighter than that of their Yorkshire counterparts, a coalition of workers prevented the city's merchants and manufacturers from expanding production. In 1803, Scipion Mourgue, an important manufacturer from Amiens and a willing accomplice in Chaptal's economic plans, was dispatched to investigate the situation. He reported: "Since 1788 and to perpetuate their vexatious influence, the cloth shearers have not permitted the manufacturers to place an apprentice with them except in proportion to natural deaths among them. For 15 years, the number of cloth shearers has fallen. Those who remain easily maintain the dependence of the manufacturers. From the moment when a manufacturer tries to train an apprentice they have not approved, all the cloth shearers make what they call a *cloque* [blister] (in other places it is referred to as "damning" or "outlawing" an establishment) and stop work. They receive from the other workers enough money to survive until the refractory manufacturer voluntarily agrees to remove the apprentice they did not wish to accept. The misfortunes that have already occurred, stemming from the general spirit of licentiousness that has prevailed in society for the last 14 years and plus the negligence of the administration has resulted in a fear of even greater problems that has kept the manufacturers in a firm dependence." Appalled, Mourgue explored how the new legal framework established by the labor laws of Germinal, Year XI enabled workers to escape their "bondage." He reported to Chaptal: "I can tell you freely that whatever hopes there have been to see private interest stimulate the commercial possibilities enough for machines to be introduced, the government ought to abandon them because private interests face greater obstacles [to mechanization] than the advantages they would derive. He has tried in vain to convince them of the power the government would use to punish even the smallest sedition on the part of the workers. All the manufacturers responded with a profound sentiment of terror: 'no doubt the government will punish them, but it is our families who will suffer and our workshops that will burn!'" Mourgue had high hopes that stern measures by the central government, overseen by a soon-to-be-established Consultative Chamber of Arts and Manufacturing, would overawe the workers and permit the introduction of the carding machines introduced by Guillaume Ternaux in Reims. Given that 29 prominent woollen manufacturers had been guillotined during the Terror and that the association of cloth shearers had enjoyed their "empire" in Sedan since the middle of the eighteenth century, it was not surprising that the "insolence" of the *tondeurs* outlasted the Napoleonic regime. As was noted in chapter 4, only in the transformed economic and political environment of the early 1820s did the cloth shearers lose control of the shop floors of Sedan and fail to prevent the introduction of new machinery.[118]

In light of the persistence of local traditions and challenges, evaluating the concrete effects of Chaptal's activities is difficult. Certainly, he and those who directed the institutions created during this period were far from infallible. For example, with Chaptal's full agreement, the Society for the Encouragement of the National Industry refused to support the improvement of Nicolas-Louis Robert's breakthrough paper-making machine—on two separate occasions. Neither Bonaparte nor Chaptal was much impressed by Robert Fulton's steamboat experiments until Fulton left France for North America.[119]

But from an institutional standpoint, Chaptal was the father of the nineteenth-century French economy. The distinctive deployment of state dirigisme melding theory and practice to further a liberal agenda, yet leaving the government in a stronger position than before, remained characteristic of the French industrial landscape for decades. It is this enduring mix of practicality, deregulation, and attention to containing the threat from below with a longer-term understanding of how to play industrial "catch up" that those most interested in exploring the "liberalism" of the French Revolution overlook. Chaptal's influence far outlasted his tenure as Minister of the Interior. For more than 25 years, his successors continued most of his policies and regularly asked his advice.[120] There is substantial contemporary evidence of an impressive increase in public interest in scientific and technological matters at every social level during the Napoleonic era.[121] France's industrial competitiveness also revived dramatically during the Consulate before declining under the Empire.

The rest of this book explores the contradictions and paradoxes embedded in the Chaptalian industrial framework. Chaptal's most important legacy was a generation in the making; it took that long for a coterie of entrepreneurs, tinkerers, and scientists trained in both theory and practice and accustomed to following the state's leadership yet oriented toward the market to germinate. These beneficiaries of the industrial system initiated during the transitional period 1795–1804 competed successfully with England not only in the theoretical sciences, but also in developing economically feasible industrial technology.

Chaptal's heritage must also be understood in the context of economic liberalism. Chaptal recognized the new context of the relationship among entrepreneurs, the laboring classes, and the state after 1789. Some French voiced a deep commitment to liberal ideas, but many more recognized the political and social drawbacks of this conception of economic decision making. As a result, the French and British policy paths to industrial society continued to diverge. The search for institutional means of implementing state industrial policy that began early in the Revolutionary decade was resolved concretely by Chaptal and his collaborators. Chaptal's reforms widened the gap between France's industrial practice of the nineteenth century and that of the eighteenth, as well as the gap in industrial practices between France and Britain. The Chaptalian approach to industrial development meant that for France the British model of economic transformation was a path not taken.

7

Facing Up to English Industrial Dominance: Industrial Policy from the Empire to the July Revolution, 1805–1830

In terms of industrial productivity and output, the ascendance of Great Britain in the first half of the nineteenth century is a common assumption. The overall trend may be evident, but there were significant peaks and valleys that should be incorporated into longer-term evaluations of economic change. Avoiding such an historical bulldozing of the twists and turns of economic change complicates our picture of the Industrial Revolution both in Britain and on the Continent and more accurately reflects the topography.[1] The highs and lows were not reciprocal. This chapter investigates economic competitiveness of both sides of the Channel at the height of the Empire's political and military sway and then explores how France coped with Napoleon's two defeats to achieve the high levels of growth discussed in chapter 1. It also argues that Britain's greatest benefit from victory in the revolutionary and Napoleonic wars was the defeat of a French industrial hegemony on the Continent created by force of arms.

Jean-Antoine Chaptal dominated early-nineteenth-century French industrial policy, but his policies were interrupted after Napoleon crowned himself Emperor in December 1804. Although Chaptal's disciples remained prominent among French economic policy makers, his approach to economic development, industrial expansion, and technological change was overshadowed by the Emperor's mercantilism. Under the Continental System (1806–1813), Chaptal's state-directed laissez-faire was abandoned in favor of political and military imperatives: Napoleon's campaign to dominate Europe economically was not rooted in French advantages in technology or production.

Napoleon's two defeats guaranteed Great Britain's industrial ascendancy. With the Emperor exiled to Saint-Helena, the approach to industrial development cobbled together under the Consulate returned to favor as a means of playing economic "catch up." The final section of the chapter depicts the changed circumstances of peacetime economic competition and how they affected French industrialization after 1815.

For political reasons, it was impossible for French policy makers to follow either of the two most effective approaches to economic growth at their disposal. Above all, their task was to narrow the productive and technological gap between Britain and France, which had widened considerably under the Empire. The continuity of economic policy makers from the Consulate to the Restoration (1815–1830) strongly influenced the path followed after 1815. For this long-serving coterie, the resumption of Chaptalian industrial policies was the logical alternative to politically impossible approaches. However, in contrast with the halcyon days of the Consulate, those directing French industrialization after 1815 confronted narrowed export possibilities in many sectors because Britain had both widened its technological lead and consolidated its political and imperial primacy. The final section of the chapter portrays how French policy makers stimulated technological innovation during the Restoration to enable a twice-defeated nation to recover while laying the foundations for greater industrial competitiveness later in the century.

The policy break with Chaptal's market-minded yet dirigiste approach was apparent soon after the proclamation of the Empire. The planning and implementation of the industrial exposition of 1806 were steps along the path that led to the Berlin and Milan Decrees. When the victory over Austria and Russia at Austerlitz in December 1805 did not convince the British to make peace, Napoleon turned to economic warfare to bring "Perfidious Albion" to its knees.[2] He linked imperial pretensions—military and economic—by using a Parisian industrial exposition to celebrate his latest crushing victory and the opportunities springing from it. Jean-Baptiste-Nompère Champagny, Chaptal's protégé and successor as Minister of the Interior, wrote: "This occasion will be consecrated to the two types of glory most precious to France. One is the immortal glory so justly obtained by the triumphs of French arms. The second is the peaceful glory which promises that our industry will be able to compete with our rival nations."[3] Each département's Consultative Chamber for Manufacturing, Industry, and Crafts was to choose the finest examples of its industry to send to Paris. In 1806, instead of Chaptal's market-oriented display of technologically sophisticated, well-made, or competitively priced manufactured goods, Napoleon wanted to present a "complete catalogue" of French products to the consuming public. This "catalogue" would complement the state's increasingly thorough efforts to survey French economic activity. From each département, the preliminary judging reports for this exposition were to be accompanied by a complete set of industrial statistics. Thus, the exposition's commercial and industrial goals were tied to the French state's growing emphasis on quantification.[4]

This grand objective, which was not related to an entrepreneurs' balance sheet, complicated the task of the prefects who were responsible for encouraging merchants and manufacturers to submit their goods to the departmental juries. Despite the success of the Expositions of the Years IX and X in generating orders for award winners, many artisans and manufacturers were leery of participating in the Exposition of 1806. They worried that the presence of exhibitors from the enlarged French nation and the invitations extended to other countries in France's economic orbit (Spain, Holland, and the Italian and German lands) would lead to "a lowering of prices" and thus a lowering of profits.[5]

These fears were accentuated in 1805 by a serious fiscal crisis. Government finances were in such dire straits that military suppliers and even the troops went unpaid. Farmers were not compensated for military requisitions. This instability drastically diminished the demand for manufactured goods.[6] To present a "complete catalogue" of French industry required broad and deep participation. Many merchants and artisans were uninterested in technical innovation and uninvolved in long-range or international commerce, which were the chief attractions of participating in a Parisian exposition. The recession left them even more timid. The trade environment also contributed to entrepreneurial uncertainty: an imperial decree of February 22, 1806 had already raised the tariff duties to extremely high levels for raw cotton and yarn while prohibiting the importation of white cloth, cotton cloth, or muslin. Anxieties about the effectiveness of such prohibitions and about the availability of raw cotton contributed to entrepreneurial ambivalence. These anxieties forced the government to postpone the beginning of the fourth Parisian industrial exposition from May 25 until late September of 1806.

To assuage the fears of timorous merchants, manufacturers, and artisans, local administrators had to take an even more active role in promoting the Exposition.[7] From Amiens, Nicolas Quinette suggested that, at a national level, participation in the Exposition be a requirement to receive military contracts and that the central government subsidize the transport and travel expenses of manufacturers who had previously won a medal.[8] When Champagny ignored his ideas, Quinette focused on convincing local entrepreneurs that the Exposition would help them forge new economic ties in the expanding empire. He also argued that Paris' "enormous population will be increased further by the victorious army and by other Frenchmen and strangers. It is also at the moment of peace . . . when new commercial relations will be established with the conquered peoples."[9] Quinette was so persuaded of the Exposition's potential benefits to local industrialists that he exhausted the departmental budget to pay the transportation costs of those chosen by the Somme's jury to go to Paris.

On September 26 the Exposition opened at the Esplanade of the Invalides. It was scheduled to run for three weeks. The inspired efforts of local and regional institutions led by the Consultative Chambers and the Chambers of Commerce increased the overall number of exhibitors from 540 in the Year X to 1,422 in 1806. This increase came from merchants, manufacturers, artisans, and inventors who had been thoroughly uninterested in attending previous expositions. Le Creusot first sent goods to an exposition in 1806.[10] Exhibitors also came from several foreign countries along with 104 French départements including 81 of the original 83. The Exposition reflected Bonaparte's imperial dream that a unified Europe-wide market that would allow the French to focus on exports to earn the profits to pay for industrial and scientific development. France could then achieve its manifest destiny—Continental industrial hegemony. The other parts of Europe would then function as the near-exclusive supplier and market for French industry in the same manner as England's colonies and former colonies overseas.

To judge more than 4,300 products, Champagny created four separate juries: one for mechanical arts, one for chemical arts, one for fine arts, and one for textiles.[11] Huge crowds attended the Exposition; contemporaries estimated, quite generously, that 10 percent of the population of Paris paid a visit.[12] Luxurious textiles made with high-quality thread that imitated English goods such as cotton calicos, muslins and percales along with cashmere and woollen velours, silk satins or novel mixed-fabric goods (blended silk-cashmere shawls, Welsh-style woollen flannels with a cotton weft) drew the most attention. Previous medal winners—including Abraham-Louis Breguet of Paris, who displayed machine-milled, hand-finished self-winding watches with second hands—attracted crowds of the curious, as did James Douglas, who displayed nine machines for preparing and spinning wool. The juries chose a wide range of goods and machines to be placed on permanent display in the Conservatoire to encourage entrepreneurs wary of their more conservative work forces to shift to mechanized production.[13] In the case of the merchant community of Cholet in the Maine-et-Loire, these hopes appear to have succeeded.[14] The judges' report claimed that France had made sufficient technical progress in recent years to be competitive with Britain: "[Under the Directory] it had become . . . urgent and essential [for French industrialists] to restore quality and to encourage a return to producing superior merchandise. . . . Today, this objective has been completely fulfilled. Our textiles are no longer inferior to those fabricated before 1789 and, in fact, they surpass them in many ways."[15]

The jury report of 1806 revealed a grave miscalculation by Napoleon and those charged with overseeing the industrial economy. Under the Consulate, the French had concentrated, with notable success, on the production of low-cost everyday products. The extensive support given to technical innovation and mechanization under Chaptal's ministry facilitated this focus.[16] The Empire, however, channeled state technical assistance toward luxury goods. In addition, whereas the Directory and the Consulate realized that most technical innovation came from artisans or small-scale manufacturers and used institutions such as the expositions and the Consultative Chambers to spread knowledge of new inventions or processes, under the Empire the government only favored products capable of large-scale commerce.[17] The Empire's territorial growth and the effects of the British blockade meant that only relatively expensive, generally higher-quality goods could be profitably sent long distances overland or shipped via inland waterways. This shift was displayed at the Exposition of 1806, which was more a trade fair than a showcase for inventive machines and industrial processes.

Publicized as a triumph, the shift visible at the Exposition of 1806 was almost completely politically motivated. Political considerations had always played a major role in French economic policy and in the course of industrialization, but the boundless ambitions of the Napoleonic regime distorted the process of industrialization under the Empire.[18] The most notoriously twisted warp on the French loom was the Continental System.

The Continental System: Mercantilism Revived

Napoleon's declaration of the French Empire in December 1804 began a new era in European politics; it also touched off a more intense effort to dominate Europe economically. Buoyed by the spectacular victories over Austria and Russia at Ulm and Austerlitz in late 1805, Napoleon and his collaborators progressively shifted French industrial policy in a more imperial direction to burnish the splendid image produced by Napoleon's military and political achievements. The crushing Franco-Spanish naval defeat at Trafalgar on October 21, 1805 clouded that vision for French commerce. Inability to defeat Britain at sea forced Napoleon to fight his enemies where he thought he could hurt them most—in their wallets. Napoleon's off-again, on-again disdain for the "nation of shopkeepers" led him to issue the Berlin Decree of November 21, 1806, which inaugurated the "Continental System." To consolidate political and military hegemony, the Continental System was intended to facilitate France's replacement of its rival Britain as Europe's industrial powerhouse.

The Berlin Decree placed Britain under a state of blockade to be enforced by France, its satellites, and its allies. In addition, European ports were closed not only to British shipping but also to any ship that put into Britain or its colonies on its way to a Continental destination. Of course, because of the eclipse of the French navy, Napoleon's ability of prevent ships from stopping at a British port was almost nil. The Empire could, however, prevent a ship from ever leaving a French-controlled port if it had docked in a British harbor. Thus, the French blockade of Britain amounted to a "self-blockade" to keep enemy goods out of Continental markets.

François Crouzet delineated three chief threats to the British economy in the fall of 1806. The occupation of northern Germany portended the confiscation of huge amounts of British goods and monumental losses for exporters. Second, the Berlin Decree might panic British public opinion, endangering public borrowing. Finally, if the Decree was strictly enforced, British exports to the Continent could fall drastically.[19] These measures responded directly to Britain's violent reaction to the humiliation of Russia and Austria at Austerlitz when the British declared a formal blockade of France on May 16, 1806.[20]

With mutual blockades in place, these long-time adversaries—depicted by contemporaries and by historians as a tiger and a shark—spent 1807 sewing up holes in their nets and sinking their hooks. The most conspicuous loophole in the blockades was the treatment of neutrals. This lacuna included both the transit function of sending goods across neutral territory and the use of neutral shipping which had undermined earlier blockades.[21] In January and November, the British issued Orders in Council that obliged neutral ships to get permission to trade in all non-national Continental ports. No ship could pass from one French controlled port to another. Neutral ships were supposed to deposit their cargoes in Britain, pay a tax, and get a license to trade. Obviously, continuing on to trade with France was prohibited, yet over time a significant number of licenses were issued allowing precisely that.[22]

The French returned the favor in November and December of 1807 in a series of measures known collectively as the Milan Decrees.[23] The Milan Decrees stemmed from Napoleon's incensed reaction to the British Orders in Council. By targeting the trade of neutrals, Napoleon intended to turn the blockade into a more effective instrument of war.[24] The Milan Decrees mandated the arrest of all British subjects on French-controlled soil and the confiscation of their possessions. Colonial goods and a number of other products were henceforth to be considered as "English" unless a contrary certificate of origin existed. All ships that complied with the Orders in Council by calling at British or British colonial ports were to be confiscated along with their cargos. These regulations applied not only to belligerents but also to neutrals. A British port of call effectively denationalized all ships and made them lawful prizes in port or at sea. Neutrals could no longer escape either the tiger or the shark.

Ultimately, the enforcement of the blockade led Napoleon into wasteful military adventures and contributed to the expansion of the French

Empire beyond the bounds of what could be ruled effectively by a bureaucratic state in the early years of the nineteenth century.[25] As in 1793–94, political and military requirements thoroughly distorted economic development and derailed certain types of technological improvements. After the Year II, such distortions did not doom France to perpetual second-class economic status. In the case of the Continental System, after 1815, however, French policy makers recognized in hindsight the opportunities missed in pursuit of political hegemony and military glory.

These observations notwithstanding, Louis Bergeron argued that the Napoleonic period was a decisive moment in development of French capitalism.[26] At the policy level, the Continental System revived a different view of economic competition. François Crouzet emphasized that this era reoriented the Continental economies away from production for export within the Atlantic trading network toward an axis demarcated by the Rhine. As a result, European industries looked inward and focused more on the national market.[27] Thus, for Continental Europe, the blockade reinstated mercantilism. Imperial France paid lip service to the need for scientific and technological advance; however, by avoiding direct industrial competition and distorting market relations, the Napoleonic economic warfare distorted earlier policies. A revived mercantilism permitted substantial numbers of French entrepreneurs to earn considerable profits without committing to technological improvement, thereby undermining the culture of industrial competitiveness fostered by Chaptal.

Bertrand de Jouvenel asserted that Napoleon envisaged increased French exports as the best means of avoiding industrial downturns.[28] Barring British goods and replacing them with French goods depended on finding opportunities in central and eastern Europe to replace colonial raw materials and market outlets. A decree issued on April 30, 1806 established a "drawback," imitating the English measure of the same name. In the French version, manufacturers who paid the increased import duty on raw cotton could recoup what they exported at a rate of 50 francs per metric quintal.[29] If Trafalgar ultimately permitted Britain to become "the workshop of the world," Imperial France aimed to be "the workshop of Europe." Michael Broers stated that this abandonment of the Revolutionary principle of the equality of peoples in the Empire to set up an "uncommon market" or rather a one-way common market was perhaps the major reason for Napoleon's ultimate failure.[30] Geoffrey Ellis depicted these policies as born both of military and political strength and industrial vulnerability; in other words, they emerged from a unique set of circumstances at a critical moment in French industrial development.[31]

Many of these long-term consequences were not intended either by the state or by producers. Supporters of economic warfare, including Minister of Finance Martin-Michel-Charles Gaudin (soon to become Duke de Gaète), believed that the blockade would force Britain to "succumb soon."[32] Imperial policy makers did not imagine that the blockade would endure for nearly seven years or that the war with England would continue uninterrupted until April 1814. The French also had high hopes for developing trade with the United States—especially after April of 1806, when the U.S. Congress passed a ban on British imports that was to take effect six months later.[33] These measures bolstered the confidence of French entrepreneurs hard-hit by British competition and encouraged them to expand production and make new investments.[34] Thus, policy makers who hoped that the blockade would provide a breathing space for French industry did not anticipate the depth or duration of depression experienced by the ports.[35]

When Napoleon and his ministers recognized the extent of economic dislocation, they took action. They provided loans to manufacturers with extensive inventories of unsold goods and ensured that needy large-scale firms interested in technological innovation had sufficient credit to keep their work force fully employed. The French government devoted more than 7 million francs to these two programs in 1807–08. To acquire useful goods and raw materials while eliminating backlogs of agricultural products, Napoleon initiated the system of "licenses," which permitted a certain level of exchange with Great Britain. These measures and the industrial dislocation that spawned them were not foreseen when the Continental System was inaugurated.[36]

The blockades were the most remarked-upon diplomatic aspect of the new approach to economic warfare. They drew England into conflict with the United States, and they drew France into unwanted commitments in Italy, the German lands, the Netherlands, and the Iberian Peninsula. However, the blockades were built on earlier measures and earlier policies.[37] For example, Napoleon's Decree of Jena (October 15, 1806) declaring that all English merchandise found in conquered territories would be seized and turned over to the army had been tried during the revolutionary decade.[38] In Britain, the Order in Council of January 7, 1807, which throttled European coastal shipping, extended measures passed in 1756 at the height of the Seven Years' War.[39] As Ellis reminded us, emphasizing the overseas trade aspect of the blockade and its effects overly privileged the maritime element of the system.[40] All the trade provisions of the Continental System had been implemented earlier, though never as systematically.[41]

Napoleon's reasoning considered international politics first, but there were substantive reasons to believe that France might be able to pick up the industrial slack left by what French policy makers hoped was the ebb tide of British exports to the Continent. An important component of this hope was the enlargement of French territory. Many contemporary historians depict the expansion of French borders solely in terms of the political unrest that resulted from imperial conscription, smuggling, and religious policies. But through the prism of the economy, territorial expansion refracted very differently. The addition of northwestern Italy, the Rhineland, Belgium, and other areas provided 14 million more consumers for French producers, lengthening the French demographic lead over Great Britain. Secondly, the annexed territories provided reservoirs of mechanical and/or technical knowledge in metallurgy, machine building, coal mining, and other sectors. The established industrial centers of Piedmont, Belgium, and the Rhineland could be tapped to complement the needs of the French core in ways that presaged a promising synergy. These areas also contributed another form of human capital in their profusion of entrepreneurs, many of whom were accustomed to facing up to British competition and had experience in satisfying diverse sets of customers in long-distance trade. Finally, the expanding frontier encompassed important reserves of raw materials that France had either lacked or had insufficient stocks of, notably Italian silk, Belgian coal, and Rhenish iron.[42]

The département of the Roër (Ruhr) highlighted the economic significance of annexation. At incorporation, the Roër had a minimum of

3,000 industrial enterprises, nearly half of them "large." These enterprises employed more than 120,000 men, women, and children and exported at least 100 million francs' worth of goods outside the département. An additional 334 mining companies, employing more than 10,000, provided, among other things, more than 180,000 metric quintals of iron suitable for making cementation steel. While part of France, inhabitants of the Roër obtained fourteen brevets of invention. Advanced machines such as mule jennies, flying shuttles, and steam engines were installed in a variety of industries at a multitude of sites.[43] Access to imperial scientific and mechanical expertise developed these technological capacities. If France began the revolutionary era as a demographic colossus, the addition of such productive areas allowed the Empire to extend those advantages in impressive fashion.

The increased emphasis on quantification and on record keeping provided more grist for those who hoped that France could replace Britain in the markets of Europe. In March 1807, François Bardel, a large-scale manufacturer in Paris and a member of the Consultative Bureau of Arts and Manufactures charged with collating imperial industrial statistics for the Ministry of the Interior, reported: "Since the law of February 22, 1806, calicos are manufactured in nearly every part of France. All the formerly motionless machines have been started up again. Despite this, all cotton goods are selling very well with demand proportionate to the number of existing machines. There is no fear that there will not be enough [machines for all the entrepreneurs who want them] and the price [of cotton goods] has stabilized. . . ."[44] Bardel was not exaggerating. An industrial inquiry conducted in 1806 revealed that the French economy had made impressive advances: more than half of all French cotton enterprises had been founded since 1800.[45] Reliable data on national growth or production for the Consulate are non-existent, but the fragmentary documentary record and the autobiographies of important policy makers attesting to significant growth in this period[46] has been confirmed by local studies of important industrial regions.[47] Foreign trade increased from 553 million francs in the Year VIII (1799–1800) to 790 million in the Year X (1801–02) with further expansion in 1802–1806.[48] Trade statistics also demonstrated that French imports of crucial raw materials (such as tin) and common industrial products (such as pottery) were shrinking.[49] Impersonal statistical aggregates were not the only cause for optimism.

Many important and innovative entrepreneurs believed that France could furnish the goods demanded by Continental Europeans. Lièwen Bauwens, one of the largest cotton spinners in France, engaged in a well-known public debate with former Minister of the Interior and now Senator François de Neufchâteau about whether the French cotton industry could withstand British competition. They also disputed whether state industrial policy should concentrate on sectors like linen where the British technological edge was less significant. Bauwens argued that the French cotton industry was an important weapon in the fight against the hated foe: England depended on the demand for its cotton goods to force its way into markets while maintaining employment and the national balance of payments. In peacetime, Bauwens contended that France would be able to compete without great difficulty because it could import raw cotton from the colonies. Until then, however, the state must continue to support the industry to set the stage for future prosperity.[50] Bauwens had won medals at the Expositions of the Years IX, X, and 1806. His reputation stemmed from using the flying shuttle to spin cotton. He spirited the device out of England one step ahead of the authorities. The confidence of Bauwens and other manufacturers rested on powerful economic incentives to produce rather than on improvements in technological capacity.

In 1807, Guillaume Ternaux, a technologically innovative and award-winning entrepreneur with large-scale operations in woollens at Sedan, Reims, Louviers, and Verviers that employed 12,000 workers, declared: "And I also will make war on England!"[51] Eighty-four Norman cotton manufacturers petitioned Napoleon to bar British goods from the Continent. They "pledged themselves to furnish 600,000 bolts." In view of the impressive enumeration of productive capacity by the signers of the pledge, this figure, equivalent to 60 percent of British imports, carried considerable weight.[52] François Richard, known as Richard-Lenoir, was the pre-eminent cotton manufacturer of Imperial France. He went even further than the Normans; he claimed to have asked Napoleon for many of the measures incorporated into the Continental System. Richard assured the Emperor that "in less than two years, France would be able [to produce enough cotton goods] to suffice for domestic consumption." He depicted the French industrial situation in simple yet evocative terms: "We have been born, but as children we cannot fight against grown men in the fullness of their strength, but in a few years we will be able to do so."[53] Many such voices reassured Napoleon that French industry was maturing quickly.

The French entrepreneurial elite was as good as its word. The output of machine-spun cotton yarn doubled between 1806 and 1808 and again (at a minimum) from 1808 to 1810. Total production increased at an even greater rate: from 2 million kilograms to more than 4.5 million in 1808. Exported cotton goods increased tenfold in the same four-year span. Not only did cottons do well; silks, woollens, mixed-material textiles, and hardware also experienced export booms. At the same time, French industry retooled, often with the latest technology, and mechanized enough to finally surpass the production levels of the 1780s.[54] At the height of the Empire, French industrial production was approximately 50 percent greater than under the ancien régime. A symbol of this growth was the increase in the number of large-scale mechanized cotton mills. In 1789, there were six in France. By 1814, there were 272. The consumption of raw cotton (which is a rough index of output) increased about 300 percent between 1790 and 1810.

This is not to say that France caught up to Great Britain. If, at the end of the Empire, the Continental cotton industry boasted 1.5 million spindles, of which about 1 million were in France proper with smaller concentrations in Saxony (250,000) and Switzerland (150,000), Great Britain still had an overwhelming advantage with 4.9 million spindles in 1811. However, over a ten-year span, Britain's lead clearly diminished as France's protected industries developed even more rapidly thanks to the hot-house environment of the Continental System. The rapidity and the extent of French expansion in this period have been ignored or distorted by longer-term evaluations of relative growth.[55]

With vast market possibilities to exploit, and with the working classes relatively quiescent because of the high-wage environment created by conscription, the French utilized the protection granted by the self-blockade to mechanize, develop and/or adopt new technologies. Reassured mightily by the mercantilist policies and even more so by the powerful repressive apparatus deployed successfully by the Napoleonic state, investors poured money into the industrial sector under the Consulate and the Empire. This investment foreshadowed the capital-intensive economic development of the middle third of the century.[56] In Alsace's vibrant calico industry, cylinder printing first appeared at Mulhouse in 1805 and spread rapidly. Spinning received the bulk of the other technical improvements, but the quality of artificial dyes and methods of bleaching also made notable progress.[57] Other textile innovations of the period were the Jacquard loom, the new cotton-printing machines developed by Samuel Widmer, and Philippe de Girard's wet flax-spinning process.[58] Both Alsace and Normandy witnessed widespread replacement of obsolete spinning wheels by mule jennies and water frames. Large factories began to be built after 1806, and steam power was first used to run them in 1812.[59] Another novel feature of production under the Empire was the increased formation of large vertically integrated firms, usually family operated or directed by limited partnerships. The various imperial trade restrictions favoring France encouraged investment in mechanization and in new technology. By 1811, 6 million francs' worth of Swiss capital had been invested in the Alsatian textile industry. Changes in management styles, in the manner of capital infusion, and in production techniques were noted in many of the textile regions of northern France, especially in the North (Lille-Roubaix-Tourcoing, Saint-Quentin, Valenciennes), in Normandy (Rouen, Louviers, Elbeuf), in Champagne-Ardennes (Reims, Sedan, Troyes), and in Alsace (Mulhouse).[60]

Technological innovation in Normandy accelerated under the Directory and the Consulate, stimulated by the arrival of a new wave of English entrepreneurs. Valentin Rawle built a hydraulic cotton spinning mill at Déville near Rouen in 1798 and constructed another on the opposite bank in 1804 at a cost of nearly a million francs. In 1801 he had mule jennies (brought from Manchester) installed by an English technician named John Ford. Five years later, his two mills had 18,000 spindles, employed 950, and produced 162,000 kilograms of cotton thread per year.[61] Led by Ezéchiel-Prosper Bélanger and other noted technicians who specialized in designing and building hydraulic motors, the major development of the Imperial era in Normandy was the widespread utilization of the network of streams flowing into the lower reaches of the

Seine River to power increasing numbers of textile machines. Although French industrialization has often been denigrated for being less dependent on steam engines than its British rival,[62] steam was not the only or even the best way to power industrial machinery, particularly before 1820. France had abundant sites suitable for the use of water power. By 1823, in the département of the Seine-Inférieure alone, there were 121 hydraulic spinning mills. Water power was also used by the pioneering copper foundry at Romilly and by other hardware establishments.[63] That France had less coal than Britain should not be overestimated as a factor in French industrial or technical power or potential. The French focus on water power was eminently rational. Many entrepreneurs in Britain, the champion of steam power, would have focused on hydraulic mills if sites had been available.[64]

The Continental System stimulated the continued evolution of French industrial geography. For the revolutionary era as a whole, northern France benefited while southern France and the hinterlands of the Atlantic ports experienced de-industrialization. The cotton boom and the expansion of French borders to the east and the north only accentuated the shift. In Saint-Quentin and in the Lille-Roubaix-Tourcoing region, cotton production replaced or complemented less dynamic textile sectors. In Mulhouse, and particularly in Paris (which contemporaries recognized as an emerging center of large-scale industry), textile manufacture sprouted in the sheltered environment of the Continental System.[65] By 1807, Paris boasted 12,000 cotton workers, almost two-thirds of whom labored in Richard-Lenoir's workshops. The cotton industry set down enduring roots that weathered the prolonged economic difficulties of 1810–1818 and provided the basis for long-term growth after 1815. The large-scale industry that survived in the southeast even after maritime trade rekindled—in Marseille, Clermont, Carcassonne, Lodève, and Bédarieux—focused on specialty production, various forms of government support, or unique resources.[66]

The Continental System was conceived as a weapon of war; it was perhaps the most effective offensive tactic available to France in the conflict with Great Britain. It was grounded in mercantilist conceptions of economic development that eschewed market competition as a means of channeling productive resources. By annexing some of Europe's most significant industrial regions and attempting to reduce the rest of the Continent to furnishing raw materials and markets for French industry, Napoleonic policy makers envisioned a mercantilist Continental System. The Continental System was the industrial complement to the general agricultural prosperity of the period. It was also to substitute for France's lost colonies and for its former access to the Spanish and Portuguese colonial empires.[67]

Continental colonialism benefited inland transit points, including Lyon and (especially) Strasbourg, that lay on the main overland routes to the east and the north. But the self-blockade meant that Bordeaux, Nantes, Le Havre, and other Atlantic ports that had maintained a reasonable level of colonial trade via neutrals would wither on the vine along with their hinterlands. The Atlantic-oriented commercial and industrial networks that had been so profitable and so productive in the eighteenth century did not survive into the nineteenth.[68] Napoleonic economic policies contributed greatly to their decline.

Despite the deep depression of the ports and the industrial slump suffered by southern France, the Continental System enabled many industrial areas of the Empire to flourish. Impressive gains in total production and in technology occurred in the cotton, silk, and chemical industries. Yet Napoleonic mercantilism could not overcome a basis economic fact of the period: land transport was slower and vastly more expensive than portage by water. As a result, the flows of raw materials and manufactured goods were retarded. France, therefore, suffered a perpetual competitive disadvantage vis-à-vis Great Britain.[69] Only when the European railroad network developed, around the middle of the nineteenth century, did a system of interior communications as envisioned by imperial policy makers become a viable economic alternative to the sea routes closed off by Britain or lost to the self-blockade. The Napoleonic resort to mercantilism failed, at least in part, because of insufficient development of the inland transportation network,[70] yet the impressive industrial growth of the first years of the Continental System revealed that, although the attempt failed, imperial evaluations of French potential were not essentially misplaced.

The mercantilist orientation of French policy under the Empire departed from earlier initiatives undertaken by Chaptal and his collaborators. The imperial state prodded French entrepreneurs to take advantage of trade opportunities rather than compete industrially with Britain through technological improvement. This short-sighted conception of trade failed because of the length of the war. It also halted French technical amelioration in its tracks and ensured English industrial domination after the imperial eagle had fallen for the last time.

The Continental System II: European Reactions and Wartime Depression

Most commentators note the disdain with which the British treated both the French attempt to keep their goods out of Continental markets and the French ambition to replace them. This disdain stemmed in part from confidence, but a liberal dose of hubris was involved. We must not be blinded by statements like Sir Francis Ivernois' well-known polemical poem: "Your blockade doesn't block/And thanks to our nautical skill/The folk you think you're starving/Can more than eat their fill." Compelling though much less egotistical data from the era of the Continental System uncovered by economic historians shed a penetrating light on the long-standing economic duel between traditional rivals.[71]

François Crouzet wrote the definitive study of the British economy during the Continental System.[72] He found that that "business as usual" conditions prevailed in the months after the French declaration of blockade. The occupation of the Hanseatic ports jolted confidence, but it was only at the end of the year that the recession threatened to deepen into a depression. January–February and September–December of 1807 witnessed slumps, but different industries, sectors, and regions enjoyed uneven fortunes. The ongoing plight of the woollens industry and the deep decline in exports of both cotton yarn and iron and steel goods were balanced by constant exports of finished cotton stuffs, and partly compensated for by increased shipments abroad of copperware and brassware. These strengths were complemented by some often-forgotten industrial products (refined sugar and salt) and supplemented by greater quantities of re-exported colonial products. Historians who seek to downplay the stakes of the Anglo-French economic rivalry often cite official export statistics depicting only a mild decline in British manufactured exports. These statistics, which show a British decline from £27,403,000 in 1806 to £25,171,000 to 1807 while total exports dropped from £36,527,000 to £34,567,000 in the same period, minimized the crisis facing some sectors, some industries, and some regions in the immediate aftermath of the declaration of the Continental System.[73]

Although there was a 5 percent total decline in official British exports in 1806–07, the percentage of exports going to Europe fell much more drastically, reaching a low (in real terms) of 28.0 percent in 1808 from a usual share in the low forties. Crouzet argued that exports to the Continent had been in decline for some time, and that the Continental System only accentuated the trend. Yet exports of colonial products and other agricultural goods to Europe increased overall, in real value and as a percentage of total exports, thereby highlighting the difficulties of the manufacturing sector.

The notorious Orders in Council of November 11 and December 18, 1807 were Britain's response to the growing efficacy of the Continental self-blockade. Crouzet asserted that the revocation of the right of neutral shipping to pass through the blockade marked a turning point in the economic history of the period. Although provoked by the Milan Decrees, the British Orders in Council sought to eliminate competition for Continental commerce and to increase demand for both colonial and manufactured goods.

Cutting off the booming trade conducted by neutrals on the part of both the tiger and the shark seemingly had drastic consequences for

Table 7.1
Real values of British exports to Europe, 1805–1811, in pounds sterling. Source: Crouzet 1958, II: 884, 886, 888.

	Manufactured goods	(Colonial) re-exports	Total exports
1805	41,069,000 (33.2%)	10,040,000 (67.8%)	51,109,000 (40%)
1806	43,242,000 (26.3%)	9,787,000 (63.2%)	53,029,000 (33.1%)
1807	40,480,000 (22.2%)	10,003,000 (64.2%)	50,483,000 (30.5%)
1808	40,883,000 (22.1%)	9,088,000 (54.7%)	49,970,000 (28.0%)
1809	50,242,000 (31.5%)	15,775,000 (71.9%)	66,018,000 (41.2%)
1810	49,976,000 (31.3%)	12,727,000 (67.5%)	62,702,000 (38.6%)
1811	34,917,000 (36.8%)	9,022,000 (63.2%)	43,940,000 (42.2%)

Britain. Having antagonized the United States, Britain saw its exports to the U.S. decline by 66.2 percent from 1807 to 1808. Exports to Europe continued to drop, but less dramatically than the previously year (12.1 percent vs. 9.3 percent). British exports to the US and Europe reached their lows for the period 1802–1811. The British economy—particularly export-dependent industries—was kept afloat by smuggling and by the much-remarked-upon burgeoning of markets elsewhere in the Americas. Expanded trade came on the heels of repeated British military intervention, notably the occupation of Buenos Aires in 1806 and of Montevideo the following year.[74] Britain also benefited greatly from the subjugation of French and Dutch colonies and the definitive independence of Haiti (the former Saint-Domingue). Looking back, contemporary commentator J. R. McCullogh described the speculative fervor of British entrepreneurs: ". . . more Manchester goods were sent out [to Latin America] in the course of a few weeks that had been consumed in the twenty years preceding; and the quantity of English goods of all sorts . . . was so great that warehouses could not . . . contain them. . . . Some speculators even went so far as to send *skates* to Rio [de] Janeiro."[75] Thus, the Continental System struck a severe blow to the British manufacturing sector. The decline of exports by approximately one-fourth (between the first semesters of 1808 and 1806) was, however temporary.

The difficulties of enforcing the French self-blockade allowed Heligoland, Malta, Portugal, and the Netherlands to become the chief conduits for both licit and illicit British trade. The Continental System depended both on European peace and on the opportunities afforded by the unique diplomatic situation after the Treaty of Tilsit. (See below.) With the return of war on the Continent, British exports to Europe recovered spectacularly in 1809–10, reaching commanding new heights. The recovery was due in part to the system of licenses instituted by Napoleon. Such trade seems to have benefited both economies. Official British exports to France reached £1 million in the first six months of 1810. This boom did not come at the expense of France, rather British exports complemented Continental growth. Thus, conjuncture rather than zero-sum competition describes the European economy during the first half of the Empire.

A conjuncture was evident not only in the demand-driven export boom of 1809–10 but also in the bust of 1811. The depression of 1811 rocked the French Empire but had even more profound repercussions in Britain. Most accounts link this depression to the financial crisis that occurred at the end of 1810 as a result of industrial overproduction, excessive speculation in goods destined for the European market, and the collapse of trade with the United States. French licenses temporarily ended, and there was a major crackdown on smuggling dating from the Fontainebleau Decree. The North German duchy of Oldenburg was occupied. This tightening of the self-blockade was made possible by the Austrian defeat at Wagram in 1809. This trade imbroglio contributed substantially to British woes[76] by accentuating a prolonged and generalized slump in demand for British manufactured goods and colonial products. In the wake of these developments, a cascade of bankruptcies threatened the stability of the English economy and left tens of thousands of workers unemployed.

The economic crisis coincided with historically high bread prices. A currency shortage following heavy grain purchases from France in 1809–10 to make up domestic shortfalls accentuated the financial plight. According to some informed contemporary observers, an earlier monetary devaluation of 30 percent seemed to have caught up to Britain at this time.[77] Britain suffered its worst economic setback in a generation. It affected even the boom sectors of cotton goods and colonial raw materials. In the environs of Manchester, one-third of the cotton weavers had no regular work because of the collapse of the U.S. export trade. The pain was even more widespread among the 12,000 cotton spinners. Only 3,000 were fully employed, while 2,500 worked part-time. Overall, Britain's exports fell by 30 percent and its total industrial output by about 25 percent in 1811.

This era also witnessed the steady increase in British state repression discussed in chapter 4. With the custom protecting workers and guaranteeing their economic independence sabotaged, the outbreak of Luddism in segments of the working classes disgruntled by technological change took on new meaning.[78] Prodded and protected by public support stemming from the economic situation, the Luddites may or may not have been a "quasi-revolutionary movement," as E. P. Thompson argued, but the Luddites certainly posed a threat to public order and to Britain's capacity to wage war. They may even have affected the "First Industrial Nation's" ability to maintain the momentum of the nascent Industrial "Revolution."[79]

Despite the best efforts of the Luddites, however, 1811–12 was not 1789. Wartime Britain was not the impotent, tottering France of Louis XVI's failing years. Although the British government was divided over Catholic emancipation and how to handle George III's declining physical health and tenuous mental equilibrium, even such shocks as the assassination of Prime Minister Spencer Percival in April 1812 could not shake the foundations of Hanoverian rule. Underlying fears about the possible fragility of the political control exercised by the British elite could be seen in the heightened military reaction of the weak Liverpool ministry to the Luddites. This policy continued for decades. Repression permitted the continued mechanization of crucial sectors. The British state also committed impressive resources to maintaining industrial employment. This intervention was conducted in an economic environment punctuated by plummeting customs revenues, burgeoning debt, and a shortage of ready cash. In March 1811, in a stunning but not-remarked-upon parallel to Napoleon's 1807 actions, Parliament agreed to lend up to £6 million to manufacturers at 5 percent interest if they kept their workers employed. Only 119 entrepreneurs took the British government up on its generous offer. The rest appear to have preferred to lower their prices to clear inventory, even if it meant taking a loss or forcing workers onto the parish dole.

The economic crisis of 1811 was short-lived. By the end of that year, and more visibly in 1812, economic recovery began on both sides of the Channel. In Britain, textiles paced the revival. In 1812, an across-the-board improvement in total output, industrial production, and the re-export trade followed. This recovery was aided by the progressive disintegration of the self-blockade after the French invasion of Russia in June 1812.[80] The process picked up speed in 1813 and continued until the definitive end of the Napoleonic Wars in June 1815. Yet it is worth emphasizing that, although the cotton industry and some other industries (linen, silk, coal, iron, steel) registered impressive growth in the succeeding decade, in constant prices Britain's industrial output in was less in 1821 than in 1811, a notably depressed year, despite an 8 percent increase in the percentage of national wages and salaries paid to the industrial sector.[81]

For Britain the outcome of the conflict with Napoleonic France was victory, but it was by no means the unchallenged triumph so often depicted by historians. Nor did peace bring either domestic social peace or rapid industrial growth. If Britain emerged from nearly a quarter-century of warfare in better competitive shape than France, as it indubitably did, the basis of this success requires careful examination. British economic growth was certainly reduced by the Revolutionary and Napoleonic wars. "No economic historian would now say that the French wars were good for the British economy," Crouzet asserts.[82] The chief difference was that structural changes in British industrial production accelerated rapidly during this period, while the pace of French mechanization and technological improvement stagnated under the Empire thanks to the discovery and employment of alternate means of profit taking,[83] contributing mightily to the British industrial edge after 1815.[84] The significance of the Luddites' failure to impede mechanization could not be plainer.

It was during the Continental System that Britain lengthened its industrial lead over its chief Continental rival. If France, with a stable population, had 20 percent more industrial output in 1815 than under the ancien régime, British industry and population growth spurted in the same period. Britain's population nearly doubled between 1781 and 1831, and its industrial output expanded by 250–300 percent between 1790 and 1821.[85] The Continental System helps to explain why French industry did not keep pace with Britain, despite the high quality of its scientific establishment, its huge demographic weight, and the vast potential markets won by imperial arms. During the Restoration and later, French industrial practice and policy had to play "catch up" even more desperately than before. Thus, the conclusion that the defeat of Napoleon set the stage for English economic domination is accurate, but it obscures the actual sources of that industrial success.

English economic hegemony could not be predicted when a victorious Napoleon and Tsar Alexander I of Russia signed the Treaty of Tilsit in July 1807. At least up until that moment, French industry seemed to be making rapid strides. From 1808, French cotton manufacture was queezed by the limited availability of raw cotton and the high price and relatively low quality of what was smuggled in or arrived from the Levant. Even mechanized spinners concentrated on lower-quality threads. Hand spinners were hard pressed to compete. Dyers suffered from steep price increases in or the unavailability of key raw materials.[86] The war in Spain curtailed the supply of merino wool, particularly after 1810, while limiting access to the Latin American market.[87] It is unsurprising that the loss of neutral trade affected imperial France more than Great Britain.

Loopholes undermined the self-blockade. Throughout the revolutionary era, high consumer demand for British manufactures and inexpensive colonial products encouraged smugglers hoping for windfall profits, particularly along the Baltic coast and in the Netherlands. Another major difficulty faced by the French stemmed from the very nature of the Continental System. By subordinating so drastically the needs of the allied and satellite nations to those of France, Napoleon made it impossible for even his firmest allies or his neediest dependents to uphold his policies. The Grand Duke of Frankfurt, the King of Saxony, and the Pope openly allowed the transit of contraband across their territories. Field Marshal Joachim Murat, made King of Naples after Joseph's switch to Spain, circumvented the restrictions on maritime trade. But perhaps most nettlesome was Napoleon's younger brother Louis, since 1806 the king of Holland. Louis not only closed his eyes to an extensive contraband trade, but he also openly flaunted the self-blockade by opening Holland's ports to U.S. ships in June 1809. Only the establishment of an interior customs cordon from the Rhine to the Trave forced Louis to back down. Similar flouting of the customs in 1810 led to Louis' dethronement and the annexation of the Netherlands. Holes in the Blockade led France to expand into northwestern Germany, the Netherlands, Illyria, and the Papal States, exacerbating the political discontent that ultimately helped to bring down the Empire.[88]

Clear evidence of a mounting and highly lucrative contraband trade accompanied falling French exports. Although these figures were far less dire than they may seem (because areas that had been important export markets were now part of the Empire), they do reveal an area of concern to French policy makers influenced by mercantilism. Exports fell from 456 million francs in 1806 to about 330 million in both 1808 and 1809. Yet imports fell even more, leaving France with a favorable balance of accounts—a glaring counterpoint to the deficit of 83 million francs in 1803. Such an economic picture warmed Napoleon's mercantilist heart, but it did not fill the treasury. With customs revenues a significant part of the state's income, the drastic decline in the customs from 60 million francs to 11.5 million francs in 1808–09 was worrisome, especially in wartime. These concerns contributed to a major deviation from the self-blockade—the institution of a system of licenses.[89]

The project of trade licenses came directly from Napoleon in March 1809. He hoped to seize some commercial advantages at British expense and to help the ports to recover some of their former trade.[90] In its first incarnation, after paying a significant fee for the privilege, the licensee could export wine, brandy, fruits, vegetables, grain, and salt to Britain in exchange for wood, hemp, iron, and cinchona bark. Cold, hard cash (that is, no paper money) was also accepted. Approximately 250 licenses were issued in 1809. Pushed by Mollien, Champagny, and Montalivet, a second type of license was authorized in December 1809 which favored the needs of industry. Oils and textile raw materials including wool and cotton were added to what could be imported while, in this version, only three-fourths of a cargo had to consist of agricultural products. Regularized by the decree of February 14, 1810, licenses rapidly became big business.[91]

The official report submitted to the Emperor in June noted that 351 licenses had been issued, authorizing 10 million francs' worth of exports and 6 million francs' worth of imports. Nor do these figures represent the true volume of imperial trade with Britain. In addition to outright smuggling, there was extensive unreported trade, particularly in grain,[92] and, with the customs barriers cracked opened, licensees often added clandestine shipments.[93] Since agriculture benefited more than industry from the license system, Napoleon responded to complaints from manufacturers by allowing the sale of prize goods taken at sea or smuggled on land. Prizes were subject to an imperial tax of 40 percent. Only certain types of raw cotton were exempted from these seizures. Official ties to networks of British smugglers, first through the Channel port of Dunkerque and later through Gravelines, supplemented these actions. In consultation with a specially convened council of merchants and manufacturers,[94] Napoleon and his aides reorganized the self-blockade in the summer of 1810.[95]

The decree of Saint-Cloud, issued on July 3, 1810, formally and publicly affirmed the licenses. Only a French citizen could hold a license, but they were later made available to the residents of former Hanseatic cities, to Italians whose homelands had not been incorporated into the Empire, and to the inhabitants of Danzig. Another decree, dated July 25, placed all maritime trade under state control, highlighting the mercantilist views imposed by Napoleon over the objections of the entrepreneurial community, the Council of State, and the General Council on Trade and Manufacturing.[96] Every ship leaving a French-controlled port had to pay a license fee to the Emperor of 1,000 francs. The measure also stipulated that imports had to be balanced by exports—and exports had to come from a list of certain "favored" goods. These goods varied according to the port of departure, but a constant element was silks, which had to constitute from one-third to one-half of the goods exported—a sign of the special favor enjoyed by Lyon.

These Napoleonic measures constituted a French navigation act similar to the one enacted by the National Convention in 1794.[97] Since imperial shipping was largely kept in port by the British blockade, Napoleon could not control trade as rigidly as he would have liked.[98] The British Navy's command of the seas forced Napoleon to make concessions on licenses to the United States to take advantage of this important neutral's ships. Napoleon's political maneuvering pushed the U.S. into a war with Britain but did not help to revive French overseas trade using American intermediaries. If the terms of the Continental System as embodied by the Berlin and Milan Decrees had been enforced, the licenses would only have allowed in foodstuffs and raw materials from the United States or from other parts of Europe. In fact, however, the decrees of July 1810 let in a flood of colonial goods and manufactures even though they were clearly of British origin. Napoleon permitted the entry of sugar, coffee and cotton, but at a high price: the Trianon decree of August 1 more than tripled the duties on coffee and increased the levy on cotton from the high level of 60 francs per quintal imposed by the decree of February 22, 1806 to 800.[99] Indigo went from 15 francs per quintal to 900. If increased revenue was a major factor in Napoleon's thinking, then his plans were reasonably successful. By the end of 1810, the new trade decrees generated additional revenue of 150 million francs, not counting proceeds from sales of confiscated goods.

These measures did not favor manufacturing. But, as in 1806, Napoleon thought that French manufactures would shift to other fabrics if duties were raised to new heights. Since a trickle of legal cotton goods would enter at enormously high prices, the Emperor also hoped to deflect consumer demand and lessen smuggling.[100] It seems that he even believed that the British would lower their prices in response to ensure that their cotton goods remained within reach of the middle classes' purchasing power.[101] Mollien noted that, on the contrary, the licenses increased smuggling and "increased bitterness against industrial regulations. . . . and were therefore more detrimental to the commerce of the Continent than to that of England." Until the licenses, French industrialists had believed themselves able to satisfy Continental consumer demand, but because of the high tariff placed on raw cotton, "by an inexplicable contradiction, these taxes gave the products of English industry more advantages than absolute prohibition had done."[102]

Under the Empire, Napoleon's economic maladroitness took on staggering proportions and revealed the limitations of mercantilist thinking in the new industrial era. This master of battlefield maneuver clearly did not understand the extent of Britain's dominance of the European market, nor did he comprehend Continental consumers' almost passionate demand for British cotton goods. To give French producers a fighting chance, Napoleon resorted to highly unpopular repressive measures against smuggling. The Fontainebleau decree of October 18 punished smugglers with 10 years of penal servitude and with branding. Confiscated colonial goods were to be sold and captured manufactures destroyed, which led to numerous bankruptcies. An enormous police apparatus backed by new customs tribunals was created to enforce these draconian measures. Imperial heavy-handedness further alienated French consumers and allied and/or subject peoples, especially in the German lands, in Switzerland, and in Prussia. Russia refused to apply these policies, as did Austria, thereby contributing to the atrophy of the French dominance won at Wagram. On December 31, 1810, Alexander I opened Russian ports to neutrals, severely compromising the self-blockade.

Even more maladroitly, Napoleon's policies maintained the regulations that hampered the exchange of goods in Continental Europe.[103] Closing the borders of Prussia, those of several allies in the Confederation of the Rhine, and those of Switzerland to enforce the Trianon decree inhibited trade, as did the long lines at imperial customs barriers.

Napoleon also permitted the payment of dues in kind for imported colonial products. In combination with every country's habit of collecting customs dues on goods sent long distances, prices rose and there were long holdups in getting goods to market. These trade barriers vastly increased the number of bankruptcies, made cash money ever more dear, and rendered commerce increasingly difficult. The supply of Napoleon's new licenses exceeded the demand for them. Of 1,153 licenses approved through November 25, 1811, only 494 were actually issued. The trade generated came to 45 million francs' worth of exports and nearly 28 million francs' worth of imports. In short, trade was strangled and most of the opportunities for profit or the exercise of entrepreneurial skill that existed after the Decrees of Fontainebleau and Trianon were illicit. Legal trade was asphyxiated by high prices on industrial raw materials and other staples, poor grain harvests, a lack of specie, and a wave of bankruptcies stemming from overspeculation in colonial goods. These were the major factors underlying the industrial collapse of 1811 felt on both sides of the Channel.

In France as in Britain, a financial crisis made worse by high grain prices constricted demand for manufactured goods. Markets collapsed and industrial unemployment skyrocketed. Despite the mistaken emphasis by later commentators on the importance of laissez-faire polices, in fact, the British floated loans to pump money into the economy by propping up beleaguered firms while cracking down on labor unrest with heavy-handed military repression that long outlasted the Napoleonic wars. The response on the Continent was similar, but surprisingly the resort to the military was far less within "old" French borders than in Britain. During the late Imperial era, only smuggling required a forceful response from the French state.[104]

Napoleon, like the British, supported large banks and loaned huge sums to large employers, including Richard-Lenoir, Bauwens frères, and Gros-Davillier in Alsace. These expenditures came to nearly 13 million francs. The Emperor also stepped up his personal expenditures and encouraged his court and his military entourage to do the same. He bought 2 million francs' worth of silks and granted them to merchants to export. He also arranged for troubled banks, mostly in Rouen and Amiens, to receive 2 million francs of additional credit to purchase accumulated stock. All in all, Napoleon authorized the loan of 18 million francs to industrialists in 1811–12, of which only half was repaid. Mollien opined that "Napoleon did not perceive the strange contradiction into which he had fallen by maintaining the Continental blockade and then recognizing the damage it caused through his financial support."[105]

Troubled economic conditions led the implacable political enemies to relax trade restrictions in 1811–12. The British Board of Trade lifted the embargo on certain colonial products while allowing merchants trading under license to be insured. After March 25, 1812, the French and British licenses were issued along almost identical lines. The French disbursed approximately 800 licenses in 1812. The balance of trade strongly favored them with 58 million francs' worth of exports to Britain and 22 million francs' worth of imports. This reawakening of cross-Channel trade contributed to a general revival of manufacturing both on the Continent and in Britain.[106] As late as 1813, Montalivet reported that the legal cotton trade from Great Britain had been confined to raw materials rather than finished goods while, since 1807, France had exported an average of 17 million francs' worth of cotton products annually. In 1812, the industry employed at least 233,000 workers; its output was valued at 290 million francs, with the raw cotton costing 155 million.[107] The manufacturing revival of 1812 illustrated that French manufacturing could survive even in a deeply antagonistic productive international environment.

Contemporaries recognized that the environment was adverse. Chaptal, from his post on the trade council (established in 1810), pushed for the creation of a separate Ministry of Manufacturing and Commerce to conserve, nurture, and perpetuate institutions devoted to scientific advance, technological improvement, and industrial expansion. Napoleon adopted the idea in 1812 and entrusted the new institution to 1812 to one of Chaptal's protégés: Jean-Baptiste Collin, Count de Sussy.[108]

The revival of 1812 did not survive the disaster in Russia. Shortfalls in grain deliveries to the cities led the state to stipulate that farmers must bring their grain to market and that merchants must disclose the extent of their stocks. As in 1793–1795, variable local "maximums" on grain prices were established in May 1812. These maximums were ignored at harvest, but once again France had to mobilize fully for war to replace the vanished Grande Armée.[109] War production boomed, but the large-scale requisition of men, materiel, and animals left little to sustain the manufacturing economy. Demand slumped and long-distance trade diminished just as stocks reached new heights because of the increase in production in 1812.[110] The tide did not finish receding until after the first Restoration (April 1814). Trade and manufacturing recovered in the last half of that year, but slowly and fitfully, in part because of the initially "liberal" trade policies enacted by the Bourbons on April 23. Although Louis XVIII poked holes in the French tariff wall to placate his British patrons, the results were far from devastating.[111] In fact, there were a number of positive industrial signals. In Rouen, cotton production in the second half of 1814 equaled the first half of 1810.[112] The shock of Napoleon's escape and the episode of the Hundred Days traumatized the incipient recovery and lost France additional money and territory.

In terms of the supply of natural resources and in terms of entrepreneurialism, market size, and access, the France of 1815 was a very different place than the France of 1812, that of 1807, that of 1801, or even that of 1795. The challenges and possibilities of the industrial economy of Restoration France were much more constrained than their imperial predecessors, a fact recognized by contemporaries, who also understood the magnitude of the losses of Belgium and the Rhineland for French industrial competitiveness.[113] This new environment underscored the connection between trade policies and industrial realities. Napoleon's two defeats ensured that all pre-1815 French efforts to mechanize or industrialize on the English pattern were false starts. Wartime Imperial policies that had helped to create vast industries and to extend earlier industrial successes faced bleak prospects in the new era.

This interpretation of the course of economic development does not challenge the long-term conclusions of the existing literature. An investigation of the Continental System, however, does undermine the myth of an inevitable British industrial dominance based on a laissez-faire economic system. Without question, Britain contributed mightily to the defeat of Napoleon. In all likelihood, British growth would have been even more rapid, but for the wars that wracked Europe and undermined the nascent world economy in the generation from 1792 to 1815. The greatest benefit derived by Britain from the revolutionary and Napoleonic wars was not, however, territorial, but rather competitive—the wars ruined the French drive for industrial hegemony built upon military success.

The final section of the chapter traces how the Chaptalian framework initiated under the Consulate was dusted off and given new impetus by the Restoration. The goal was to permit—over the long term—a defeated and diminished nation to mechanize, to innovate, and to compete with a victorious Britain. In the next chapter, I will go beyond delineating the French divergence to examine alternative opportunities for gain. It was not entrepreneurship that France lacked, or an interest in deriving profit or opportunities to do so. French mechanization and technological innovation stagnated for rational economic reasons in the same way that Britain took advantage of its competitive strengths to forge ahead by accelerating the structural transformation of its economy. Neither course was inevitable or foreordained. They must be understood in context. By looking forward to the Restoration and then turning back to the Empire, I will examine why Britain furthered its technological/industrial lead. As a consequence of that lead, a modified mercantilism rather than laissez-faire remained the dominant industrial policy employed in western Europe during the first third of the nineteenth century.

Trade, Technology, and the Role of the State in Restoration France

In the formulation and implementation of its economic policies, the Restored Bourbon monarchy depended on former servants of the Empire. Although a few former revolutionaries remained beyond the pale and a number of former Imperial enthusiasts were initially suspect, many administrators from previous regimes remained in place during the early years of the Restoration. Even after the political purge that followed the Hundred Days the industrial decision makers remained in place. These dedicated bureaucrats maintained that scientific advance and technological innovation was the best means of improving French industrial competitiveness. Their previous experience increased their awareness of the reduced circumstances—territorial, financial, and political—facing the Bourbon state.[114] This nexus of common perspectives reinforced the prestige and attraction of the Chaptalian approach to industrial development during the Restoration.

A laissez-faire approach to trade was never an option under the Restoration, just as the Corn Laws prevented its adoption in Great Britain.[115] The bureaucratic reliance on duties levied on specific items inherited from the Napoleonic era persisted. With most commodities and many manufactured products, notably textiles, falling in price, deflation increased the real cost of fixed duties as much as 1,000 percent between 1815 and 1848.[116] Because of the Bourbon regime's narrow basis of support, policy makers could not avoid the outright prohibitions of certain imports demanded by special interests. Landowners with forested holdings, port cities, and textile manufacturers lobbied to garner an absolute monopoly of the domestic market.[117] The demands of different factions of protectionists such as between the sheep grazers and the manufacturers of woollens were resolved by imposing high duties and then providing a drawback for exporters.

The Restoration tariff of April 28, 1816 was enacted while parts of the country were still occupied by foreign troops and just as the first inclinations of a poor harvest became apparent. This law revealed how far French policy had drifted from that of the first restoration and the Napoleonic system of licenses. The relatively light duties set in 1816 (40–55 francs per quintal, depending on whether the goods arrived on a French ship) increased during the reign of Louis XVIII.[118] High French tariffs led to reprisals by other countries that drastically curtailed exports of many important products such as wine.

Market fragmentation meant that exports could not function as the chief engine of growth for French industry although the makers of certain articles or in certain sectors exported high proportions of their output.[119] In terms of market access, the France of the Restoration lagged behind the France of the ancien régime, which took advantage of the "family compact" with Spain to reach customers in Latin American. Nor did Britain's return of France's colonies compensate. Saint-Domingue was lost to France forever, and independent Haiti was neither as rich nor as good a customer. The end result of these trade policies was that industrial growth had to rely on indigenous resources—if export or import duties were factored in, Bourbon tariff policy ensured that France could not compete. High tariffs did stimulate French metalworking and machine building by successfully preserving the home market for domestic manufactures; nascent industries survived and, in many cases, thrived.[120] Here again trade policy stood for industrial policy because it necessitated a particular approach to technological change and manufacturing born of political needs not economic considerations.[121]

Market fragmentation also, paradoxically, helped French industry. Although the end of the Revolutionary and Napoleonic wars again permitted coastal shipping, and long-distance maritime French trade could penetrate the Baltic and the eastern Mediterranean, the redistribution of European commerce and industry away from the coasts (most notably the Atlantic, but also the Mediterranean and the Baltic) meant that oceangoing transport was no longer as much of an economic advantage. Europe's rivers had always been commercial highways. The age of steam accelerated and accentuated that trend in preparation for the emergence of the railroads. Railroads initiated a vastly different economics of transportation with broadly beneficial results for France. But before steam ships and later railroads were common, high transportation costs to inland markets not readily accessible by sea helped France to compete in some niche markets for mass-produced goods.

The French emphasis on niche markets was broad, deliberate, and important. A transitory factor like transport costs was not the major influence on this approach to competition. Historians of industrialization have long noted the French concentration on luxury goods. Some even assert that this specialty was part a long-standing aspect of the French national character. Yet, for a country attempting to compete with a dominant industrial power with higher technologically based productivity and impressive advantages in the "invisible" areas of shipping, banking, and insurance, what other economic strategy made sense? Since Britain's lead in these areas dated from the late seventeenth century (if not before), was it any wonder that France closed off the national market and developed export specialization in niche markets and/or luxury production? The eighteenth-century boom associated with the development of the Atlantic economy was an aberration in the long-term development of French industry. In the early nineteenth century, niche specialization and luxury goods were more important than ever to French industry because of British ascendance.

This trend was particularly visible in textiles. If the cotton spinning industry could not compete with Britain in international markets, other sectors of French textiles could. These included cotton printing,[122] high-quality woollens,[123] and France's most lucrative textile product: silks. From the nadir of the Maximum in 1793–94, Lyon recovered under the Empire, in part, because of the Jacquard loom. This important technology which automated the transfer of patterns made on punched cardboard forms to cloth, put brocades within reach of middle-class purchasers. Invented by Joseph-Marie Jacquard around 1801 and improved continually by a number of talented workers, this loom enabled the French silk industry to recover its earlier dominance within the international economy. The role of preferential government treatment was essential. Such treatment included but was not confined to the monopoly accorded by the Continental System, the tariff schedules of 1808 and 1811, and the export regulations of the licenses that mandated trade in silks. By 1814, the annual value of Lyon's silk production totaled 40 million francs. By 1830, that figure had doubled and Lyonnais silk made up one-third of all French exports. The silk industry employed about 37,500 in and around Lyon.[124]

Nîmes,[125] Saint-Étienne, and Saint-Chamond also experienced booms in silks as workers ceaselessly improved first the *métier à la zurichoise* and then the Jacquard loom, which was fully automated in 1819 by Burgin and applied to the ribbon and lace industries. The adaptation of the Jacquard loom capped a process, begun during the Revolutionary decade, of mechanizing each of the six steps in ribbon manufacturing. At the height of the Empire, the département of the Loire produced 10–12 million francs' worth of silk ribbons. This relatively troubled industry only emerged from its imperial cocoon around 1818. By 1828, the industry employed 27,500 and produced 37 million francs' worth of ribbons. Inhabitants of Saint-Étienne received 22 brevets of invention during the Restoration in this industry and left France (in 1826) in the unusual position of being able to refuse to export the most up-to-date Jacquard looms to England.[126]

As in Lyon, the efforts of entrepreneurs, tinkers, and workers in Saint-Étienne were seconded by the government, which provided institutional support through the Chambers of Commerce, the Consultative Chambers of Manufacturing, and the Conseils des prud'hommes established under

Napoleon. Jacquard worked at the Conservatoire and acquired grants from the Society for the Encouragement of the National Industry, the Academy of Lyon, the municipality of Lyon, and the First Consul. In April 1805, he received a subsidy of 50 francs for each machine he installed over the next six years. Saint-Étienne required more institutional assistance because the city lacked any means of providing a formal education (until 1807) or informal technical training. A Design School for ribbon making was established in 1804 and expanded in 1809 to provide a free course in mathematics. Another course focused on machines and their mechanics. Even more importantly, a Mining School was established in 1816, executing an imperial plan.[127] The government also learned from past mistakes and assisted development by passing regulations to ensure that entrepreneurs with mining concessions had the capital and technical knowledge necessary to take advantage of their opportunities.[128]

The region's raw materials and the wealth generated by silk, hardware, and arms was channeled by these institutions toward developing a better transportation network of roads and canals, attracting new industries, and realizing the potential of existing manufactures through technological innovation. Factories making plaited cloth and elastic cotton fabrics were highly mechanized processes brought from Paris and established at Saint-Chamond and Saint-Étienne under the Empire. From 1818, steam power was the basic power source in these industries. Steadily increasing production of these textiles complemented silks for the rest of the century.

In addition to assisting niche and luxury industries, the French government understood that, over the longer term, international competition with Great Britain entailed building capacity in steel making, in machine building, and in the use of coal as a fuel despite Britain's seemingly insurmountable advantages. The département of the Loire, centered on Saint-Étienne, Saint-Chamond, and Rives-le-gier, was rich in natural resources but poor in technological innovation. Here the major strands of French industrial policy converged.[129] New educational and advisory institutions established and supported by the central state provided training and linked the formerly isolated population to national and international developments. The School of Mines in Saint-Étienne gave locals enough mathematical training to use geometry in excavation and even taught many pit directors to read. The first directors, notably Louis-Antoine Beaunier and Louis de Gallois-Lachapelle (graduates of the National School of Mines and the Polytechnique respectively), definitively established the use of coke in steel making at Rives-le-gier. Both were also entrepreneurial pioneers who set up technologically advanced firms based on their own first-hand investigation of the latest English practice. Entrepreneurialism complemented their educational and official activities on behalf of the government. They regulated the various mining concessions to facilitate economies of scale in production. They also encouraged the construction of a tramway of horse-drawn cars on rails. Intended to transport coal between the Rhône and Loire Rivers, the tramway opened in 1822. The railroad linkage from Saint-Étienne to the Loire was finished in 1827. Secondly, the French state supported the efforts of local entrepreneurs to develop or preserve niche markets in textiles or luxury goods either by installing known machinery or designing labor-saving devices. Thirdly, the state provided administrative direction, as well as material and technological support to permit this département to capitalize on its vast mineral endowment and bring French metallurgical practice into the nineteenth century.

Recognizing the backwardness of the steel industry and the many failed attempts to naturalize making coke steel in France, the first restoration sought out an Englishman named James Jackson with broad entrepreneurial experience and different sorts of metallurgical skills. He brought his family to France in October 1814, hoping to take advantage of the lack of technologically equivalent competition. The British state proscribed Jackson and sequestered his goods, worth an estimated 180,000 francs.[130] At the urging of Chaptal, the Society for the Encouragement of the National Industry, and the local Consultative Chamber, Jackson and his four sons began construction at Saint-Étienne of a coke-burning blast furnace that would use the puddling process common in England. Delayed by the Hundred Days, Jackson began producing coke steel in the fall of 1815, assisted by his sons and four British metal workers from Birmingham brought to France by his 15-year-old daughter Anne. In 1816, his mill at Trablaine made cementation steel at 1.50 francs per kilogram, using charcoal and Swedish iron. He gradually reduced the price to 1.20 francs in 1818, in part by making use of local coal. This price was roughly equivalent to British output and far lower than the 12 francs charged by other French producers.

Yet this achievement was only a start, since the vast investments in raw materials, skilled workers, machines, and buildings that steel production required made it an uncertain business. Although the government provided 3,000 francs a year to rent the factory, sales were disappointing because of overproduction. A special grant of 6,000 francs from the Ministry of the Interior in October 1819 recognized the importance of "service that [Jackson] had rendered in propagating knowledge of the process of steel making in the arrondissement of Saint-Étienne." Established at a new site, for four years (1822–1825) Jackson received a generous subsidy of 20 francs per metric quintal sold (not produced) as well as a no-interest loan of 38,330 francs for new equipment. Contracts from the Ministry of War supplied the essential financial underpinning of the enterprise in its early years. The case of Jackson and coke steel showed that technology transfer in a vital sector often necessitated a long-standing state financial commitment.

In this instance, the French investment paid impressive dividends for national industrial performance and international competitiveness. Jackson made steel until his death in 1829. His sons moved the enterprise to Rives-le-gier the following year and prospered fabulously for another generation. Generations of stéphanois steel workers passed through their factories and spread their techniques far and wide. Joseph-Étienne-Marcellin Bessy and other entrepreneurs followed in Jackson's footsteps, combining workers brought from England with local commercial capital to found profitable steel-making enterprises. Bessy, like Jackson's sons and many other stéphanois, found that heightened competition necessitated making substantial investments in new technologies that lowered production costs. Taking advantage of close financial and commercial contacts with England, Bessy, followed by his brother, employed English steel-making techniques, notably a blast furnace designed in Birmingham, beginning in 1820. Coke replaced charcoal in 1827 and, in 1831, the firm patented an improved rolling mill to make gun barrels. The successful revival of Le Creusot in 1816 also followed this pattern.

The emerging steel-making complex in the Loire facilitated the rapid expansion of the commercial arms industry. Steel production and the need for innovation stimulated the use of locally built steam engines. Although steam power had been introduced to the region in 1789, it had spread slowly. By 1816 there were only 12 Watt-type engines, and they were employed exclusively in mining. After 1818, steam power was increasingly employed in textile factories, in arms making, and in other forms of metalworking. By 1828, there were eight local workshops making steam engines. In 1835, there were 162 steam engines in the arrondissement of Saint-Étienne, deploying 4,000 horsepower (20 percent of the national total). In turn, these tinkerers contributed to the development of such varied machines as the hydraulic turbine, chrome steel, the sewing machine and the locomotive. In the period 1840–1880, the Loire produced one-third of all French steel, about the same percentage of French coal, and nearly all French ribbons, elastic fabrics, and plaited cloth. Assisted by Restoration industrial policy, Saint-Étienne and the département of the Loire emerged from long-standing technological backwardness to become a hotbed of technological innovation and industrial production rivaling the English Black Country.[131]

During the Restoration, the French state confronted British technological supremacy. Technical "inferiority" should not mask the significant strengths of the French industrial economy. Despite its reduced financial and political resources, the government developed these strengths whenever possible. In the short term, the Bourbon administration recognized that several different sectors required encouragement if France was to improve its international competitive position. These sectors included the manufacture of high-quality and/or high-value-added luxury goods (such as silks) and niche products (such as ribbons). It also included important and basic industrial technologies, most notably steel making and the use of coal as fuel. These various industries received impressive levels of financial support and technical assistance from the central state, particularly in départements (such as the Loire) that had long been sites of industrial production but not of technological innovation. To widen the range of possible innovators, the state founded institutions at the local and/or regional levels to encourage the exploitation of existing opportunities and natural resources.[132]

Such policies resumed the model of institution-based innovation. The wide-ranging emphasis on education, technical training, and interaction among scientists, tinkers, bureaucrats, entrepreneurs, and workers that Chaptal developed was disregarded by the Emperor. The resumption in 1819 of industrial expositions (organized on the model of those held in the Years IX and X rather than the one held in 1806) signaled that the Restoration's economic decision makers recognized the missed opportunities of the Empire.[133] They understood Chaptal's policies as a means of playing industrial "catch up." The policy makers of the Restoration also understood that such approaches required time to work. Once centers of innovation were established, they could become self-perpetuating, as in Saint-Étienne.

The Restoration made the most of desperate straits. The grasping, ambitious mistakes that led ultimately to the fall of the Empire fundamentally damaged the international competitiveness of French industry. Louis XVIII knew that the slow economy enabled Bonaparte to return in 1815, but for political reasons he was unable to follow the British model of industrialization and equally unwilling to imitate the French Republic's approach. The difficult productive environment inherited from Napoleon was made worse by the Hundred Days and by occupation of 1815–1818. Along with these burdens, however, the Empire also deeded a group of experienced administrators to the Restoration. They revived the Chaptalian framework as the best means of charting a course between Scylla and Charybdis. Despite the constraints imposed by English political, technological, and industrial dominance, France's economic performance under Bourbon stewardship was impressive. Sustained by the industrial revival after 1815, the Bourbon regime postponed revolution until 1830.

8

Coalitions and Competition: Entrepreneurs and Workers React to the New Industrial Environment

State industrial policy under the Empire and Restoration was remarkably stable in one major area: the government's relationship to groups of workers and—somewhat surprisingly—to confederations of small-scale producers. Napoleon Bonaparte, Louis XVIII, and Charles X disagreed about the proper role of economic competition and the function of government intervention in the industrial economy. Despite their differences on fundamental issues, however, these monarchs concurred that neither entrepreneurs nor the laboring classes had a place in the policy-making process unless specifically invited by the central state. Industrial policy during this crucial era was essentially top-down; the government believed or acted as if it believed that only through such sponsorship could unfettered competition among domestic entrepreneurs and within the laboring classes improve the French international industrial position. Although the trade environment of the early nineteenth century was mercantilist, domestically Adam Smith reigned supreme.

Creating an unfettered and essentially competitive economic environment may have been the goal of a long line of state bureaucrats and policy makers, but increased competition in the domestic industrial sector did not mesh with the conceptions held by a vast number of entrepreneurs or the laboring classes. Both groups responded to state efforts to foster greater competition by banding together or extending enduring bonds to impose some sort of limitation on the freedom of the market. The lingering reliance by French entrepreneurs and workers on illegal coalitions to maintain their standard of living and/or way of life during the first decades of the nineteenth century reveals why the policies initiated at the state level took so long to bear fruit. Although such coalitions had characterized the ancien régime, during the Revolution, these models of sociability were seen as economic impediments. They were also imbued with a political resonance that made them anathema to Empire and Restoration alike.

The French state, like its British counterpart, outlawed combinations of workers.[1] The 1749 labor law, the Le Chapelier Law of 1791, and articles 1790 and 1791 of the Civil Code of 1804 all forbade association. Severe penalties for the laboring classes were specified by articles 291–294 and 414–416 of the Penal Code. Not only were workers forbidden to meet or engage in any form of collective action, with the law of 22 Germinal, Year XI (April 12, 1803), a laborer was obliged to present a livret to his employer to ensure that he fulfilled his work contract and paid back advances. The *livret* detailed the employee's work history and was retained by the employer.[2] When a worker arrived in a new town, he had to present his livret to the mayor or another municipal authority before searching for employment. Interestingly, women retained the right to organize because of the wording of the law. Beginning in January 1812, the Minister of the Interior authorized the prefects to issue parallel decrees in places, such as Lyon, where the concentration of women workers in the silk industry made them prone to organize.[3] Draconian attempts to rein in the "threat from below" and eradicate the memory of the Year II failed to convince French workers to follow the government's lead in industrial matters.[4]

Although the Foucauldian implications of these measures are clear, it must not be forgotten that the Napoleonic state, like the ancien régime, subjected employers and employees alike to its discipline. Perhaps the clearest indication of the state's genuine intent was the aforementioned law of 22 Germinal, Year XI. Although the government created the *livret* in response to complaints of entrepreneurs from all over France about the unruliness of their work forces, the law outlawing coalitions also applied to employers. The penalties for combination were, however, far less severe for entrepreneurs. The very same measure that outlawed coalitions of entrepreneurs established the Consultative Chambers to collect their views on industrial topics. The nineteenth-century state closely guarded the economic initiative.

The legal subordination of workers to employers was an important principle of nineteenth-century French administration, but masters as well as men were to be reduced to isolated individuals competing in the marketplace. The state sought to dominate not only potentially unruly workers, but also the properties classes who supposedly directed these regimes. Entrepreneurs and workers flouted industrial policy through coalitions and combinations which reveals the fundamental incongruities of French society and politics in the first half of the nineteenth century.

The repressive regulatory environment drove both worker and entrepreneurial coalitions underground, but did not eliminate them. Creative organizational responses by masters and men underscored the limitations of any top-down industrial policy. These confederations also help to explain how and why France, a seemingly stable country with a notoriously powerful repressive apparatus, faced revolution, revolt, or civil war in 1830, 1832, 1848, 1851, and 1871. Exploring both the successes and the failures of these coalitions demonstrates the continuing resonance of the Revolutionary legacy in the early nineteenth century. Industrial policies handed down from on high were difficult to enforce, a fact often overlooked by focusing on the actions of the state. This chapter redresses the balance by delineating the various forms of resistance to state-imposed regulation and/or discipline and then illustrating how resistance to competition tempered and distorted state industrial policies. French capitalism was farther from laissez-faire in the early nineteenth century than it had been in the eighteenth.

Worker Coalitions in the Early Nineteenth Century: Riot and Rebellion Redux

Under the Consulate, the laboring classes claimed their fair share of the improved economy. In many different cities and trades, workers banded together to win higher salaries or better working conditions and to control access to employment.[5] The relative dearth of adult male laborers thanks to Napoleonic conscription generally increased wages and made workers far less willing to accede to the demands of their employers. The coalitions formed by laborers to protect their interests and to limit the free play of the labor market spanned a revival of *compagnonnage*, strikes by skilled workers, and the kind of pressure tactics so common under the ancien régime. The continuity in tactics, both formal and informal, used by the laboring classes, should not, however, be overemphasized. It was precisely in this period that larger groupings of workers coalesced beyond the confines of a single trade, sector, or region paving the way for a new era in worker organization after the Revolution of 1830. The newer means of organization by workers that dawned under the Restoration highlighted the political lessons learned from the French Revolution.

In the first decades of the nineteenth century, compagnonnage once again became the journeymen's best organizational weapon to resist a hostile legal environment. The practice of compagnonnage was usually based on the work requirements and traditions of a specific city or trade. The brotherhoods experienced a major revival under the Consulate after hibernating for much of the revolutionary decade.[6] As under the ancien régime, through *compagnonnage*, journeymen, alienated from an increasingly repressive regime, expressed collective grievances that had no other legal or political outlet.[7]

An incident in Bordeaux illustrated both the continuity in the journeymen's tactics from before 1789 and the efficacy of compagnonnage. In May 1805, Prefect General Charles Delacroix of the département of the Gironde reported that "compagnonnage exists in Bordeaux in all its force and it holds the owners of workshops in a ruinous and humiliating subjection."[8] The locksmiths of Bordeaux had formed a "seditious organization" to prevent Jean-Louis Gaudet from establishing a workshop. Gaudet was an accomplished metal worker from Paris and a member of the capital's Lycée des Arts. He was placed under a ban after he refused to return the livret of a compagnon who wanted to search for a better-paying position, demanding that the worker finish the job he had contracted to do. The worker "laughed in his face" when Gaudet voiced this idea. Delacroix stated that compagnons often refused to permit masters to hold their livrets and enforced their demands for higher wages, control over hiring, and immediate release from labor contracts either through violence or by placing a shop under "damnation." Gaudet had also alienated the compagnons by seeking to choose his own apprentices rather than allowing the syndics of the compagnonnage to furnish them.

Delacroix jailed Jean-Baptiste Lachouille (called Bourguignon), the syndic of the locksmiths for 15 days, but his fellow compagnons refused to work for Gaudet until their leader was released. A number of other compagnons went on the tramp rather than continue to work in Bordeaux for unsympathetic masters. The locksmiths' resolve had been stiffened the year before when they battled another group of compagnons from a rival trade; they had even fought the police when they tried to stop the fracas. The locksmiths' victory led them to re-establish a confrérie in a local church overcoming an explicit rebuke from the Archbishop. Delacroix concluded his report on the incident with these words to the Councilor of State responsible for the police in his district: ". . . this description will show you the powerlessness of our careful measures taken up until this time against seditious associations of workers."

Nor were the locksmiths alone in their militancy. Delacroix noted that other brotherhoods had also become increasingly proactive in dealing with their employers. For 1805, I have found evidence in Bordeaux of revived compagnonnages or other forms of organization in at least thirteen trades: stone cutters, carpenters, wheelwrights, tailors, hat makers, shoe makers, joiners, blacksmiths, curriers, cabinet makers, farriers, bakers, and locksmiths.[9] These brotherhoods and/or informal organizations were in contact not only with each other but also with like-minded groups in other cities. According to Delacroix, employers who complained about the brotherhoods' demands were forced to pay a fine. If the master continued to protest, they "were lost, without resource." The locksmiths' conflict with Gaudet led compagnons in other trades, including farriers, cabinet makers, blacksmiths, and joiners, to demonstrate in support and to advocate openly the practice of compagnonnage. Rather than engage in a protracted conflict with his labor force or earn the hatred of other employers, Gaudet returned to Paris, depriving the economically troubled city of Bordeaux of a practitioner of a mechanical process in metalworking that would not be re-introduced for decades.[10]

Repressive laws emanating from Paris did not stop the practice of compagnonnage in Bordeaux during the Napoleonic era and the Restoration.[11] Inability to destroy labor organizations exhibited continuity with the ancien régime. In fact, in December 1806, Fauchet, Delacroix's successor as prefect, reported that the mayor of Bordeaux met with "shopowners, workers in each profession and the police to name inspectors and deputy inspectors" to oversee labor relations and the quality of the goods being produced. The first set was named by the mayor, but the following year they were to be elected by the workers.[12] After a promising start, this cooperative initiative was disbanded by order of the Minister of Police to avoid seeming to re-establish the *corporations*. Recognizing the resistance and persistence of corporate ideals avoids overestimating either the power of the state or the passivity of the laboring classes. In Bordeaux, up to the Revolution of 1830, informal organizations and/or brotherhoods of journeymen maintained an impressive degree of control over hiring and firing along with other aspects of working conditions. They also had some influence on wages.

Large-scale street violence between different groups of journeymen advertised the government's inability to stamp out compagnonnage in Bordeaux during the Restoration. The compagnons of Bordeaux had ties to their peers in other French cities. The third major assembly of workers to unite and codify the "rite" of compagnons known as "du devoir" was held in Bordeaux in November 1821 to make peace among the fractious brotherhoods there and to end divisive violence. This *Congrès compagnonnique* failed to unify the brotherhoods, but it marked a high point in this phase of the national movement.[13] Compagnonnage in Bordeaux peaked in 1824–1826 before declining under the July Monarchy as different labor tactics came increasingly to the fore.[14]

Nor was southwestern France unique in its labor militancy. In the first three decades of the nineteenth century, journeymen in various trades organized in at least 23 départements, including the cities of Paris, Bordeaux, Orléans, Tours, Toulon, Nantes, Lyon, Marseille, and Toulouse.[15] These brotherhoods were concentrated in certain artisanal specialties, including construction, the manipulation of leather, wood, metal, and foodstuffs, and the hand-crafting of clothes, shoes, and hats. These trades were not yet at the center of nineteenth-century industrialization, but these organizations strongly influenced the actions of the state, the perceptions of entrepreneurs, and the activities of other workers.[16] Traditions of organization influenced the formation of mutual-aid societies that, because they were permitted by the state, became so much a part of the associational experience of the working classes in the early part of the century.[17] After 1830, such societies provided institutional wombs for nascent labor organizations.[18]

If certain trades were particularly prone to labor militancy, so were certain cities, particularly if they were on the route of the *tour de France* followed by the compagnonnages. The silk workers of Lyon were notoriously aggressive in defense of their wages and prerogatives, as their rebellions in 1831 and 1834 suggest, but the hatters of that city were no less active. In 1817, and again in 1819, the hatters of Lyon struck to restore a wage cut. For a decade after these strikes, they prevented the diminution of their wages, the introduction of unorganized workers onto the shop floor, and changes in the pace of production, all while raising money to support strikes in other trades.[19] Successful action by compagnons to restore wage cuts or slashed piece rates, maintain control over access to the trade, or improve working conditions took place in Nantes (1809), Paris (1813, 1821), Marseille (1823, 1826), and Toulon (1825–26), in addition to Lyon and Bordeaux.[20]

Organized groups of workers won similar prerogatives or concessions in places outside the confines of the *tour de France* and in trades that did not have compagnonnages. The blanket makers and cotton weavers of Paris (1821) the cotton spinners and tinsmiths of Rouen (1820, 1822, 1825), the paper workers and cotton spinners of Amiens (1812, 1822), the cotton weavers of Roubaix (1819), the paper workers of Bar-sur-Seine (1812), the miners of Saint-Étienne (1823), and the Loire River transport workers at Roanne (1819) all combined successfully against their employers.[21] Such movements, rooted in working-class communities of long standing, built on older traditions to enjoy new successes.

As in Great Britain, the shearmen of the woollens industry attempted to use their monopoly of skill to dominate the shop floor and prevent either entrepreneurs or the state from infringing on their hegemony. If machine-breaking in 1817–1821 or other crowd action could not prevent mechanization in the southern towns of Carcassonne, Castres, Limoux, Lodève, and Vienne, it did slow down the pace or bring it to a standstill in other woollens centers, including Olóron, Abbeville, and Sedan. Elbeuf and other production centers that were more congenial to mechanization and technological innovation benefited.[22]

Neither the customary demands of compagnon journeymen nor the strikes organized by other groups of urban workers evoked as much fear as combinations of factory workers in the small villages and towns of the growing industrialized districts. Entrepreneurs recognized that they could easily find themselves at the mercy of disgruntled workers. The small size of individual villages and the close proximity of other factories in a nascent "red belt" meant that the concentration of workers with similar concerns was overwhelming.[23] During the summer of 1825, in the valleys of the Cailly and Austrebethe, northwest of Rouen, the frustrations of disgruntled factory workers in the cotton industry boiled over. Widespread anxiety about unemployment and dislike for profit-maximizing business practices by wealthy factory owners gave rise to a militant worker coalition capable of mobilizing thousands of workers from a wide swath of nearby villages.[24] This spectacle unnerved anxious entrepreneurs and exposed officials to the dangers of the new industrial geography.

The early 1820s were generally prosperous for the French cotton textile industry. Production expanded outside Rouen as entrepreneurs took advantage of the fast-running streams and rivers, especially the Cailly and Austrebethe, close to their junctures with the Seine. By 1823, there were 53 spinning mills with tens of thousands of spindles in these two valleys. When sales slackened, the manufacturers of Normandy began to cut wages, claiming that they had to make up for the rising price of cotton if they were to maintain prices (rather than raise them) in a weakening market. Throughout the eighteenth and nineteenth centuries, lowering wages was common business practice among French industrial entrepreneurs. They decreased the earnings of their workers to maintain their comparative advantage rather than exploring the possibilities of technological innovation or searching for new outlets. That French manufacturers were able to employ this tactic heightened the significance of instances where they could not.

Throughout the "little valley of Manchester," the wage cuts imposed by manufacturers prompted workers to organize to defend their interests. Similar drives had taken place in the summer and fall of 1822 and again in 1823 in response to entrepreneurial "speculation."[25] Early in 1825, workers began to set aside money for a strike fund. By early summer, with the price of bread rising, the police noted an increase in after-hours cabaret conversations reflecting worker resentment. Anonymous letters threatened violence if newly purchased cotton spinning machines were not dismantled to maintain employment. Particularly disliked foremen were beaten. Choruses of the Revolutionary anthems "La Marseillaise" and "Le Réveil du Peuple" were heard in the back alleys of the factory villages around Rouen. The police took note of the increased level of activity, but their preliminary efforts failed to uncover the "ringleaders" of the "cabal."

The workers made their move in the second half of July 1825 when raw cotton prices returned to previous levels. In the valleys of Cailly and Austrebethe, the cotton spinners threatened to leave their looms en masse unless their employers restored their wages. Beginning in the villages of Pavilly and Barentin, the strike threat forced one employer after another to give in to the workers' demands. Those who did not accede faced not only a strike, but a boycott by other workers. Among the most recalcitrant entrepreneurs was Jacques Levavasseur (1767–1842), a former president of the Commercial Court of Rouen. In 1811, he had established a large hydraulic cotton mill along the Cailly at Le Houlme, 8 kilometers from Rouen. Despite his liberal politics, Levavasseur was ennobled in 1815—probably because of his marriage to the daughter of the Marquis de Chapais de Marivaux. He increased his fortune by importing American cotton in his own ships and set up other large spinning mills along the Andelle River in the département of the Eure. At Le Houlme, Levavasseur employed 100–110 workers of whom about 15 percent were children under the age of 12. Everyone toiled 14 or 14 1/2 hours a day. On Thursday, August 4, this determined entrepreneur, who ultimately left a fortune valued at 14–15 million francs, finally caved in to workers' demands and increased wages. However, "refusing to be dictated to by workers," Levavasseur exempted a number of unskilled laborers. This set the stage for a confrontation the following Saturday (which was payday), August 6.

A large crowd of workers from nearby factories gathered outside the factory as the slighted workers demanded the same increase received elsewhere in the valley and by their fellows in Levavasseur's mill. Failing that, the workers asked for their livrets. Levavasseur refused to return them or to increase wages. Fearing just such an event, he had requested a police presence. Seven mounted policemen and twelve on foot were on hand. When "the rabble" would not disperse peaceably, Levavasseur convinced the sergeant heading the mounted police to charge the crowd with sabers. He later claimed that the workers wanted to destroy his factory. Six workers were seized and detained within the factory before the policemen withdrew before a hail of cobblestones that slightly wounded one trooper. News of the confrontation spread like wildfire through the forewarned valleys of Cailly and Austrebethe, bringing groups of workers armed with cudgels to Le Houlme. A guard post near Rouen manned by four policemen and eight soldiers was overrun; the guards were disarmed and beaten. The mayor of Le Houlme, Benjamin Adeline, himself the owner of several mills, intervened to prevent further bloodshed. Adeline convinced Levavasseur to free the imprisoned workers. Disgusted by the lack of official support and realizing that his factory would be safer if he were not there, Levavasseur left for Rouen, escorted by four policemen. En route, his coach was attacked with sticks and stones. Some workers were wounded, a few severely, by the policemen. Back in Le Houlme, the crowd in front of Levavasseur's factory disbanded only when 50 soldiers arrived to reinforce the police picket. Thirty-four workers, including both men and women, were arrested for carrying firearms and sent to jail in Rouen. The workers blamed Levavasseur for the strife and for the wounds suffered by their comrades. That evening a crowd of 700–800 returned to the factory and threatened the two squads of policemen on guard. Taking advantage of the distraction, a few people sneaked over the back wall and broke all the windows on the first floor. The following day, a crowd threw rocks at the police picket after mass.

On Monday, August 8, angry and determined crowds of striking workers converged on Levavasseur's factory at Le Houlme. Their quarrel was with the form of capitalism practiced by Levavasseur in the shelter of the state, not with the machine or with mechanization itself. Throughout the valleys, others entered factories and tried both to stop work and to enlist supporters. After several tentative approaches, the swelling crowd (estimated by a policeman at 2,000) emboldened the workers. They decided to avenge their comrades and punish Levavasseur's greed. Women carried stones; the men carried cudgels, pitchforks, and axes, and a few had pistols and muskets. An advance on the picket line around 11 A.M. was blocked by Adeline and the police commander, who tried to negotiate a solution. This fruitless attempt lasted 5 hours, during which a squad of troops arrived. Around 4 P.M., the tocsin rang in Le Houlme and the workers surged forward. Bayonets and musket balls greeted them. Worker casualties were impossible to estimate, but among the many wounded soldiers and policemen, only one later died (from a gunshot wound to the head). The crowd kept trying to force its way into the factory until the prefect, Baron de Vanssay, sent in 400 troops to protect the factories of the valley of Cailly from the crowd's fury.

In response to the events in Le Houlme, the prefect deployed massive force. Troops guided by policemen interrogated numerous "suspects." Hundreds were arrested in the search for "les mutins" and "les rebelles." The town was plastered with posters bearing the Penal Code articles forbidding "coalitions of workers and seditious meetings." Cafés and cabarets closed at 7 P.M., 2 hours before sundown. Hordes of police spies infiltrated the region. Circulars were distributed to other industrial centers, including Amiens and Strasbourg. It was feared that workers on the run from prosecution might seek jobs near the frontiers. Dozens of tinsmiths who had formed a coalition in Rouen three years earlier were also detained. In spite of official precautions, a mysterious fire at a house owned by Levavasseur and two attempts to set the police stables ablaze troubled the peace. Unlike the unrest itself, the fires were swiftly snuffed out. In the days after the riot, nearly all the workers returned to their jobs, but manufacturers in the two valleys reported renewed threats from their workers if they did not raise wages, even for those who had not faced earlier cuts.

By the end of August, only fifteen were still imprisoned. Four additional "leaders" of the coalition were charged for their role in the riots of August 6–8: one for killing the policeman, one for wounding a supply officer, and two for delivering speeches exhorting the crowd to attack the picket line. The trial, conducted by the Assizes Court of Rouen on September 20–23, was notable for two developments. The first was the conflict between Adeline and Levavasseur. The mayor, supported by most of the policemen who testified, accused Levavasseur of hard-hearted measures: his greed provoked the workers and incited the confrontation. The presiding judge, however, was uninterested in questions of "provocation." Instead, he emphasized the similarity to the politics of the Year II. If workers were allowed to combine to confront their employers with impunity, anarchy threatened. The martial strains of "Le Réveil du Peuple" almost drowned out the instructions he gave the jury. The threat from below prompted harsh penalties: death for the worker who killed the police officer, and sentences of 12 years of forced labor, 10 years of forced labor, and 8 years of confinement for those convicted of lesser crimes. A police tribunal convicted fourteen other workers of "having taken part in a coalition of workers" and sentenced each of them to two months in prison.

The events of August 6 and 8, 1825 in Le Houlme confirmed the need for co-optation and conciliation. Because of the animosity and influence of Levavasseur's father-in-law, the Marquis de Chapais de Marivaux, Adeline lost the position of mayor, which he had held since 1819, for not "keeping the peace." His criticism of Levavasseur and the implied crack in elite solidarity could not be borne. But once the immediate crisis died down, the prefect urged the manufacturers of the affected valleys to increase wages and to provide additional "encouragements" to "good workers." By early 1826, the wage demands sought by the coalition had generally been met. Levavasseur was so disgruntled that he brought in workers from Brittany willing to work for lower wages. For the next several years, groups of local workers harassed his business.

The severity of the crackdown in the fall of 1825 did not prevent the formation of other clandestine worker coalitions in the factory towns dotting the valleys of Cailly and Austreberthe. Nor were these workers deterred from engaging in strikes. Repression was not prevention. The manufacturers of the region recognized the formidable power of coalitions of workers by giving into their demands in July 1825 and again six months later. When another group of textile workers was brought before the Assize Court in December 1825 on charges of coalition, the judges refused to prosecute even though the workers had fought with the police. Wages in France might have been low, but, as Jacques Levavasseur learned to his chagrin, the collective wishes of the working classes could be ignored only at great risk.

Although interesting as a piece of local and/or labor history, the events in Le Houlme must be considered in their regional, national, and international contexts. From across the Channel, British revocation of the law forbidding coalitions of workers in June 1825 was reported in newspapers and spread by word of mouth. This news may have emboldened the cotton workers of Normandy. A powerful coalition of paper workers in Thiers was also under indictment that fall. Rival rites of compagnons fought in the streets of Bordeaux in June. The police were powerless to stop them.[26] An even more bloody confrontation between compagnons occurred on September 1 at Tournus, near Mâcon.[27] Strikes by compagnon bakers rocked the cities of Marseille and Toulon.[28] New textile machines were being introduced into rural areas around Abbeville and Amiens, regions that had resisted them successfully up to that point.[29] Groups of bakers in the département of the Nord met regularly and openly in cabarets that summer. Persistent rumors of spinners from Manchester organizing in the département led the prefect to write to the Minister of the Interior on more than one occasion to assure him that there was "no committee of workers in Lille."[30] Chillingly to those with long memories, the summer of 1825 also brought the first power looms to Normandy.[31] In such a moment of general ferment, with a new and more radically conservative monarch newly come to the throne, developments in the industrialized valleys near Rouen seemed part of a broad threat to the social and economic order.

French laborers refused to passively accept the disciplinary dictates of industrial capitalism. They adapted traditional organizational forms and developed new ones to meet the challenges of industrialization. These organizations had significant though often intermittent successes in preserving shop-floor control and wage levels. Despite the repeated efforts of both the state and of entrepreneurs to realize a "liberal" (read: individualized) work environment, the French laboring classes overcame repression through organization to achieve their own goals.[32]

Producer Coalitions: Limiting Industrial Competition

Workers organized in response to a troubled, even hostile, economic and political environment dominated by large-scale commercial and manufacturing interests. They were not alone. Collective action by both large- and small-scale entrepreneurs was an important survival strategy in the early decades of the nineteenth century. Producer coalitions demonstrated that the transition to an increasingly competitive industrial economy enforced by the central state affected entrepreneurs and the working classes alike. Thus, the path pursued by the French government in the first 30 years of the nineteenth century illustrated the limitations of laissez-faire principles. The tactics of combination used by entrepreneurs to protect their collective interests paralleled those developed by the laboring classes. Entrepreneurs sought to convince the government to revive limitations on competition from the ancien régime. When these attempts failed, the efforts of entrepreneurs to circumscribe competition shifted underground. Coalitions provide a unique insight into how and why technological innovation assumed a different role in the French industrial economy than across the Channel.

The first major effort to limit competition by forming coalitions of entrepreneurs concerned the revival of the *corporations*.[33] This drive reflected the same organizational impetus that led to the formation of the Chambers of Commerce and the Consultative Chambers of Manufacturing.[34] In the name of international competitiveness, groups in places as widely separated as Thiers and Lille sought to revive some form of producer cooperative that would be licensed and/or regulated by either the central state or organs of local government.[35] The fundamental goal of these efforts, however, was not the restoration of the sort of *corporations* abolished by the Le Chapelier Law. Instead, local institutions hoped to eliminate illegal underground coalitions, which persisted long after 1791. A clandestine coalition among the iron masters of Bordeaux avoided repeated efforts by the police to break its power. It also curtailed production and raised prices, strangling an already troubled industry. The Chamber of Commerce of Bordeaux hoped to co-opt these producers to prevent them from undermining or avoiding the existing regulatory structure.[36]

Nineteenth-century supporters of coalitions focused on their role in improving the training of laborers, a goal compatible with the Chaptalian approach to industrial development. In the Year X (1802), the prefect of the Seine reported to Paris' Chamber of Commerce that "to leave apprenticeship voluntary or optional contributes to the propagation of charlatanism of all types [in manufacturing]." From his vantage point as overseer of the largest center of industrial production on the Continent, he noted that entrepreneurs rarely took the time to train their workers beyond the most minimal level. Unfettered competition gave small-scale producers no incentive to teach a worker to grasp the entire production process (which would enable workers to strike out on their own). The prefect observed that inadequate training had plagued French industry since the abolition of the corporations.[37] Other highly placed bureaucrats, including Paris' Prefect of Police Louis Nicolas Dubois and Councilor of State Antoine Thibadeau, made similar arguments.[38] For political reasons, the Napoleonic government and its successors rejected the recreation of the *corporations* and formal apprenticeships as means of improving worker training.[39] Chaptal clearly thought that educating workers more broadly in mathematical and mechanical principles, either privately or under the aegis of state institutions, was enough to encourage the development of a productive and innovative industrial labor force.[40]

An overlapping group of French administrators and policy makers emphasized the disciplinary benefits of reviving *corporations*. This argument, based essentially on social fear, was predicated on the notion that if given the proper protections and incentives, the masters—as good industrial entrepreneurs—would control the restive working classes in their employ and prevent workplace conflict among competing groups of laborers. Taming recalcitrant port and dock workers who already had a kind of de facto *corporation* was a recurrent problem. Under Napoleon, such organizations existed among the port workers of Bordeaux, Rouen and Toulon and the river haulers of Saint-Quentin and in the Morvan area of the département of the Nièvre. These coalitions aimed to maintain wage rates, to improve working conditions, and to exclude "outside" workers. In Bordeaux, the police delegated certain powers to employers taken straight from old-regime statutes that tacitly recognized the administration's inability to control these workers. Both administrators and police officials saw a reimposition of social hierarchy in all commercial enterprises as particularly needful because the revolutionary decade had undermined industrial discipline.[41] Rather than revive the *corporations*, however, the Bonapartist state led by Chaptal instituted a more direct means of enforcing worker discipline: the livret. This approach was intended to bridle the revolutionary potential of the French laboring classes in a manner unnecessary in the solely rebellious British context. The institution of the livret was made possible by the popularity and repressive power of the Napoleonic state; after 1815, the threat of the guillotine wielded by recalcitrant workers once again haunted the nightmares of French entrepreneurs.

Experiences gleaned from the expanding nation during the revolutionary/Napoleonic wars reminded French administrators of the two-edged disciplinary challenges posed by *corporations*. In Holland, an area with *corporations* before its annexation in July 1810, reports of "a tyrannical power exercised by the heads of the *corporations* that arrogated some powers belonging only to the judicial authorities" both in Amsterdam and in other Dutch cities led to police crackdowns. The Council of State was responding to a request from the Minister of Manufacturing and Commerce. The crackdown was directed against formerly privileged masters who "eluded legal measures and continued their abuses."[42]

A third major argument made by proponents of reestablishing the *corporations* was that only by privileging producers could the high quality standards necessary to French exports be maintained. This argument was advanced by those involved in large-scale commerce or by "experts" on the subject in government. Except for the protectionist-minded manufacturers of Lille, it was not an argument heard from actual producers.[43]

To ensure quality and to meet the needs of the capital, food suppliers required financial backing. A government decree instituted a quasi-*corporate* organization under elected syndics for the bakers of Paris on October 11, 1801.[44] The butchers followed on September 30, 1802. But instead of the candlestick makers, Prefect of Police Dubois authorized the establishment of placement offices for twenty different trades in 1804. Michael Sibalis suspected that this measure was a test project for the re-establishment for the *corporations des arts et métiers*. The creation of trade organizations called syndicats or chambres syndicals followed. Despite the staunch opposition of the Ministry of Commerce and of the large-scale commercial interests represented by Paris'

Chamber of Commerce, fourteen such employers' associations were approved by Dubois between 1805 and 1810, ranging from the wine merchants to the renters of carts in those areas where it is a "profession whose exercise is of special concern to public health or public welfare." Five more syndicats were in operation by 1815, and the administrators of Paris under the Restoration founded an additional five. Although these associations were dim reflections of the ancien régime's *corporate* system, they did help to structure the labor market and the competitive environment in Paris. The utility to the state of the syndicats can be seen clearly in the survival of these ad hoc creations of the Napoleonic police until their formal legalization in 1884 by the Third Republic.[45]

Chaptal and his successors were firmly committed to eliminating vestiges of the ancien régime that restricted industrial development. After 1815, however, the prospect of restoring institutions that had existed before 1789 improved. From 1815 to 1818, a flood of petitions from entrepreneurs all over France inundated the central state.[46] A special commission was convened to study the question. A report in November 1817 was followed up by another submitted in April 1819. The commission walked the fine line between license and liberty in French state oversight of the industrial economy:

If complete liberty is the soul of commerce and manufacturing, then license is the scourge. . . . From all over, a storm of reclamations against the abuses of an unlimited liberty of commerce and industrial manufacturing has arisen.

Such a career, indistinctly open to all, has, all too often been taken up by greedy men without capital, credit, and knowledge who are strangers to the honorable and fastidious sentiments that have distinguished French commerce for so long; instead, they only see the profession as a means of rapidly acquiring a fortune.

From that situation comes fraudulent manufacturing, the deterioration of quality, the adulteration of dye colors: all of which lose for our manufactures their advantages through the theft of established names or processes, all to the determent of their inventors or owners.

This leads to deceitful announcements to retailers which rebound to the consumer. And because of this practice, hazardous speculations ruin not only the entrepreneur but the investor who had confidence in them.

These habits lead to the multiplication of thefts by workers as well as even more culpable embezzlements.

This leads to coalitions of manufacturers, of masters, of artisans or of workers, either to raise or to decrease wages arbitrarily or to limit the number of apprentices to a very small number compared to the needs of commerce or of consumers.

What follows is the illegal introduction of foreign merchandise to encumber the stores and paralyze domestic manufacture. This leads to bankruptcies that are more or less fraudulent, but that are so common and so scandalous that they become a new way of making a fortune.

To the commission, these widespread problems stemmed from too much commercial liberty. The reestablishment of the *corporations des arts et métiers* was, for many, the remedy.

Despite the dangers of the commercial and industrial practices decried above, in a stunning rejection of the ancien régime, the commission refused to turn back the clock. They found that no major overhaul of the existing system was needed. The best means of maintaining or even improving French industrial progress was "merely the full and entire execution of the existing laws and regulations." The commission endorsed the Chaptalian framework and the need for government oversight to prevent liberty from running rampant in the industrial economy. They suggested that in suitable locales, a new set of officials should coordinate the activities of the various institutions created to succor French industrialization and ensure the enforcement of the laws.[47] Financial and political considerations combined to scuttle this proposal. For the remaining decade of the Restoration, the economic initiative remained with the entrepreneur especially with large-scale interests.[48]

For those who believed that cooperation and coalition among small producers could mitigate the harsh individualizing demands of a liberalized market economy, the return to the past had been rebuffed. At the same time, both government oversight and an acceptance of the concentration of capital were reaffirmed. This rejection of custom and cooperation in favor of a state-enforced legal individualism drove coalitions even further underground. As in so many other areas, during the first decades of the nineteenth century, calm waters did not reflect the deep currents of the economy. In the 1820s, entrepreneurial coalitions retained many eighteenth-century features.

Because the police focused primarily on the threat from below, and because of the hidden and sometimes precarious existence of illegal combinations, documentary evidence of formal coalitions by producers is scanty. Much of what does exist comes from the state's investigation of worker complaints about short-term or temporary actions by employers. In the course of a long-standing wage dispute with their employers, in May 1808, the master carpenters of Paris, "under the pretext of the unwillingness of workers to accept a rejection of their demands for higher pay, have claimed indemnities much greater than the original salary demand."[49] In mid July 1811, Lyonnais silk merchants unilaterally reduced the wages of their workers 50–66 percent. The Council of Prud'hommes and the Chamber of Commerce did nothing.[50] The "weavers of several Parisian workshops" wrote twice to the Emperor during the Hundred Days to accuse "Boudard, manufacturer, of trying to form a coalition to lower salaries."[51] Other masters, faced with long-standing worker militancy, particularly in cities and trades on the *tour de France*, organized periodically when challenged by compagnons. In Nîmes, in 1827, the ringleaders of a masters' coalition were themselves former compagnons who took advantage of their ties to the workers to exploit them.[52] In Paris, in 1820–21, an informal coalition of master chair makers attempted to prevent Jacques-Étienne Bédé from finding a job to punish him for acting as the workers' negotiator in a labor dispute over the preservation of customary practices. The employers' unity fractured and did not survive the conflict.[53]

My findings suggest that petitions for redress were more numerous under the Bonapartist regime, which seems to have been perceived by workers as more even-handed than the Bourbons. Although such short-term combinations were an important and consistent element of entrepreneurial tactics, longer-term coalitions had even greater potential to protect the interests of small-scale producers, particularly in industries, sectors, regions, or towns where a coherent group of masters held a stranglehold over a manufacturing bottleneck. Such a coalition existed—at least intermittently—among the dyers of Amiens for more than a decade despite concerted efforts to destroy it on the part of the government and by the city's mercantile elite.

The uncertainties of the cotton industry of Amiens elicited a common response from both workers and manufacturers: combine. This tactic was particularly effective in Amiens because vertical integration in production was non-existent. For each stage of the production process, "work on the various products was performed successively. If one step in the process was too expensive, the consumer would not purchase the finished product and the entire industry would be paralyzed."[54] Beginning in the early 1820s, the cotton spinners formed a coalition that united significant numbers of workers. The entrepreneurs who dominated the dyeing and cotton printing industries that were essential elements of "value added" in the French sale of textiles abroad also formed their own organizations in June and July 1821. Both groups defended essential interests, i.e. maintaining control over access to jobs, customary work practices and ensuring continuity of production and thus employment. Producers' coalitions excited just as much concern from authorities as combinations of workers. And like the workers, the dyers and cotton printers used a variety of tactics to thwart official demands that they disband their illegal *corporation* to allow the market free rein.

Producer coalitions first attracted public notice on October 8, 1821, when the cotton printers unilaterally raised prices 20 percent. At the same time, 29 of Amiens' 30 dyers initiated a new means of payment so that the up-front costs were not borne by these industrial middlemen, but rather by the merchants who reaped the greatest profits. At the same time, the dyers refashioned the complex calculation between the amount of labor that went into a particular industrial process and its price, all in exactly the same way, raising the final cost of the manufactured article from 2–2.5 francs to 4 francs. The same day that the increases were announced, the leading cotton merchants, gathered together in the city's Chamber of Commerce, began investigating how such concerted action was possible. Almost immediately, they discovered that 23 master dyers had formed a formal coalition. It was created in June after nearly a year of discussions culminated in a three-day meeting (June 27, 28, and 30) that resolved the details. The dyers then deposed their act of association before a notary on July 6. Six dyers joined after the fact.

In addition to fostering mutual aid, this association aimed to preserve the financial independence and work standards of the practitioners of the trade. The dyers blamed the problems of Amiens' cotton industry on those merchants who tried "to lower the price of fabrication below reasonable bounds to the detriment of the quality of the goods being manufactured and consequently to the detriment of the consumer." To protect themselves, each week, the dyers agree to pay two francs into a "common fund" for each "velverette et cannelet" [the basic cotton goods produced in Amiens] and three francs for each "velvetine" that entered their workshops to be either dyed or bleached. To oversee its execution, three commissioners were to be elected monthly by at least half the signatories. The commissioners were obliged to inspect each dyer's establishment at least once a week and to keep detailed registers on the goods coming into the city's dye shops. In addition, the coalition was to have a syndic and a treasurer, elected for one-year terms by two-thirds of the members. The syndic would keep track of the money and could impose stiff financial penalties on those who circumvented the agreement. The treasurer deposed 15,000 francs in cash with a notary to guarantee financial probity. Every four months, the treasurer would disburse the funds collected "so that each of the interested parties would receive a portion perfectly equal to those of the others," no matter how much a dyer had paid into the fund.

This ultimate expression of Jacobin egalitarianism varied enormously from the wartime exigencies of the Year II. It was clearly designed to benefit smaller producers and to remove incentives to expand production beyond a certain level—all in the name of maintaining quality standards and employment.[55] At least three annual meetings restricted to signatories were to be held and a register of deliberations kept. Existing firms could join simply by a written declaration of adherence to the statutes. The heads of new firms had to be approved at a general assembly. One-thirtieth of the deposits were to be retained as a reserve fund to assist those who had accidents "that did not stem from negligence." All the signatories were obliged to acquire and retain fire insurance at their own expense. Each year, the city's poor were to receive 10 percent of any unused funds. The same basic formula was followed by the cotton printers. They even used ancien régime measurements long out of ordinary usage when determining how much they were to deposit in the "common fund." By combining aspects of mutual aid with small-producer collectivism, both the dyers and the cotton printers of Amiens hoped to preserve their status and livelihoods in an uncertain economic environment.[56]

The large-scale interests represented in the Chamber of Commerce orchestrated a massive administrative response to these small producers. Accusing both the dyers and the cotton printers of using "combination" to exercise "monopolistic control" over the industry, the Chamber of Commerce demanded that the mayor of Amiens and the prefect break up the coalitions.[57] The mayor played down the existence of the combination and tried to get the masters to disband voluntarily. The prefect sought to wash his hands of the whole matter by leaving the issue to the judicial system. This lack of support did not reassure the merchants even though the Tribunal of Commerce acted the very day the coalition was discovered—even before any evidence was submitted to it. Both coalitions were legally dissolved on October 8.[58]

Both the dyers and the cotton printers refused to accept the government's dictates. The cotton printers reestablished their coalition on October 10; the dyers followed on November 29. These small producers claimed that they were involved in a "guerre mutuel" to protect their "daily bread" against the merchants who were trying to get the government to do their dirty work for them. If successful the state would guarantee that they "were left with no resources to resist them [the merchants]." They rejected the Chamber of Commerce's charge that they had formed a pre-1791-style "*corporation*"—their association was strictly voluntary.[59] In this incarnation, the coalitions emphasized mutual aid, at least in the organizational documents deposed with the notaries. Both groups even attempted to register with the Council of State as charitable organizations. The Chamber of Commerce, however, complained to various authorities that the intent remained the same: to raise prices and protect the independence of small producers, thereby distorting the market and rendering the textile industry of Amiens less competitive. The activism of the large-scale merchants through the Chamber of Commerce was another form of collective action available only to a more eminent level of entrepreneurs.

Despite its recourse to the power structure, the Chamber was disappointed again. The mayor reported the affair to the prefect and then "resolved to stay out of it." The Councilor of State who oversaw charity and mutual aid found that the new statutes for the coalition no longer contained obviously illegal provisions and that it would be necessary to prove the economic effects of the coalition conclusively in order to outlaw it. Government sanction was unnecessary, said the Count de Capelle. The courts agreed.[60]

Instead of undertaking the difficult task of proving the deleterious economic effects, the Chamber petitioned the Minister of the Interior to punish the members of the coalition and remove the threat to local industry. The Minister, Count Joseph-Jérôme Siméon, was incensed that the dyers and cotton printers used the "cloak of mutual aid to shield clauses that seem abusive and tainted by illegality," notably the naming of syndics and commissioners along with regular assemblies, both of which contravened the Le Chapelier Law. He believed that these statutes violated the penal code because the syndic could levy penalties, "a power which should belong only to the courts." He also noted that the dyers' and cotton printers' control over prices clearly marked these associations as a legally prohibited form of coalition.[61] The Chamber of Commerce claimed that only Siméon's spirited intervention pushed the courts to take action. In April 1822, the coalitions of the dyers and cotton printers were again dissolved.

The cotton industry of Amiens was troubled. The region had taken advantage of the Continental System to mechanize cotton production and to expand output, particularly of velours. From a zenith in 1810, production fell by half in 1816; it recovered only slightly in the succeeding decade. Yet the number of dyers and cotton finishers remained the same from 1805 to 1828. Amiens' producers conserved traditional methods and resisted technological change. At every step, they depended on cheap labor rather than machinery. In part, this approach resulted from the squeeze on profits exercised by merchants divorced from the production process. Despite the willingness of the laboring classes to accept starvation wages, velours from Manchester were approximately 45 percent cheaper than those from Amiens, severely retarding the regional cotton industry. A new crisis caused by technological obsolescence began after the Revolution of 1830.[62]

At least 26 of the 30 master dyers formed what was nominally a mutual-aid society in January 1831 and deposed statutes with a notary.

Their goal seems to have been to freeze out a few newly formed dyeing establishments using more modern methods. As in 1821–22, they established a system of paying into a common fund according to output, but receiving an equal share of the proceeds. The prefect noted that what made judicial prosecution difficult was that "the dyers had thought ahead to avoid having any regulations that are in opposition either to established rules or to the laws. It is more than probable that the true intent of the association is to dictate to the manufacturers, at least to some degree. But by careful means they have hid these goals. They are the subject of secret conventions for which the public dispositions deposed with the notary are only a preparation and a means of execution."[63] In March 1832, a large manufacturer planned to patronize a new independent workshop. Two coalition members sprang into action. According to the Chamber of Commerce, they pressured the owner of the new shop to form a "bridge of gold." They would perform some of the work in exchange for promises to share orders. Subcontracting diminished the sums deposed in the common fund, but co-opting potential innovators kept the monopoly intact.

Such maneuvers could not hold back technological progress and the reconfiguration of the division of labor forever or permanently stymie the interests of the large-scale merchants. In 1833, the Jacquard loom was introduced in Amiens. Over the next five years, the number of cotton workers fell from 6,000 to 1,500. Coalitions had ensured the decline of an industry. Cotton production survived only through the development of bombazine, a new specialty that was mechanized from the start. The ability of small producers to sidestep the dictates of the state was more impressive than their economic success. At best, the dyers and cotton printers of Amiens only delayed the destruction of their livelihoods.[64]

The persistence of coalitions in Amiens in the face of opposition from both big interests and the state illuminated several important lacunae in current accounts of industrialization. First, many small producers did not believe that technological improvement was the only or even the best competitive strategy. In industries that were not vertically integrated, producer coalitions were preferred. Second, the role of the state was not always as effective as it has been depicted. Neither the French government's laissez-faire principles nor repression succeeded in Amiens where cooperation rather than competition was the rule. That the dyers and cotton printers of Amiens could maintain technically illegal combinations over a period of years and sometimes decades demonstrated the practical limitations on French state authority.

Alternatives to Competition: Counterfeits, Knock-offs, and Smuggling

Combination and coalition limited the free play of the market in the French industrial economy. These tactics were used by small producers and by workers. They were, however, far from the only illegal methods employed to ensure the survival of less efficient or less technologically advanced enterprises. From the theft of raw materials by workers, to systematic short-changing in the size, quality, or thread-counts of textiles, to counterfeiting, to smuggling, cheating was an endemic means of economic survival for French producers.[65] Such survival tactics both predated and outlasted the Industrial Revolution. Deployed on the shop floor and in the counting house, cheating took advantage of the loopholes in the nascent international market economy in ways that had very little to do with increasing efficiencies in production. Alternatives to market-oriented competition complicated the process of industrialization in the age of revolution.

A frequent complaint of nineteenth-century textile entrepreneurs was that workers stole enough raw materials to undermine their profit margins.[66] Unable to stop this practice, industrialists demanded government intervention to curtail workplace theft of raw materials, notably wool and cotton. Between 1803 and 1811, with a peak in 1807–1809, petitions to this effect flooded in from all over France, usually from Chambers of Commerce, Consultative Chambers, or departmental agricultural councils. On the request of the Chambers of Carcassonne and Lyon, the central state empowered the Councils des Prud'hommes to oversee a campaign to eradicate or at least limit workplace theft.[67]

Officials of the Ministry of the Interior believed that state action could not resolve this long-standing problem. Chaptal himself argued that, at best, strict police enforcement of the regulations "could contain and diminish the number [of thefts]," not prevent them altogether.[68] In Sedan, where entrepreneurs complained bitterly about the widespread theft of valuable and hard-to-get merino wool, in 1803, Scipion Mourgue described the lack of punishment as a "vice of local police," but he was more concerned about "the notorious bad faith of certain manufacturers who make use of the raw material from theft, thereby enshrining the habit and aggravating the crime."[69] Claude-Anthelme Costaz, reporting for the Ministry, feared to poison the relationship of entrepreneurs and the working classes. Troubled by memories of the revolutionary committees' enforcement of the Maximum in the Year II, Costaz was particularly worried about potentially heightened militancy by laborers at the margins. They were most likely both to steal or to express their grievances with action in the street. A better and more appropriate response to labor militancy, opined Costaz, was for entrepreneurs to act rather than relying on the government to solve their problems. He suggested that industrialists precisely account for raw materials entrusted to workers whether they labored at home or at a centralized manufactory. The existing regulations, particularly the livret, were sufficient legal leverage for entrepreneurs to enforce more stringent quality and/or quantity control over raw materials. A few weeks later, Champagny sent a circular to every prefect on the "need to repress the theft of rawmaterials committed by workers." The best means of doing so was lifted directly from the 1673 regulations obliging entrepreneurs to maintain a ledger containing a strict accounting of what came in and went out of the shop.[70] Despite spirited local resistance to this infringement of "freedom of enterprise," at least some officials enforced Champagny's directive.[71]

Costaz noted that the entrepreneurs' demands for greater regulation were linked to their hope to recreate the *corporations des arts et métiers* or to restore special regulations to govern centers of high-quality production.[72] The woollen manufacturers of Reims petitioned the government in October 1802 to that effect. In March 1813, the cotton entrepreneurs of Amiens weighed in with a similar request, and 14 months later the Council of Manufactures reached conclusions similar to Costaz's. The Consultative Chamber of Louviers was still demanding specialized regulations and diverse government protections that would enable them to maintain their reputation for quality woollens in December 1820.[73] At the same time, however, the Chambers of Commerce or Consultative Chambers of Amiens, Carcassonne, Elbeuf, Paris, and Sedan explicitly rejected greater state regulation.[74]

Theft of raw materials by workers mirrored long-standing productive practices of their employers. Nicolas Quinette, a particularly knowledgeable prefect as well as a former Minister of the Interior, found systemic and systematic cheating in all phases of textile production in the département of the Somme. In May 1809, Quinette reported that local entrepreneurs in Amiens and Abbeville used poor sales or any economic difficulty as a pretext to lower the salaries of their workers. As in Britain, French entrepreneurs employed increasing numbers of women and children at lower wages to tend the machines.[75] In times of lax demand, Picard textile manufacturers lowered the quality of their wares using time-honored methods of cheating, said Quinette. The possibilities included cutting the thread count, using lower-quality cotton or wool, short-cheating on the length or width, and applying less expensive dyes.[76]

Since prices were rarely lowered at the production end, counterfeiting was also a favored tactic. The manufacturers of the Somme would buy their competitor's goods and then bundle them with their own in an attempt to pass their goods off as the product of more highly regarded, better-quality producers, either by region or, in the case of a well-known manufacturer, by the individual. Even some of France's best-known textile firms, including Richard-Lenoir, engaged in these practices.[77] Counterfeiting the marks of other manufacturers and selling smuggled goods as their own were also widespread tactics.[78]

Inspectors of manufacturing and intendants had reported these tactics since 1750; it is not surprising that Amiens had trouble finding customers when the lucrative colonial market for clothing slaves disappeared and military orders dried up.[79] Producers condemned workers for stealing, but they were at least equally guilty of cheating on quality. Both the French reputation in luxury goods and the many administrative attempts to foster reasonably priced everyday goods were endangered in ways unrelated to technological capacity. Given the remunerative potential of these widespread practices, the lack of "progress" in French textiles noted by many observers through the 1830s can be explained in a non-technological way that does not always reflect glory on the legend of the heroic entrepreneur.[80] The economic rationality of highly profitable alternatives to laissez-faire competition took advantage of contradictions and limitations of the market. These calculations underlay manufacturers' "market culture" during the Industrial Revolution in France.

Despite the wartime conditions under which the Napoleonic government operated, administrators formulated a systematic response to these problems. A first tentative step was taken on 3 Fructidor, Year IX (August 20, 1801) when the central administration decreed that all textiles should bear a "national" trademark to reduce smuggling. In January 1806, the Chamber of Commerce of Lyon recommended establishing a municipal trademark for its principal fabrics. All counterfeits would be seized. As in the early 1780s, manufacturers claimed that it was impossible to implement. The Chambers of Commerce of Carcassonne, Marseille, and Montpellier solicited government decrees (of September 21, 1807 and December 9, 1810) that prescribed the number of threads, the length and breadth, and the color of the stamps for each type of fabric involved in Levantine trade. Offices in designated ports were established to affix and verify the national stamp on woollens. Chaptal denigrated these measures as a return to the pre-1791 system of regulation, terming them "useless and vexatious."[81]

But in the chemical industry, in which he had a vested interest, Chaptal recognized the potential benefit of increased regulation to eliminate fraud. Three decrees were issued (one on April 1, 1811, another on September 18, 1811, and a third on December 22, 1812) to force soap manufacturers to affix an attestation of the contents of their product. For soaps made from olive oil, Marseille was to have a unique trademark that included the name of the manufacturer and the place of fabrication. According to Chaptal, this left the initiative with the manufacturer and did not interpose the state between the producer and the consumer. The Councils of Prud'hommes were charged with inspecting workshops and stores to ensure the fulfillment of the decrees by manufacturers of "good faith" while furnishing "a guarantee to the consumer that he has not been tricked by the manufacturer."[82] Despite Chaptal's ambivalent reaction to administrative efforts to ensure quality production, the clear trend under the Empire was in the direction of greater regulation.

Under the Restoration, the government revived some ancien régime administrative practices like the *marque des étoffes*. First resuscitated on

April 28, 1816, the stated goal of the measure was to limit imports and to ensure that was what sold as French-made was not, in fact, of British manufacture and smuggled into France. A new set of officials, most of whom were placed at the frontiers executed this measure. Manufacturers were also to regulate themselves. In good Chaptalian fashion, Councilor of State Louis Becquier, charged with the Ministry of the Interior, recognized in August 1816 that "the administration can, without doubt, intervene more directly in the trademarking of cloth. . . . But, fearing to constrain industry and limit commerce, the administration has preferred the guarantees of self-interest on the part of the manufacturers and merchants rather than relying on more direct action of the authorities." Becquier concluded that "that this ordinance is one of the first acts of an entirely new type of legislation." Perhaps this young engineer did not realize that he was going back to the future.[83]

Five further instructions on trademarking were promulgated over the next three years revealing how complicated and contested implementation was.[84] In Amiens, manufacturers worried that only someone with local knowledge could differentiate among the various types of goods. They wanted to ensure that any trademark could be affixed within the département. They did not want their goods inspected by their rivals in Lille.[85] Laissez-faire administrative procedure deliberately left merchants and manufacturers to oversee trademarking. This approach guaranteed that cheating on quality in one form or another would remain a viable and potentially profitable option for entrepreneurs throughout the first half of the nineteenth century and beyond. In absolute terms, however, such fraudulent activity was overshadowed by an even more clearly illegal maneuver: smuggling.

Smuggling was another persistent problem for French industrial development that was difficult to stop through state action.[86] During the eighteenth century, endemic smuggling on both the land and sea frontiers was common both in war and peace.[87] It spiked in the mid to late 1780s (around the time of the Anglo-French Commercial Treaty) thanks to an enormous demand for British goods, and it was accentuated by widespread English dumping in France. By 1790, the French state employed 19,000 officials to oversee the burgeoning trade between the two rival nations and to clamp down on smuggling. The outbreak of continental war in 1792, followed the next year by a worldwide maritime conflict diminished smuggling somewhat, but rampant inflation and the vigilance of the local revolutionary committees enforcing the Maximums drastically curtailed smuggling in the Year II. After 1794, smuggling once again became an accepted, widespread, and sometimes even patriotic entrepreneurial activity.

The policy makers of the Consulate established a regulatory framework to crackdown on this illicit activity. The Directorial administration had imposed stiff penalties for smuggling in 1797: confiscation of all goods involved and of the means of transport, from horses and wagons to ships. A financial penalty of 3 times the value of the goods and a mandatory prison sentence that ranged from 5 days to 3 months for unarmed civilians was also mandated. Members of the armed forces or the administration, or anyone taken under arms, received obligatory sentences of 5–15 years in irons. A high-level conference of government ministers, local officials, and judges recommended establishing a special criminal court solely to judge fraud cases. They decided that "seizures are received with a great deal of disfavor by the majority of courts. They forget that the [illegal] introduction of English goods ruins our manufactures . . . in the eyes of most judges smuggling seems to be only a minor infraction that can safely be overlooked." General Jean-Henri-Becays Ferrand, prefect of the Meuse-Inférieure—one of the chief avenues of smuggled English goods into France—reported that the "long wait in allotting the shares to soldiers and customs officers from any seizures" had been "the principal cause of declining zeal among them [these agents]." The court was created in September 1801.[88]

The customs service was reorganized and increasingly professionalized in July of that year. The *douanniers* received substantially augmented salaries of up to 4,000 francs. During the brief interval of open maritime trade after the Peace of Amiens, the Consulate upped the regulatory ante by detailing half-brigades of infantry (1,300 men) and regiments of cavalry (220 men) to crack down on smuggling in each of six frontier military divisions (5, 6, 7, 24, 25, and 26). If armed while engaging in smuggling, if either a soldier or a customs official was killed or wounded, the suspect "was considered to have taken part in an armed gathering." The resumption of war in May 1803 made these crimes punishable by death. During the spring and summer of 1805, mobile columns of troops and/or national guardsmen were deployed to catch bands of smugglers.[89]

Smuggling reached new heights after the declaration of the Continental System in 1806. From October 1810, new courts dealt with minor cases in the vastly expanded area under the French customs umbrella. Thirty-four customs tribunals spanned western, central, and southern Europe, from Rome to Gröningen and from Brest to Parma. These courts could impose up to four years of forced labor. Prevotal or military police courts handled appeals. They could levy up to 10 years of forced labor. Any and all British goods were to be burned publicly. It was understood, however, that this vast administrative apparatus, which employed about 35,000 customs officials, could only slow the epidemic of smuggling plaguing Napoleonic France.[90]

"Smuggling" referred to a host of illegal activities. Grain was commonly smuggled during times of dearth, particularly when either the central state or local administrators banned its export either from a département or from the nation. The prefect at Anvers, in the département of the Deux-Nèthes, reported that grain was exported illicitly from and through the region toward Cologne, Liège, and Louvain down the Meuse and the Rhine. It was almost impossible to stop this trade: "It is the day laborers, the artisans, the workers, the men that one would refer to as of 'the people' who execute this contraband. They earn up to 432 francs for every five quintaux exported. All you need is 18 francs in your pocket to start and you can double, triple, or even increase your capital ten-fold. A share-cropper who earns only 24–30 sols a day, finds it more profitable to save up to buy a sack of wheat and then carry it himself on his shoulders into Dutch territory. Such examples became contagious and soon fraud had become widespread despite the watchfulness of the customs and the military and the severity of the laws."[91] Grain smuggling was particularly widespread during the early Consulate and the late Empire. Eleven generally good harvests (from 1802 through 1812) meant that food was less scarce and grain smuggling less lucrative during that fortunate era, but localized shortages and overly heavy military requisitions in some areas led to illicit trade in grain throughout the period.

Grain smuggling was a major concern of the Napoleonic administration, but it paled in comparison to illicit commerce in colonial goods which included not only luxuries such as coffee, chocolate, and sugar but also such industrial raw materials as indigo, cochineal, and (most important) cotton. Manufactured goods, overwhelmingly British in origin, constituted another category of contraband. Bonapartist officials often winked at the clandestine grain trade (a policy tacitly sanctioned by the system of licenses instituted by Napoleon himself) or let in a few shipments of colonial products because, with prices absurdly high because of tariffs, they could only get their "fix" of coffee, chocolate, or sugar from the British. These same officials, however, actively combated any influx of manufactured goods, even before the formal declaration of economic war to the death in 1806.

The British state fostered the smuggling trade in manufactured goods. In May 1805, a special agent of Fouché described the process. Whenever sales for the manufactures of Manchester or Birmingham slackened, the government purchased the unsold goods and sent them to the Continent. The agent observed that this process was not as expensive as it might seem because they forced the manufacturers to sell at a low price. Nor was there any drawback to pay. The loss on these initial transactions was, at most, 5 percent. The goods were distributed to British commercial agents who contracted with merchants in Emden, Amsterdam, Hamburg, or Frankfurt to bring the merchandise to the Continent. These locals found people to smuggle British products into the port or across the frontier. This agent concluded: "Fraudulent imports of this sort, are, at present, one of the principal outlets for English manufactures. Repression is very difficult because the profits are so vast and so much greater than any other form of speculation. Even if the British manufactures have a wholesale value of three-fourths of their nominal value when they arrive in the hands of French merchants, these latter can earn a profit of 10–12 percent and still undersell French-made goods by a considerable margin."[92]

The British government and a significant number of enterprising middlemen purposely handed French merchants—whose interests suffered heavily from the war and from other government policies—opportunities to profit that they could not and did not ignore, despite the damage done to domestic industrial interests. Jean-Baptiste Gaudoit and a female associate named Manchot set up a workshop in October 1806 to make lace in Caen, which Manchot would oversee while Gaudoit found markets for their product. Gaudoit made several trips to southern English ports and later to Denmark to find clandestine ways of exchanging goods. Gaudoit and Manchot employed up to 3,000 men, women, and children and exported their entire production to Britain despite official prohibitions. Over the period 1804–1808, they sold 1.358 million francs' worth of goods on the other side of the Channel. When he was arrested for the second time in 1808, Gaudoit revealed that he had been involved deeply in illicit commerce since 1801, sending silks, bronzes, watches, wine, shawls, wigs, clover seeds, and lace to Britain. He imported cotton cloth, coffee, tobacco, and gin. Initially, Gaudoit brought his goods in—via local merchants—through Anvers and Rotterdam, but he often shipped his merchandise by way of Hamburg and a number of small Dutch ports. After 1806, he increasingly went through Frankfurt and Düsseldorf. His account books reveal that trade skyrocketed *after* the inauguration of the Continental System in 1806. Seizures either by customs or by the military amounted to a paltry 89,000 francs (2.65 percent of sales) of which 36,000 was at the hands of the British. While Gaudoit was in police hands, six respected merchants offered to pay a "caution" of 50,000 francs to free him, and Manchot agreed to help pay any fine levied against him. The authorities were also swayed by the claim that imprisoning Gaudoit permanently would "rob" 3,000 workers of their jobs. Gaudoit was released in February 1809 after paying a large but relatively insignificant fine of 15,000 francs. Gaudoit appears to have been one of the largest smugglers in France, but, as the police records indicate clearly, he was far from alone.[93]

Newly incorporated areas were notoriously lax in repressing smuggling. In October 1801, the Marquis de Sémonville, the French minister to the Batavian Republic (the Netherlands), wrote to Foreign Minister Talleyrand that "since the outbreak of maritime war, all the cities of Belgium and Batavia have turned to a double contraband trade, equally prejudicial to both nations." To bring coffee, sugar, tobacco, and other valuable goods from the British colonies across the Belgian-Dutch border, "they have corrupted some lower level customs agents and fought with others." In the Belgian départements alone, he estimated that there were at least 2,000 full-time smugglers. "Nor do they have to rely on the obscurity of the night, the usual path taken by fraud: it takes place in the full light of day, under the eyes of customs agents who they have compelled to keep silent."[94] Herbouville, the prefect at Anvers, reported that several merchants employed gangs of porters to bring goods across the frontiers, protected by bands of supposed "outlaws." The merchandise was warehoused in homes on the outskirts of the city or even in the city itself before being divided up so as to avoid municipal tax collectors. "Friday, market day, is the most favorable for fraud."

In April 1808, the police commissioner for the Deux-Nèthes stated flatly that smuggling was too widespread to repress. He offered an amnesty to try to shift commerce from illicit to legal channels. This tactic earned him a sharply worded reprimand from the Council of State.[95] The various Imperial initiatives to restart trade with Britain discussed in the previous chapter were not the only mixed messages sent to the entrepreneurs of the annexed territories. Later in 1808, the police reported from Anvers that "the director of the customs gave written orders to officials that from this moment onward they were to stop all smugglers. But at the same time, he told these same people verbally to disregard his commands."[96] In November, customs officers recovered a machine used to stamp bales of tobacco after inspection; it had been stolen two years earlier. Evidence was accumulated by opening mail (a power recently granted to the local constabulary by the Minister of Police), and a significant number of arrests were made in several cities. "What was surprising is that the Flemish and especially the people of Anvers whose faith teaches them that to trick or cheat the government is a 'good work' are astonished to see the smugglers and the forgers [of inspection marks] stopped."[97] Such equivocal actions, even in an area generally favorable to annexation by France, were revealing. Despite Napoleon's reputation for nascent authoritarianism, the inability of the Napoleonic state to crack down effectively on smuggling in the expanding Empire should not surprise.

To the north, the Dutch dominated sea-borne smuggling between Britain and the Continent.[98] From the Consulate onward, Fouché's police network described the scope of this trade and the deep involvement of both French and native officials. Packet-boats operated between various British ports (particularly Gravesend and London) and Rotterdam even after the resumption of war in 1803. In October 1804, letters found aboard a ship captured by a privateer and then turned over to the police reported that "the English are multiplying their maneuvers to introduce their merchandise into France." A number of Dutch merchants in different cities helped the smugglers by providing false certificates of origin. "The French commissioner in Rotterdam received 5 percent of the value for providing the certificates. The commissioner in Amsterdam receives 80,000 florins a year for the same service, which is paid to him by the Chamber of Commerce."[99] British merchandise also entered Dutch ports through Walcheren Island. From there, the goods were brought into the Escault River estuary by groups of English and American merchants.[100]

Unwillingness by Louis Bonaparte, King of Holland, to compromise Dutch commercial interests by cracking down on smuggling led to the annexation of the kingdom by France in 1810.[101] A staggering volume of smuggled goods was found by French agents. In June, 800,000 francs' worth of British merchandise was confiscated from Breda's shops and warehouses alone. More than one-third of it was colonial goods. The customs agents hoped to ruin the smugglers, most of whom were important merchants, but the prefect believed that by saving them from bankruptcy the new administration could win their support.[102] Such gestures and the increased French repression only seem to have stimulated the Dutch to find more obscure islands to trans-ship goods from British hulls before bringing them onto the Continent.[103]

Although the ports of Belgium and Holland were the preferred entry points for smuggling British goods into French markets, a number of new avenues surfaced once surveillance there increased. Hamburg, Lübeck, and other former Hanseatic ports trans-shipped vast amounts of British merchandise particularly colonial goods, thanks to loopholes in the "self-blockade." By 1810, French authorities estimated that one-sixth of the population of Hamburg relied on income from smuggling. Imperial administrators decided that the only possible means of slowing down, much less eliminating the flood tide of British goods smuggled into this region was to place French officials in charge of the customs. Like Holland, northwestern Germany was incorporated directly into France in 1810.[104] In this region, Heligoland, and, to a lesser extent,

Norderney Island, functioned as trans-shipment and distribution points similar to Walcheren Island. From August 9 to November 20 of 1808, 120 British ships carrying merchandise worth more than £80,000 docked just at Heligoland. The following year, the re-export of British goods was valued at almost £3.7 million.[105] Most commerce, whether licit or illicit, did not stop; it simply moved further east into the Baltic, especially after the French burned more than 5 million francs' worth of British goods found in Hamburg.[106] Even in France proper, commercial networks built on smuggled British goods were widespread; Gavin Daly asserted that a majority of Rouen's merchants and manufacturers participated in this illegal trade.[107]

Numerous smuggled British cargos, including colonial goods and manufactured products, entered France directly.[108] Prussia, neutral until 1805, was the "fake" flag of choice for ships bringing in British sugar for the Paris market. According to the police, "Emden sugar" was a joke well understood by French wholesalers and grocers. To close the Dutch route, General (later Field Marshal) Marmont ordered the army to help the customs officials. "Prussian" ships—using papers provided by Klose, the Prussian customs agent in Emden—docked at French ports, notably Cherbourg. In 1805 the local police agent reported that there were more than 4 million francs' worth of British goods currently in the city that had arrived in American hulls.[109]

Other Atlantic ports received a fair share of smuggled goods. In addition to Cherbourg, Morlaix and other small Breton ports reaped some benefits from close ties with England and the Channel Islands. Around Nantes, the clandestine article of choice was salt. Conducted mostly by military deserters, smugglers revived ancien-régime tactics of avoiding collectors of the salt tax. They gathered in groups of up to 150 to lead as many as 300 horses laden with salt through the hedge-divided maze of the bocage.[110] Smuggled goods brought across the Pyrenées deluged Bordeaux. This trade, conducted by armed bands, steadily increased after 1808. Prompted by requests from local merchants, British ships also landed goods at various small docks along the Gironde estuary, with the full connivance of administrators and customs officials.

Fraudulent licenses and illicit transactions with neutral shippers were common in Bordeaux, where support for Bonaparte was relatively low and sympathy for Britain relatively high. Some of the tricks used to avoid paying customs included trans-shipping goods to different ships multiple times in hopes of claiming that British goods were actually French-made, using real licenses to leave restricted ports, but then calling on unscheduled ports to load and unload contraband. This and counterfeiting licenses may have been less violent than the more blatant smuggling common in other parts of the Empire, but they still contributed to the flood of British goods (both colonial and manufactured) that inundated France.[111] Such tactics were also used along the Mediterranean border between Spain and France and in Italy.[112]

In discussions of the Continental System or smuggling, maritime questions often unjustly take pride of place. In eastern France, and especially along the Rhine (which formed the border with Switzerland and the German states), smuggling was a response to artificial barriers imposed administratively blocking long-standing commercial relationships rather than blockades or self-blockades. François Crouzet argued that inland smuggling outweighed its maritime counterpart. With Strasbourg's emergence as the main transportation hub of the Empire, Frankfurt, Heidelberg, Darmstadt, Mannheim, Rastatt, Kehl, Offenburg, Lahr, and especially Basel could add smuggled goods to merchandise in transit through the area.[113]

Switzerland was a particularly important entry point for British cotton goods, which could pass through customs as local manufactures. Swiss and German officials, from the King of Bavaria and the Grand Duke of Berg on down, obstructed French efforts to find British merchandise on their territory. With French administrators either turning a blind eye or actively participating in smuggling, illicit trade conducted by local merchant elites through this region proved almost impossible for the Napoleonic state to curtail, much less eliminate.[114]

A favored tactic along the Swiss frontier was to have two wagons that were "exactly the same: the same number of bales, the same weight, the same configuration. The customs duties would be paid for one and the other would enter fraudulently. One would enter [France at Mulhouse] openly and deposit merchandise which would remain there. The other would promptly disappear with the merchandise being sent to diverse places, but now complete 'denationalized' so that the origin cannot be determined and there is no trace in the books." In August 1805, customs agents found seven merchant houses in Lyon, Versailles, Basel, Mulhouse, and Cernay engaged in this practice. Five years later, similar tactics brought bales of British raw cotton across the Rhine at Kehl disguised as legal but lower-quality Levantine cotton from Macedonia.[115] Having reached the eastern border of France, merchants on both sides of the Rhenish frontier employed smugglers to bypass the customs barriers. A nighttime barge crossing delivered the goods to gangs of porters. Other porters, in groups of 50–60, received a princely 18 francs a day to carry contraband British goods over the mountain passes from Basel to the départements of the Doubs and Jura. Prefects, policemen, and customs agents complained that once the frontier had been breached little could be done other than exercising strict passport controls.[116]

Official corruption was the greatest barrier to cracking down effectively on smuggling. Corruption extended into the military, into the civil administration, and, most distressing, into the customs service. The enormous opportunities for gain enticed the poorly paid French officials. In certain areas, opposition to Napoleon or even resistance to French rule motivated corrupt activity, at least to some degree. The Magnier family was notoriously venal. Jean-Charles Magnier-Grandprez headed the customs in Strasbourg, while his brother and first cousin were high-ranking officials in other places. In exchange for a percentage of the proceeds, they warned the smugglers when raids were to take place and detailed what parts of the Rhine would be patrolled at a given time. The Magnier family was widely known to be corrupt, and charges were made against them to the Council of State by several prefects in eastern France, but Magnier-Grandprez was protected for mysterious reasons by the chief customs administrator himself, Colin de Sussy.[117]

The customs office in Alsace was disreputable; the customs office in Anvers was infamous. In May 1805, Blutel, the customs director at Anvers openly told his agents not to arrest smugglers, but to shake them down for bribes. A close associate, Captain Bigaune, purchased a vast tract of land for cash; a few months earlier, he had reported having no fortune to speak of. Merchants involved in smuggling declared bankruptcy, then retired in luxury to England. Verbrouck, the mayor of Anvers, and his son, the head tax collector in Ghent, were so deeply involved in smuggling "that they will even pledge bail for them. It is because of the influence of the mayor that the police are so negligent or apathetic." The judges, "none of whom support the French, liberate almost everyone accused of smuggling." The French consul took a cut of 3–5 percent on every transaction. Even after Fouché deployed military units to the Deux-Nèthes, the customs service continued these practices. Thanks to popular support, corruption in the customs service was endemic in certain areas and widespread in others.[118]

What does this litany of illegal behavior signify? Smuggling was only the best-publicized of the illicit behaviors limiting the free play of the market. To defend the Continental System by cutting down the volume of underground trade, the Emperor overextended the French Empire and severely overtaxed the administrative network. The magnitude of smuggling after 1806 reminds us that the actual behavior of entrepreneurs often had little to do with the sort of economic rationality envisioned by a Smithian laissez-faire, or perhaps that entrepreneurs had their own codes of honor and rationality. Imperial France suffered more from smuggling than Britain. The political reasons associated with military occupation were compounded by transportation problems and isolation from colonial markets. Smuggling siphoned entrepreneurial initiative from vital sectors and made it possible to fill Continental demand in areas where Britain had a clear technological lead (as in iron and cotton) without resorting to competition. This was not rent-seeking behavior, as Mokyr might argue; rather, it was a distortion of industrial specialization, whose significance was magnified by France's territorial losses and by France's diminished access to the Continental market after 1814. Take together, the illicit tactics discussed in this chapter help to explain why French industry did not begin to take off before 1815. After 25 years of war, traditional methods, niche markets, and luxury goods benefited from tariff protection to launch France's recovery in an international economy dominated by Great Britain. Eminently rational French entrepreneurs reaped significant profits without having to invest in new technologies or in re-organizing production, thereby paving the way for renewed efforts to compete internationally in areas of advantage once the occupation of eastern France came to an end.

Smuggling influenced French industrial development at two junctures: in the 1780s and during the wars of 1793–1815. In both cases, British industry benefited at France's expense. The bureaucrats directing French industrial policy failed to recognize that, given opportunities for illicit profit, entrepreneurs would take advantage of them, or band together to guarantee their livelihoods. In view of the lingering "threat from below," imitating the British mode of production and technological innovation were uncertain propositions. War and the myriad opportunities for illicit profits sidetracked Napoleonic efforts to create an industrial machine capable of beating the British at their own game. Even after the return of peace, French entrepreneurs and workers continued to limit economic competition in order to preserve their conception of proper market relations. Such alternatives to straightforward economic competition enabled France to endure the consequences of the path not taken until a new productive environment emerged after 1830.

9

Chaptal's Legacy in a Niche Industry

As a consequence of the continued fragmentation of French markets, uneven prices and local practices fueled competition, both international and domestic. In April 1826, a ton of pig iron cost 150 francs in Champagne, 265 francs in the center of the country, and 300 francs on the eastern border. Recognizing this lack of market integration, the Restored Bourbons invested heavily in communications and transport. First roads and canals and later steamboats and railroads received sustained state financial support and technical attention between 1815 and 1850. Experts trained at the *grandes écoles* championed France's scientific and technical abilities against British industrial dominance. The postal service was upgraded and its costs were lowered. The government also overhauled and extended the banking system to mobilize credit, lower interest rates, and encourage investment. These underappreciated yet important activities of the French state improved the long-term prospects for industrial competitiveness and domestic market integration.[1]

As the industrial landscape shifted, centers of production rose and fell. Some survived thanks to the quality of entrepreneurship, product specialization or technological innovation; others did so because of local or regional factor endowments. Factors ranging from sitting atop a coal field to rapid population growth sustained certain sites. The institutions created earlier in the century enabled the département of the Aube, dominated by the city of Troyes, to become a worldwide center of technological innovation in the production of cotton *bonneterie*, the making of hats and especially stockings. Technological improvement was essential to the development and consolidation of niche markets in textiles, but also more generally helped France to survive British predominance. The reaction of the laboring classes to the Troyens' cultivation of this sector and its increasing mechanization also demonstrated that the problems that had vexed Chaptal remained to trouble his successors.

In the second half of the eighteenth century, cotton spinning and *bonneterie* had grown rapidly in Troyes, Arcis, and Romilly, all in what became the Aube. This growth was based on the exceptionally low wages paid to the workers in domestic industry in southern Champagne. After the brief episode of machine-breaking in Troyes (1789–1791), the Revolutionary decade witnessed a 25 percent decline in production. The war was at fault. The fall in output equaled pre-1789 exports. Not long after 18 Brumaire, a brigade of troops was stationed in Troyes.[2] Thanks to this protection from "the threat from below," entrepreneurs rushed to mechanize. The *bonneterie* industry in the Aube rapidly joined the ranks of most technically advanced in France, a fact recognized by the jury reports from the Expositions of the Years IX, X, and 1806.[3] The high price of cotton under the Empire, however, put the industry into a tailspin, stimulating attempts to innovate. In the period 1802–1812, smuggled British goods and rigorous Swiss competition led tinkerers to patent 13 different machines to automate the manufacture of hosiery. With the return of peace, these rudimentary mechanical efforts could not compete with Englishman Marc Isambart Brunel's 1816 invention of the circular loom, the first machine to make stockings without additional stitching. Both tinkering and the inability to match British investment in advanced machinery resulted from continued reliance of Troyes' small-scale family firms on very low local wage scales and a dispersed "proto-industrial" production pattern. Only military orders and high tariffs kept the industry going under the Restoration when the falling price of cotton goods led to diminished wages nationally, thereby eliminating the Aubois' cost advantages.[4]

The French state followed a consistent policy of encouraging technological advance that began in the eighteenth century and was codified by Chaptal. The French believed that assisting innovation by providing local groups of entrepreneurs and workers with training and opportunities would facilitate tinkering and/or mechanization. This means of lowering costs and strengthening output, they believed, would make France more competitive in the international industrial arena. Chaptal asserted that second-tier industrial centers would benefit most from such an approach, but the Troyens had to wait until the Restoration for systematic institutional support.

In 1817 Troyes was given a Chamber of Commerce to replace an informal "Bureau of Commerce" formed by the prefect in 1801.[5] Its ineffectual learned society was reorganized into a "Society of Agriculture, Sciences and Arts" (in two stages: 1818 and 1828). The city's first savings bank was chartered in 1821, and in 1830 a Discount Bank aimed at assisting artisans and entrepreneurs was established by the Chamber of Commerce. Throughout the nineteenth century, these institutions played major roles in technological innovation in the region. With financial support from the municipality and the département, from 1826 to 1828, the Society reached out to workers by offering classes in practical mathematics, held three nights a week from 7 to 9 P.M., that were attended by 200 students, half of them workers. Two years later, a more rigorous class in "chemistry, geometry and mechanics applied to the industrial arts" replaced the initial offering. The course pointed out what the difficulties with existing machines were and explained what had been done, both within France and more generally, to cope with these difficulties.

Led by Joseph-August Delarothière, a worker who took the mathematics class, Troyes became a center of French innovation in textiles. Delarothière and two other workers—a mechanic and a bleacher—developed a more economical process that produced stockings at the same price as Britain's. First displayed during Charles X's peregrination through eastern France in 1828, it was patented the following year. Delarothière followed up this innovation in 1834 with an improved and automated version of Brunel's machine that attached two pieces of fabric without additional stitching. The Chamber of Commerce failed to get royal patronage for Delarothière, who could not afford to build the machine he had envisioned. In imitation of the role of private capital in funding English innovations, local manufacturers in Troyes took advantage of Delarothière's distress to pirate his design. He continued to tinker with textile machines, but, like many artisan-innovators, he died penniless in 1854.

Over the next generation, other workers from Troyes perfected the circular loom. In 1837, a watch maker named Jacquin, who had taken the mechanics course, developed a machine that speeded up the manufacture of hats. The following year, Jacquin was awarded a brevet of invention for a machine to produce pants, vests, and camisoles. Between 1838 and 1849, François-Humbert Gillet, a locksmith and another graduate of the mechanics course, pioneered four widely used machines, each of which used a mallet press, and earned the nickname of "Mallet Gillet." Other tinkerers concentrated fruitfully on incremental changes to existing looms or helped adapt factories and machines to the use of first water and then steam power. This blooming of artisanal innovation continued for at least two more generations. Between 1878 and 1888, M. A. Mortier alone earned 612 brevets of invention or certificates of addition for machinery. The machines were used not only in Troyes but also in Normandy, Picardy, and the Nord. By the late 1830s, machines designed and built in Troyes were being shipped to other parts of Europe, to Great Britain, and to the United States. By taking advantage of a long series of innovations that germinated in this formerly languishing corner of the textile industry, metal workers from Troyes, backed by local entrepreneurs, created a vibrant machine-building industry where none had existed before.

These workers/mechanics/inventors/tinkerers were the beating heart of one of France's fastest-growing textile niches. In 1846, the national production of cotton *bonneterie* was worth 55 million francs. The city of Troyes had 1,500 (mostly hydraulic) power looms and accounted for one-third of French production. As a whole, the département had nearly 11,000 looms that employed approximately 30,000 workers. Innovations saved the *bonneterie* industry and laid the foundations for the long-term prosperity of the city and the region. In developing these innovations, the hidden hands of government tariff protection and the provision of technical education by local institutions were complemented by developmental grants to workers and the 2 million franc loan for the Aube (out of 30 million francs for the entire nation) accorded by the newly installed Orléanist monarchy in October 1830.

Despite the general prosperity of *bonneterie* and the Aube, entrepreneurs were troubled by worker unrest. In the autumn of 1833, after several large merchants lowered the price they would pay for thread, 2,000 workers formed a coalition to force the wholesalers to buy thread and to coerce the manufacturers to keep them at their looms. Only a year later did the crisis relent as raw cotton prices fell. The approaching revolutionary crisis was forecast by a bread riot that rocked the city of Troyes in August 1847. The proclamation of the Second Republic on February 24, 1848 was greeted with joy by workers in the Aube and with trepidation by the département's industrial entrepreneurs. Their fears were realized in the textile town of Romilly. Hungry, underemployed workers believed that "machines, especially circular looms" threatened their jobs and would reduce them to poverty. On February 28 and 29, pushed by small proprietors who could not afford to mechanize, the town's workers took broke every circular loom. The destruction was selective. Persons and other property were left alone with a single exception: one manufacturer, who resisted, was beaten and had all his property ravaged. The national guard did nothing as 120 looms worth 40,000 francs were despoiled, "because the majority of manufacturers and inhabitants do not disapprove of the destruction of circular looms." An equally troubling riot unsettled Troyes on April 9 and 10, when workers invaded the prefecture during elections for national guard officers to press for the selection of sympathetic candidates.[6]

With the political and economic situation uncertain and faced with recrudescence of the long tradition of worker militancy in southern Champagne, it is not surprising that the Aubois, like so many other Frenchmen, looked to a familiar name to return order to the country. In the December 1848 presidential election, Louis-Napoleon Bonaparte received 92 percent of the nearly 73,000 votes cast in the Aube.[7] Karl Marx belittled Louis-Napoleon by claiming that history always repeats itself, the first time as tragedy and the second time as farce. Manufacturers may have made the first wave of machine-breaking during the French Revolution into a tragedy, but they never considered the machine-breaking in Romilly during the Revolution of 1848 a farce.

French industrialization after 1815 followed a trajectory different than British industrialization, in part because of attitudes regarding the working classes. The French industrial strategy instituted by Chaptal took advantage of domestic circumstances and minimized liabilities. Institutional support and government protection stimulated technological innovation and contributed to France's impressive long-term economic growth. France's political restiveness, however, greatly affected its process of industrial development. If Peter Mathias' question "First, and, therefore, unique?"[8] is applied to the French Revolution rather than British industrialization, then the divergent paths to industrial society seem natural rather than irrational or culturally determined as argued by David Landes and those who follow in his wake. If we take the notion of political economy seriously, then perhaps we need to refocus attention on the subject raised so despairingly by E. P. Thompson, namely how British elites came to dominate their working classes so thoroughly during early industrialization.[9] As events in the Aube demonstrated, despite the structural changes in French manufacturing and the transformation of industrial conditions, on the eve of the Crystal Palace Exposition, French entrepreneurs had no such assurance. The British model was impossible for France to follow. Political and social revolution placed France on an alternative pathway to industrial development.

Notes

The following abbreviations are used:

AC	Archives communales
ACSP	*Arrêté du Comité de la Ville de Paris*
AD	Archives Départementales
AM	Archives Municipale
AN	Archives Nationales
BHVP	Bibliothèque historique da la Ville de Paris
BM	Bibliothèque Municipale
BN	Bibliothèque Nationale
CNAM	Musée des Arts et Métiers–Archives historiques
EHESS	Éditions de l'école des hautes études en sciences sociales
FHS	*French Historical Studies*

Chapter 1

1. Landes 1969, 124–192, esp. 127–134. My first sentence paraphrases another influential work by Landes: *The Wealth and Poverty of Nations* (1999). Here and elsewhere, I follow the practice—common in the historical literature—of distinguishing between "France" (meaning the place) and "French" (meaning the people). It mattered greatly that one ruler was King of France and another was King of the French.

2. The argument that France had an economic *sonderweg* that differed in its means and timing rather than in its lasting results is usually associated with the revisionist work of Patrick O'Brien and Caglar Keyder (1978). Although some see this as a counterfactual argument, convincing economic indicators support it. See also Sabel and Zeitlin 1985. Much of this reconsideration is based on the work of Paul Bairoch, beginning with *Révolution industrielle et sous-développement* (1963). For a recent review of the literature, see Heywood 1992, 4–14.

3. For examples, see Landes 1969, 128, 132.

4. For an accessible version, see Young 1969.

5. Bairoch 1963, 279, 289, 300, 332, 335, 346; Crouzet 1990, 342–328; Crouzet 1958, 121–148; Landes 1969, 145–147; Lévy-Leboyer 1968, 794; Markovitch 1965–66, 216–217; Price 1987, 28–35.

6. Asselain 1984, 130; Verley 1997, 317; Lévy-Leboyer 1978, 267; Daviet 1997, 17.

7. For a recent economic analysis, see Louat and Servat 1995, 280–281.

8. This is even more impressive in view of the fact that French real wages declined by about 23% in the period 1820–1855. See Hoffman and Rosenthal 2000, 451; Morrisson and Snyder 2000, 72; Crouzet 1990, 342.

9. See, for example, Lazonick 1991, 267–296, esp. 281–286.

10. Daunton 1995, 564–565. Even the British triumphalist François Crouzet agrees, although he adjusts the date for the institutional shift to after 1870 (Crouzet 1996, 59–60).

11. See, for example, the lack of reference to France in Chandler 1991.

12. Mathias 1979, 3–20.

13. Crafts 1998; Rubinstein 1993; Wiener 1981.

14. Jacob 1997. For her Weberianism, see Jacob 2000. Although she does not agree with my interpretation, I am deeply grateful to Peg for introducing me to the history of science and technology and encouraging my initial forays into this fascinating field.

15. Among the important works in this field are Berg and Clifford 1999; Fairchild 1993a,b; Roche 2000.

16. Mokyr 2002, 28–77.

17. Published as "The Intellectual Origins of Modern Economic Growth" (Mokyr 2005). See especially pp. 286 and 336.

18. Pomeranz 2000.

19. See Wong 1997.

20. Rosenband 2005.

21. De Vries 1994.

22. Crafts 1983; Harley 1999, 160–205; Williamson 1984.

23. This includes the widely read works of Gerschenkron (1962) and Landes (1969, 1999).

24. MacKenzie 1998, 22–47, esp. 34–36, 38, 40–41.

25. Brewer 1989; O'Brien 1993.

26. I encountered a similar argument made in a very different way extremely late in the writing of this book. Donald MacKenzie (1998, 49–66, esp. 50–51) used the ideas of Herbert Simon, Sidney Winter, and Richard Nelson in articulating the fallacies of neo-classical economics. MacKenzie described a provocative model that can be applied to how French and British entrepreneurs actually behaved.

27. Gerschenkron 1962.

28. Macleod 2004, 111–126; Berg 2005.

29. Thompson 1963; Randall 1991, 1988, 1986.

30. Mokyr 2002, 28–77, esp. 31, 36, 52, 64, and 74. I thank Mokyr for a personal communication (February 2003) on the subject of Chaptal as one of the paradigmatic figures of the Industrial Enlightenment.

Chapter 2

1. At present, these issues are being investigated primarily through the creation and/or unification of markets. The "Forum: New Directions in Economic History" (*French Historical Studies* 23, 2000, no. 3) illustrates this trend with important articles by John V. C. Nye and by Philip T. Hoffman and Jean-Laurent Rosenthal.

2. The Physiocrats are generally understood as followers, to one degree or another, of the related ideas of Mirabeau père and François de Quesnay. These two thinkers and the members of their circles, who were often termed "economists," emphasized agriculture as the sole wellspring of prosperity and its interests as the proper foundation of all economic thinking. The Physiocrats' goal was to permit and to facilitate the actions of the state, which, they argued, should favor landed proprietors interested in increasing productivity in myriad ways, while spreading knowledge of improved agricultural techniques and developing institutions to assist agriculture. In exchange for this support, the Physiocrats asserted, landed property ought to be the basis for royal taxes. See Fox-Genovese 1976 and chapter 3 of the present volume.

3. For how the English example stimulated technological advance in France in one critical industry, see Garçon 1998, 5–17. See also Bourde 1953, 179–199.

4. Bret 2002; Dhombres 1991; Harris 1998; Hilaire-Pérez 2000; Lewis 1993; Minard 1998; Parker 1993; Shinn 1992.

5. Perhaps the most influential recent examples of such a focus are Jacob 1997 and Jacob and Stewart 2004.

6. Joel Mokyr (1990, 254, 258, 260) provided an inspiring first step in this sort of analysis.

7. Among their many important works, I wish to emphasize the following: Sewell 1980; Kaplan 2001; Sonenscher 1989b; Reddy 1984.

8. Thompson 1963, 1971; Davis 1975. See also Desan 1989.

9. For an example of this reformulation, see Johnson 1993, 37–62.

10. This is a major theme of Rosenband 2000b.

11. Hirsch and Minard 1998, 135–158.

12. Sonenscher 1989b, 287–289; Gayot 1998, 240–241.

13. Kaplan 1979, 22–25; Roche 1987, 272; Coornaert 1941, 275.

14. Chassagne 1991, 171–172. For supporting evidence, see Minard 1998, 296–300. On evasions, see Kaplow 1964, 34–35; Schmitt 1980, 361–362.

15. In Troyes, workers' coalitions faced repeated but ineffective police intervention. Regulations were passed in 1760, 1767, 1770 and 1773 and again in 1783, 1786

and 1787. *Extrait du registres des arrêtés de la mairie de Troyes*, 6 Prairial, Year XI, AM Troyes 2I9 17, *Ordonnance*, January 14, 1760, BM Troyes, Fonds Carteron 31; Colomès 1943, 48–53, and the measures taken in the 1780s by the Prêvôté of Troyes, BM Troyes FF (suppl). See also Hirsch 1991, 175–177; Truant 1994, 123.

16. See, for example, *Adresse au Monseigneur le Contrôleur Général des Finances par le commun des habitants d'Elbeuf*, 1763, AN F12 674 and the sources cited in the previous note.

17. The other four edicts attempted to improve the freedom of trade for the capital by cutting taxes on grains and meat while simultaneously abolishing some institutions that imposed fees on the same foodstuffs as they came into Paris. See Dakin 1939, 118–252, esp. 233–235. For other perspectives on Turgot's work, see Gillispie 1980, 3–50; Revel 1987, 225–242.

18. Sewell 1980, 131; Mousnier 1979, 472.

19. Sewell 1980, 131–132. For Turgot's attack on the *corporations*, see Kaplan 2001, 77–127.

20. Coornaert 1941; Hardman 1993, 50–51; Price 1995, 46–48. Dakin (1939, 236–239) emphasizes how well Turgot planned the measure.

21. Stone 1981, 121–126. The quotation is cited by Dakin (1939, 239).

22. Bossenga 1988; Sonenscher 1989b, 290; Minard 1998, 280; Gayot 1998, 279.

23. The *lit de justice* at the Parlement of Paris was not duplicated. The Parlements of Bordeaux, Toulouse, Aix, Besançon, Rennes, and Dijon refused to register the measure; that of Nancy refused to implement it (Coornaert 1941, 178). It was not implemented in Lille (Bossenga 1991, 129). In general, the Six Edicts were not implemented in the *pays d'état*, but they were in the cities of the *pays d'élection*, including Lyon.

24. Flammeront 1898, 293–356 (the quotation is from pp. 332–333). See also Sewell 1980, 72–74.

25. Sewell 1980, 72–77.

26. Kaplan 2001, 77–84; Stone 1981, 121–126, 135; Roche 1986, 204; Roche 1987, 272–273; Reddy 1984, 36.

27. Cited by Kaplan (1979, 27–29) and Kaplow (1972, 37). See also Dakin 1939, 254.

28. Kaplan 2001, 77–127, esp. 78–79.

29. Stone 1981, 94–97, 123; Palmer 1959, I, 452–453.

30. Horn 1995.

31. Kaplan 2001, 80.

32. Hardman 1995, 50–51.

33. See Kaplan 2001, 107–115; Isambert 1833, 74–75.

34. Cited by Kaplan (2001, 109).

35. Sonenscher 1987, esp. 103–104; Coornaert 1941, 173–175; Minard 1998, 316–326.

36. Kaplan 2001, 109–115.

37. The rights of seigneurs were confirmed specifically in 1779 and 1783 (Coornaert 1941, 157). On apprenticeship, see Sonenscher 1989b, 283. On privileged areas, see Horn 2006b.

38. See Kaplan's study of this issue (2001, 166–250).

39. Roche 1991, 282–293; Garrioch 1986, 113–114; Hafter 1995, 1997; Crowston 2000, 2001.

40. Kaplan 2001, 108–115.

41. Necker, *Lettre à de Crosne*, August 7, 1779, AD Seine-Maritime C137.

42. The powerful and prestigious Six Corps were revived, although the marchands de vin were added to the drapers, mercers, goldsmiths/jewelers, furriers, hosiers and grocers (Garrioch 1986, 99). There were between 117 and 120 corporations in Paris (Sonenscher 1989b, 62).

43. On the merciers/fabricants de bas, see Tolozan, *Lettre à de Crosne*, December 20, 1784, AD Seine-Maritime C137. On the metal trades, see the *Ordonnance de Police* of April 9, 1778, AD Seine-Maritime 5E 657.

44. The *corporations* were given two months to write new regulations. Outside of Paris, this was rarely done and often took years (Kaplan 2001, 112–113, 251).

45. These reforms were part of a particularly dense period of legislative activity. The Royal Council passed 57 measures in the 27 months from February 1776 to June 1778 (Kaplan 2001, 137).

46. Lyon's *corporations* never recovered from their brief abolition. Police of the city's trades was lax after 1776 (Garden 1970, 557–558, 572–582, esp. 580; Truant 1994, 131).

47. Mousnier 1979, 473.

48. Gallinato 1992, 334–335; Kaplan 2001, 135.

49. Parker 1993, 24; Hardman 1995, 51–52.

50. Bosher 1964, 103. See also Kaplan 2001, 133–142; Minard 1998, 321–322.

51. Harris 1979, 91–97. A stunning number of otherwise fine recent books on the reforms of the era and its major figures ignore the reform of the world of work. This non-exhaustive list includes the following: Egret 1975; Hardman 1993, 1995; Jones 1995; Labourdette 1990; Price 1995.

52. Hirsch 1991, 178–180. Hirsch finds that 1778–1781 was a unique moment when something new might have taken shape.

53. The answers are instructive. See the varying views of regulation in the responses of Simon Clicquot de Blervache and the Abbé Terray found in AN F12 661. See also Kaplan 2001, 105–106.

54. Harris 1979, 174; Saricks 1965, 72; Rosenband 2000a; Kaplan 2001, 29–38.

55. Minard 1998, 322–323; Kaplan 2001, 163. In English, see Minard 2000, 483–487.

56. For the views of inspector of manufacturing Jean-Marie Roland de la Platière on the need for further changes, see his letters in AD Somme C286 and his articles in the *Encyclopédie méthodique: Manufactures, arts et métiers* (1785) and in *L'Art du fabricant d'étoffes en laines rasées, et sèches, unies et croisées*, *L'Art du fabricant du velours de coton* and *Art de préparer et d'imprimer les étoffes en laines* (1780). Also see Gillispie 1980, 354–356.

57. Minard 1998, 323–325. For references to those opposed, see Reddy 1984, 36.

58. This is a major theme in Minard 1998 and in Parker 1993. For a nice summary, see *Notice du Travail que MM. les Intendants du Commerce ont adopté pour la confection totale de la Législation des Manufactures*, n.d. [1780], AN F12 657.

59. The correspondence on this subject is voluminous. For the province of Champagne: *Observations de la communauté des fabricants de la ville de Troyes pour se conformer à l'article 2 des Lettres Patentes de Sa Majesté du 5 Mai 1779*, March 28, 1780, AN F12 658B and Bruyard, *Mémoire sur les fabriques de la Champagne et de la Picardie*, December 15, 1784, AN F12 661. For the perspective of the central government: Abeille, *Observations*, March 27, 1781, AN F12 661, Crommelin, *Lettre à M. Abeille*, August 1, 1781, AN F12 654, Necker, *Lettre à MM les Intendants qui sont dans le ressort des parlements d'Aix, Dijon, Rennes, Pau, Grenoble, Rouen et Languedoc*, September 24, 1779, AN F12 676B and Montaran, *Circulaire aux Inspecteurs des Manufactures de son département*, n.d. [1781], AN F12 743.

60. Woronoff 1994, 190; Thomson 1982, 385, 388, 437–438.

61. Reddy 1984, 34–38; Smith 1775, book V.

62. See Parker 1993, 24–33, 99–105; *Instruction donné par nous, Directeur-général des finances aux inspecteurs et commis pour l'exécution des Règlements concernant les manufactures*, 1780, AN F12 1556.

63. Minard 1998, 325–326; *Observation des Inspecteurs Ambulants des Manufactures sur les Lettres-Patents des 5 Mai 1779, 1er, 4 et 28 Juin 1780*, December 23, 1788, AN F12 553. Most reports on enforcement are in AN F12 677.

64. Parker 1993, 25; Minard 1998, 325–326.

65. Kaplow 1964, 25–33.

66. Reddy 1984, 39–40.

67. Bossenga 1988, 697–698; Bossenga 1991, 151–152; Gayot 1998, 78–81, 358; Parker 1993, 24–49; Guignet 1977, 90–95. There is a thick file on Sedan in AN F12 1358.

68. This is a major theme in Thomson 1982 and in Hirsch 1991. See also Parker 1993, 26–27; Billion 1941, 131–139.

69. All references to this edict are from Isambert et al. 1833, XVIII: 78.

70. Ferguson (2000, esp. 567) situates this interest in space as focused on the body rather than geographically. Rosenband (2000a, 455–476) shows why tramping workers were problematic for entrepreneurs.

71. Hufton 1974, 69–106, 219–244.

72. Evidence of this perception is exceptionally voluminous. For a few examples from primary sources, in different industries and different places: *Avis des Députés du Commerce*, December 5, 1788, AN F12 724; the entries for October 22, 1782, August 30, 1784, and May 2, 1785 in the *Registre des délibérations des fabricants de toutes sortes de toiles en fil et en coton*, 1779–1788, AD Seine-Maritime 5EP 753; and the views of the constructeurs de navires found in the *Lettre des Jurats de Bordeaux à M. Joly de Fleury, Ministre des Finances*, September 23, 1783, AM Bordeaux BB 179 registre 152–154. Truant (1994, 120) cites a decision by the Parlement of Paris in September 1778 which was a homologation of a judgement from Lyon that prohibited worker associations. See also Sonenscher 1987, 77–109; Sonenscher 1989b, 322–327; Kaplan 1979, 22; Kaplow 1972, 37–41; Burstin 1993, 70–72; Hirsch 1991, 173–177.

73. On the fundamental mobility of rural artisans and workers, see Belmont 1998, I: 101–127. Johnson (1993, 37–62) demonstrated the converse of what happened when workers remained in place: proletarianization and the commencement of the factory system.

74. Sonenscher 1989b, 140–145, 174–209; Sonenscher 1986, 74–96; Truant 1994, 119–133; Dewerpe and Gaulupeau 1990, 156–158; Chassagne 1991, 169–176; Coornaert 1941, 204–205. Kaplan (1979, 36–51) makes the opposite point.

75. On workers' customs and how entrepreneurs sought to defeat them to improve industrial production, see Rosenband 2000b, 49–118. See also Lewis 1993, 39–48.

76. Reynard 1999: 18–25.

77. Reddy (1984) invokes "a world without entrepreneurs"—the government had to create them, partly through market forces.

78. Bossenga 1988, 702–703.

79. Pollard 1965, 189–244; Guignet 1977, 463–596, 650–679.

80. See Baczko 1992; Day 1987; Julia 1981; Palmer 1985.

81. The primary sources are in AD Seine-Maritime C135. See also Kaplow 1964, 34–35; Chassagne 1991, 173; Gullickson 1986.

82. Minard 1998, 212–218; Rémond 1946.

83. Reddy 1984, 24–34.

84. *Déclaration du Roi, concernant les communautés d'Arts et Métiers du Ressort du Parlement de Rouen*, February 6, 1783, AN F12 760.

85. On the recalcitrance of the Parlement of Rouen, see the detailed correspondence from April–August 1779 found in AN F12 786.

86. Chassagne 1991, 171–177.

87. *Requête*, July 2, 1789, AD Seine-Maritime 5E 657.

88. *Registre des délibérations de la communauté des serruriers de la ville, fauxbourgs et banlieue de Rouen (1784–1788)*, 34–35 for July 9, 1785, AD Seine-Maritime 5EP 651.

89. Froment and Hennin de Beaupré, *Lettre*, n.d. [1782?], AD Seine-Maritime 5EP 657.

90. *Registre des délibérations de la communauté des serruriers de la ville, faubourgs et banlieue de Rouen (1780–84)*, 41–45, AD Seine-Maritime 5EP 650.

91. This paragraph is based on *Registre des délibérations de la communauté des serruriers de la ville, fauxbourgs et banlieue de Rouen (1784–1788)*, 12–13, 16–17, 31–34, 97 in AD Seine-Maritime 5EP 651; *Registre des délibérations de la communauté des serruriers de la ville, fauxbourgs et banlieue de Rouen (1788–1791)*, 2 in AD Seine-Maritime 5EP 648; and Gabriel Parel, *Déclaration*, March 30, 1785, AD Seine-Maritime 5E 656. Sonenscher has done a thorough investigation of the entry registers of the locksmiths for this period. *Work and Wages*, 101–102, 108, 120–112, 151–173.

92. Truant 1994, 124–126.

93. See the entries for August 30, 1784, October 18, 1784, and May 2, 1785 in the *Registre des délibérations des fabricants de toutes sortes de toiles en fil et en coton*, 1779–1788, AD Seine-Maritime 5EP 753.

94. Reddy (1984, 56–57) believes this applied to Rouen's entire textile sector.

95. Tolozan, *Lettre à de Crosne*, July 17, 1783, AD Seine-Maritime C144; Bouloiseau 1957–58, I, 76–207 (quotation from p. 120).

96. A case which combined the issues facing sons and widows of masters and agrégés is in the correspondence between Tolozan and the new intendant Étienne-Thomas de Maussion in 1787 (AD Seine-Maritime C126).

97. *Déclaration du Roi, concernant les communautés d'Arts et Métiers du Ressort du Parlement de Rouen*, February 6, 1783, AN F12 760.

98. *Pétition*, March 20, 1782, AD Seine-Maritime 5E 657.

99. Bouloiseau 1957–58, I: 128–132.

100. *Registre des délibérations de la communauté des serruriers de la ville, faubourgs et banlieue de Rouen (1780–84)*, 39–41, 48–49, 64, 66, 71–72, 81–85, in AD Seine-Maritime 5EP 650, *Registre des délibérations de la communauté des serruriers de la ville, fauxbourgs et banlieue de Rouen (1784–1788)*, 16, 20, 51–60, 78, 83–85, 88, in AD Seine-Maritime 5EP 651, *Registre des délibérations des fabricants de toutes sortes de toiles en fil et en coton*, 1779–1788, AD Seine-Maritime 5EP 753, *Lettre à de Crosne*, January 21, 1783, AN F12 786, Tolozan, *Lettre à de Crosne*, July 17, 1783 and Maussion, *Lettre à de la Michodière*, April 29, 1789, both in AD Seine-Maritime C144.

101. *Arrêt du Conseil d'État du Roi*, February 26, 1784, AD Seine-Maritime C122. See the registers of the bonnetiers in AD Seine-Maritime 5E 165–166 and the passementiers-toiliers in 5EP 612.

102. Kaplan 2001, 141.

103 *Registre des délibérations de la communauté des serruriers de la ville, faubourgs et banlieue de Rouen (1780–84)*, 78 in AD Seine-Maritime 5EP 650.

104. See the entry for March 8, 1782 in the *Registre des délibérations des fabricants de toutes sortes de toiles en fil et en coton, 1779–1788*, AD Seine-Maritime 5EP 753.

105. Tolozan, *Lettre à de Crosne*, March 30, 1784, AD Seine-Maritime C137.

106. Tolozan, *Lettre à Maussion, intendant de Rouen*, July 16, 1787, AD Seine-Maritime C137.

107. Minard 1998, 288–290.

108. See, for example, the entries for March 8, 1782, October 22, 1782, and October 29, 1782 in the *Registre des délibérations des fabriants de toutes sortes de toiles en fil et en coton, 1779–1788*, AD Seine-Maritime 5EP 753. Kaplan (2001, 682, n. 40) claims to have evidence of more than 1,000 such seizures.

109. These registers are in AD Seine-Maritime 5EP 165–166, 500–501, 612, 616–617, 736, 743–745, 753 and 760. See also the extensive correspondence found in AD Seine-Maritime C126.

110. *Registre des délibérations de la communauté des serruriers de la ville, faubourgs et banlieue de Rouen (1780–84)*, 24 in AD Seine-Maritime 5EP 650.

111. *Registre des délibérations des teinturiers en laine, soye, fil et coton en grand et en petit teint de la ville, faubourgs et banlieue de Rouen (1780–1791)*, 5–16 in AD Seine-Maritime 5EP 736.

112. *Avis des Députés du Commerce*, August 1, 1783 and *Avis des Députés du Commerce*, May 27, 1783, both found in AN F12 721. See also the raft of complaints in AD Seine-Maritime C137 and C173.

113. Bouloiseau 1957–58, I: 202–203.

114. Reports by respected inspectors of manufacturing frame these problems over time: Huet de Vaudour, *Mémoire*, January 1762, AN F12 560; Goy, *Mémoire*, November 12, 1782, AN F12 650.

115. This did not include three *corporations* outside the purview of the reform such as the apothecaries (Bouloiseau 1957–58, I: XLV, 208). Paris went from 117 to 50. Abbeville fell from 50 to 18 (Ruhlmann 1948, 103).

116. This correspondence from 1779 is found in AD Seine-Maritime C144.

117. The correspondence of May-August 1779 is in AD Seine-Maritime C137 and C144.

118. de Crosne, *Lettre à de Montaran fils*, January 13, 1778, AN F12 1424. There are a number of letters on sales in the latter carton

119. *Arrêt du Conseil d'État du Roi*, February 14, 1780, AN F12 786.

120. See the correspondence of Tolozan and de Crosne of June 5, 1783 and August 19, 1784 in the AD Seine-Maritime C129, the *Mémoire pour les fabriquants d'Elbeuf*, May 27, 1783, Alrond?, *Lettre à Tolozan*, June 24, 1783, and Goy, *Lettre à Tolozan*, October 1, 1785 all found in AN F12 760. See also Kaplow 1964, 28–51.

121. *Mémoire pour les fabriquants d'Elbeuf*, May 27, 1783, AN F12 760.

122. De Crosne, *Lettre à Tolozan*, August 19, 1784, AD Seine-Maritime C129.

123. This ordinance was the source of their complaints: *Déclaration du Roi*, August 11, 1779, AD Seine-Maritime J719/1. Their complaints were answered in letters from Tolozan to de Crosne of September 20, 1780 in AD Seine-Maritime C126 and December 1, 1779 and June 29, 1781 in AD Seine-Maritime C136.

124. *Mémoire pour les marchands de la ville de Neufchatel en Normandie envoyé au Contrôleur Général des Finances*, n.d. [1780] and de Crosne, *Lettre à Tolozan*, April 19, 1780, both in the AD Seine-Maritime C136.

125. *Avis des Députés du Commerce sur la demande de cassation d'un arrêt du Parlement de Rouen*, July 16, 1782, AN F12 721. Parker 1993, 38.

126. On the issue of widows in the corporate world, see Kaplan 2001, 193–200. Rouen appears to have had a long-standing place for women within the *corporate* structure, not only for widows but, in general, that was rare in provincial France (Sonenscher 1989b, 66). On the gender consequences, see Hafter 1997; Crowston 2001; Lanza 1996.

127. *Arrêt du Conseil d'État du Roi*, February 26, 1784 and *Arrêt du Conseil d'État du Roi*, November 23, 1786, both in AD Seine-Maritime C122.

128. The educational intentions of the extensive plates in Denis Diderot's *Encyclopédie* or the detailed information on manufacturing from Roland de la Platière in the *Encyclopédie méthodique* are examples of this mindset. See Brose 1998), 23–50; Jacob 1997, 131–207.

129. Harris 1992, 167–182.

130. This attitude was quite widespread. For some examples, see Carra, *Lettre à Tolozan*, November 29, 1785, AN F12 1340; Jean-Baptiste Reboul, *Pétition à l'Assemblée nationale de la France en son Comité du Commerce*, n.d. [1790?], AN F12 652, Vandermonde, *Examen de la machine presenté au Conseil par le Sieur Perrin fabriquant, pour dispenser les ouvriers en étoffes façonnées de la nécessité d'employer un aide pour tirer les cordes*, October 5, 1783, AN F12 2201; Montigny, *Mémoire et avis sur les mécaniques de M. La Salle relatives aux étoffes de soye*, April 25, 1781, AN F12 642.

131. This attitude clarifies a current debate in the history of science regarding what aspects of the British model the French actually wanted to imitate. The "learning by doing" paradigm did not translate well in France because of this lack of faith in the capacities and willingness to adapt on the part of the workforce. See Clark, Golinski, and Schaffer 1999; Harris 1998; Macleod 2004; Rosenband 2000a.

132. Schmitt 1980, 365–366. Sonenscher (1989b, 91–96, 146, 190) puts this issue in the context of subcontracting. See also Colomès 1943, 29, 57–58; Thomson 1982, 380–388.

133. Chaudron 1923a, 11; Reddy 1984, 4, 10; Gullickson 1986, 65–67, 86–87. For an opposing view, see Liu 1994, 46.

134. Crouzet 1967, 139–173; Rosenband 2000b, 134–139; Thomson 1982, 383–385.

135. Thomson 1982, 417–423; Johnson 1993, 41; Schmitt 1980, 364; Garden 1970, 588–591; Kaplow 1964, 38; Parker 1993, 40–45. In Normandy this was true for the region of the *pays de caux* (Gullickson 1986, 64).

136. The standard account is Bosher 1964.

137. Vidalenc 1958a, 276–278; Mollat 1979, 239–240; Dubuc 1975, 134–137. See also *Rapport des Travaux de la Commission intermédiaire de Haute-Normandie depuis le 20 Décembre 1787 jusqu'au 27 Juillet 1790* (Rouen, 1790), AD Seine-Maritime C2114* and the correspondence of the Bureau d'Encouragement of Rouen found in AN F12 658A.

138. Kaplan 2001, 500–599; Sewell 1980, 77–99; Sonenscher 1989b, 328–362.

Chapter 3

1. The phrase of Louis-Guillaume Otto was "La révolution industrielle est comencée en France" (cited in Landes 1999a, 128–129).

2. Grenier (1996, 425) insists on this.

3. Jacob 1997, 120–130.

4. Mokyr 2002, 35–42.

5. Stewart 1992, chapter 8.

6. Mokyr 2002, 40.

7. On the academies, see Roche 1978. Jones (2003) depicts French state activism in agriculture parallel to the industrial situation. On this sort of economic partnership, see also Reynard 1999, 1–25.

8. Gillispie 1980, 9–12.

9. Hirschman 1995, chapter 17, esp. 199–202.

10. Gillispie 1980, 6–12.

11. According to Jones (1995, 205), this intellectual legacy was strongly represented on the Finance Committee of the National Assembly.

12. The theme of French views of English success was taken up in "The sources of England's wealth: Some French views in the eighteenth century," a neglected article that has been reprinted in English in Crouzet 1990, 127–148. Crouzet is, however, mostly concerned with the period up to 1760.

13. Cahen 1939; Harris 1998; Jeremy 1977; Minard 1998.

14. A major exception to this lack of attention concerns proto-industrialization. See Liu 1994, 5–9, 22–32. More generally, see Ogilvie and Cerman 1996; Sabel and Zeitlin 1985, 133–176.

15. Berg (1994, 57–59) hints at this misuse or misreading of Smith.

16. See Mokyr 1999; Harley 1999; Crafts 1985.

17. Mathias 1983, 31–32.

18. Patrick Colquhoun, *Treatise on the Wealth, Power and Resources of the British Empire* (1814), cited on p. 128 of O'Brien 1993.

19. Price 1999, 52–53.

20. The citations are from Édouard Boyetet, *Recueil de Divers Mémoires relatifs au Traité de Commerce avec l'Angleterre faits avant, pendant et après cette négociation* (Versailles: Baudouin, 1789), 14–86, esp. 37–38 and 41.

21. Roland de la Platière wrote this in *L'Art du fabricant en velours du coton*, sponsored by the Academy of Science between 1780 and 1783 (cited in Rémond 1946).

22. *Observations Préliminaires de la Chambre de Commerce de Normandie sur le traité de commerce, entre la France et l'Angleterre*, n.d. [1786], AD Seine-Maritime C1092 and *Avis des Députés du Commerce*, January 20, 1786, AN F12 723.

23. See Sickinger 1999, 101–116. Other drawbacks existed on goods such as refined sugar to facilitate exports (letter of December 14, 1786, in *Septième régistre des lettres envoyé par la Chambre de Commerce de Guyenne, 1785–1791*, AD Gironde C4266).

24. A 1563 law known as "Five Elizabeth" set the hours of work for laborers. Justices, county sheriffs and town officials were given the power to fix wage rates annually at the Easter quarter sessions (Moher 1988, 75, 77). No French authority had similar powers. The enforcement of such measures even at the local level was well beyond the power of the Bourbon state, much less the Valois state!

25. Price 1999; O'Brien 1994; Ashworth 2003; Daunton 1995.

26. Wood (1992, 33) contends that it was the *idea* of the nation that was weak in Britain, not the power of that state.

27. Crouzet 1990, 19–20.

28. MacLeod 1988; Hilaire-Pérez 2000.

29. The quotations in this paragraph are from Jean-François de Tolozan, *Mémoire sur le commerce de la France et de ses colonies* (Paris: Moutard, 1789), 99.

30. Tolozan's correspondence with the industrial spy Le Turc was a major reason why he latched on to this explanation of French technological "backwardness." See also his *Lettre à M. Barthélemy* (the French ambassador to Great Britain) dated April 7, 1789, in AN F12 994; Hilaire-Pérez 2000, 259–263. Other important officials publicly endorsed a similar view (Donaghay 1982, 208).

31. The citations in the previous three paragraphs come from a letter from Holker to Rayneval dated December 29, 1785, reprinted in Édouard Boyetet, *Seconde Partie du Recueil de divers Mémoires relatifs au Traité de Commerce avec l'Angleterre, faits avant, pendant, et après cette négociation* (Versailles: Baudouin, 1789), 86–101. See also Rémond 1946, 118–124. Harris (1998, 410–411) had a more negative perspective on Holker's views of French competitiveness and potential. Holker's view about the competitiveness of Norman cottons and the possibility of exporting them was widely shared. See, e.g., Boulle, 1975, 319–320.

32. This document is in AN F12 1342 (cited on p. 186 of Chassagne 1991).

33. Murphy 1982, 439; Murphy 1998, 67; Murphy 1966, 578.

34. Murphy 1966, 575–578.

35. Boucher le jeune of Rouen gathered some of the material, and Jean-Barthélémy Le Couteulx de Canteleu, also of Rouen, reported it to the National Assembly. This quotations in this and the following two paragraphs come from [Louis-Ézechias] Pouchet, *Essai sur les avantages locaux du département de l'Aube, et sur la prospérité nationale, ou Adresse à Mes Concitoyens du Département de l'Aube*, 1791, AN F12 652. Similar attitudes by contemporaries include Pierre Laurens Daly, *Mémoire*, September 1, 1790, AN F12 652. See also AN F12 1342.

36. Chassagne 1991, 185.

37. The views of Pouchet and Holker and nearly all the data I have collected on this issue rebut Maxine Berg's assertion (2005) that it was British style or fashion that sold goods, not their quality and price.

38. This activist vision of the role of the state is at odds with the more associational view of British industrial development emphasized by Jacob (1997, 116).

39. The date is October 16, 1788 (Young 1969, 72–73).

40. One of the most cited views emphasizing how far behind the French remained comes from François Crouzet (1967). See also Harris 1998; Rostow 1960; Landes 1969.

41. For cogent recent reviews of the literature, see Mokyr 1999, 1–127; O'Brien and Quinault 1993, 1–30, 54–78. Interestingly, historians concerned with the heydey of British technological superiority during the nineteenth century do report the views of contemporaries. See, for example, Tunzelmann 1994.

42. Mokyr 1994b, 35.

43. Musson and Robinson 1969, 190–199, 352–371; Jacob 1997, 116–130.

44. Inkster 1991, 32–59; Mokyr 1999, 36–39.

45. Rubinstein 1993, 25; Berg 1994, 169–188; O'Brien, Griffiths, and Hunt 1996, 155–176.

46. Parker 1986, 45–58. For a more negative view, see Harris 1998, 527–565.

47. Musson and Robinson 1969, 219–221, 97.

48. See the sources cited by Bowden (1919) and Rose (1908).

49. Sée 1930, 312.

50. Donaghay 1990, 386.

51. Only during the revolutionary wars did the quality of British paper equal that of the paper produced by Britain's Continental competitors (Rosenband 2000a, 474).

52. Cited on p. 209 of Donaghay 1982.

53. The quotation is from a letter, dated June 30, 1786, to William Eden, the British negotiator of the Commercial Treaty (Hogge 1861, I: 135–136).

54. Bret 2002, 373–375. This was part of a recent reversal. French cannon shortages in the Seven Years' War led the government to imitate English means and methods of building industrial capacity (Baugh 2004, 241).

55. Erdman 1986, 57.

56. Bowden 1919, 18–35.

57. Here I am arguing in the same vein as O'Brien and Keyder (1978, 176).

58. For a brief survey of the negotiations, see Murphy 1982, 432–443.

59. See, for example, Doyle 1988, 162–164.

60. Woronoff 1994, 190–191; Daviet 1993, 53.

61. Landes 1969, 139. I contend that the following eminent historians are uninterested in the Treaty and its effects, thereby demonstrating a broader trend: Crafts (1983); Crouzet (1990); Musson and Robinson (1969); Mokyr (1999); Berg (1994); Mathias (1983); Floud and McCloskey (1994).

62. Mori 2000, 186; Christie 1982, 167; Black 1994, 101–112.

63. See, for example, P. Jones 1995, 100–102; Doyle 1989, 86–87; C. Jones 2002, 360–363.

64. Doyle (1988, 164) and Bosher (1988, 76–77) still insist on this.

65. Shapiro and Markoff 1998, 260–261.

66. The phrase is Murphy's (1982, 434).

67. That peace with England was the essential objective of the Treaty is a common thread of the three foremost specialists: Labourdette (1990), Murphy (1982), and Donaghay (1978).

68. Cited on p. 268 of Cahen 1939 in the context of a letter written to an agent in Britain on December 27, 1786. See also Murphy 1998; Hardman and Price 1998.

69. Labourdette 1990, 257–258.

70. On the influence of Physiocracy on this generation of policy makers, see Meyssonnier 1989.

71. This argument echoes Wood's provocative contention (1992, 1–2) that English capitalism was born in the countryside rather than in urban areas.

72. On Vergennes' views, see Labourdette 1990, 239–241. On Dupont's role, see Murphy 1966, 569–580; Bosher 1964, 72–79.

73. Murphy 1982, 432–458, esp. 434.

74. Lacour-Gayet 1963, 163.

75. See Donaghay 1978, 1157–1184; Lacour-Gayet 1963, 150–166.

76. Payen 1969, 99–135, esp. 131–132.

77. John Holker fils (?), *Mémoire sur l'état actuel de la filature*, April 1790, AN F12 1341.

78. Harris 1998, 108.

79. Murphy 1966, 577–578; 1982, 439.

80. *Lettres-patentes*, January 19, 1786, AD Gironde 1B 56 registre, f. 191–192.

81. *Arrêt du Conseil d'État du Roi*, May 25, 1786, AD Gironde C1644.

82. *Arrêt du Conseil d'État du Roi*, December 2, 1786, AD Gironde C1644 and *Arrêt du Conseil d'État du Roi*, August 12, 1789, AD Seine-Maritime C175.

83. *Arrêt du Conseil d'État du Roi*, August 25, 1786, AD Gironde C1650.

84. Reddy 1984, 54–55.

85. Of the 5.5 million *livres* spent to improve manufacturing between 1739 and 1789, 10% was spent during the four years Calonne was Controller-General (Donaghay 1982, 206).

86. In his 1995 book dedicated to Vergennes' domestic policies, Munro Price inexplicably fails to mention either the Single Duty Project or Vergennes' commercial and industrial goals. Every other account emphasizes these objectives.

87. This paragraph is based on pp. 106–135 of Bosher 1964. See also Hirsch 1991, 172–173.

88. Bosher 1964, 78–81.

89. For a clear summary of the calculations that went into the duties set by the Treaty and why the "collateral" issues like the Single Duty Project had such a powerful effect, see Donaghay 1982, 215–217.

90. Cited by Bosher (1964, 20). This dislocation is particularly stunning since on the eve of the Revolution, France produced more than twice as much cast iron (Verley 1997, 304).

91. Hirsch 1991, 172–173; Popkin 1999, 47–48; Shapiro and Markoff 1998, 260–261.

92. Letter from Holker to Rayneval of December 29, 1785 reprinted in Boyetet, *Seconde Partie du Recueil de divers Mémoires*.

93. Cullen 2000, 591.

94. For an excellent short survey, see P. Jones 1995, 96–106.

95. Report of Jacques-Claude Beugnot in *Procès-verbal des séances du conseil général du département de l'Aube*, November-December 1790, AN F1CIII Aube 2.

96. Donaghay 1984.

97. British exports to France fell back to 23 million *livres* in 1789 (Butel 1993, 75).

98. There were analogous complaints in Britain (Cahen 1939, 277–278).

99. Cited by Cahen (1939, 279).

100. Young 1969, 408–409.

101. Pierre Deyon, cited on p. 226 of Butel 1993; Engrand 1986, 193.

102. See the correspondence in AD Gironde 7B 1228. This quotation is from Lalau Cordier, *Lettre à Cordier-Joly à Bordeaux*, March 7, 1788, AD Gironde 7B 1212.

103. Louis Villard, *Observations sur l'état actuel des Manufactures de lainages et cotons de la Picardie*, n.d. [1788], AN F12 659. The importance of the closure of the Spanish market for the crisis of the 1780s cannot be underestimated. On its significance in Languedoc, see Lewis 1993, 63–65.

104. *Lettre des députés composant la Commission intermédiaire de la Province de Champagne au Contrôleur Général*, January 18, 1788, AN F12 678.

105. Simon Clicquot de Blervache, *Avis*, March 9, 1788, AN F12 678.

106. Saricks 1965, 96.

107. This and the following two paragraphs depend on pp. 401–422 of Donaghay 1984.

108. Donaghay 1982, 218.

109. For an example of the long-time British habit of dumping in the French market, see p. 23 of Lewis 1993.

110. See the sources in AN F12 678 cited above.

111. Chassagne 1980, 153.

112. The two most famous of these pamphlets are *Les observations de la Chambre de Commerce de Normandie sur le traité de commerce entre la France et l'Angleterre* (1787) and *La réfutation des principes et assertions contenus dans une lettre qui a pour titre : Lettre à la Chambre de Commerce de Normandie sur le Mémoire qu'elle a publié relativement au Traité de commerce avec l'Angleterre par M.D.P.* (1788).

113. See his *Lettre à la Chambre de Commerce de Normandie sur le Mémoire qu'elle a publié relativement au Traité de commerce avec l'Angleterre* (1787) and *Lettre à la Chambre de Commerce de Normandie* (1788)

114. Sée 1930; Murphy 1966; Delecluse 1989.

115. A law to that effect had been repealed in 1784.

116. This paragraph is based on the recommendations in a 1788 *Mémoire* written by the Chamber of Commerce of Normandy (AN F12 658A).

117. Donaghay 1990, 395–397; Gras 1908, 396; Butel 1993, 73–75; Thomson 1982, 385.

118. Butel 1993, 73–75, 105, 237–239, Donaghay 1982, 210, 217; Gayot 1998, 317; Becchia 2000, 333–334; Bouloiseau 1957–58, I: LIV; Liu 1994, 77; Gras 1906, 123.

119. Lyon, however, suffered gravely from the effects of the disease that ravaged the cocoons of the silk worms in 1788 (Butel 1993, 238).

120. Garçon 1998, 12–13.

121. Sources for the two preceding paragraphs: Boulle 1975, 321–322; Butel 1993, 72–88, 105, 237–239; Donaghay 1990, 394–395; Liu 1994, 77–78.

122. *Avis des députés du commerce*, May 31, 1787 and *Avis des députés des commerce*, June 26, 1787, both in AN F12 723, *Avis des députés d commerce*, January 8, 1789, AN F12 724, and *Arrêt du Conseil d'État du Roi*, August 12, 1789, AD Seine-Maritime C175.

123. Jean-François Tolozan, *Mémoire*, AN F12 1340.

124. Le Turc, *Lettre à Tolozan*, March 19, 1789, AN F12 994, *Lettre à M. de Cantelou*, July 23, 1788, AN F12 651, Le Clerc père et fils, *Mémoire des Machines*, September 6, 1788, AN F12 1343A.

125. The bulk of this voluminous correspondence is found in AN F12 1340. See also Gerbaux and Schmidt 1906, II, 114–118; Jacques-François Martin, *Mémoire sur l'art des machines pour la filature*, n.d. [1792], Bibliothèque de CNAM 272, and the *Rapport fait par Charles Désaudray sur une fabrication de mousselines superfines, à l'imitation de celles des Indes au Lycée des Arts*, 30 Floréal, Year V, CNAM U546. See also pp. 92–102 of McCloy 1952 and the works of J. R. Harris, Serge Chassagne, and David Jeremy already cited.

126. This correspondence is in AN F12 1340.

127. Chassagne 1991, 198–199; *État de ce que le Conseil veut bien accorder au S. Davis et Comp.*, 1787, AN F12 675.

128. See reports in AN F12 1340 and the *État des Encouragements accordés à différents entrepreneurs et manufactures pendant un nombre d'annees fixé par les décisions*, 1791, AN F12 678.

129. Gillispie 1980, 438–144.

130. Boyetet, *Seconde Partie du Recueuil de divers Mémoire*.

131. Thermeau 2002, 262.

132. Even as fine an historian as the late Sidney Pollard (1999, 1) fell victim to Anglocentrism.

133. For an introduction to this literature, see Prados de la Escosura 2004.

134. Meyer 1991, 126–139.

135. Bret 2000, 2002. See also Gillispie 1980, 479–547.

136. Mokyr 1990, 251–255, 294–295.

137. Reynard 1999; Rosenband 2000b.

138. Brown, *État des ouvriers des manufacutres actuellement sans travail dans le généralité de Paris,* December 11, 1788, AN F12 678; Bruyard, Brown, Lepage, Lansel, and Lazowski, *État des secours accordée jusqu'au 7 fevrier 1789, aux ouvriers des manufactures,* 1789, AN F12 678, *Rapport des Travaux de la Commission intermédiaire de Haute-Normandie depuis le 20 Décembre 1787 jusqu'au 27 Juillet 1790* (Rouen: 1790), AD Seine-Maritime C 2114*.

139. Lambert, *Lettre à M. Lefebvre l'ainé syndic de la Chambre du Commerce de Rouen,* April 5, 1788, AN F12 1788.

140. Jacob 1997, 183–186.

141. Tarbé, *Lettre à M. de Tolozan,* February 7, 1789, AN F12 678.

142. *État des secours accordée jusqu'au 7 févriers 1787 aux ouvriers des manufactures,* 1789, AN F12 678.

143. Legrand 1995, 19.

144. d'Agay, *Lettre à Derveloy,* February 22, 1789, AD Somme C403, *Lettre de M. de Tolozan à M. Demaure,* October 3, 1790, AD Somme L496, and Nicolas Desmarest, *Rapport,* April 12, 1791, CNAM U543. See also the sources in the following note.

145. See the *Registre* in *Journal de la ville de Rouen 1777–1789,* AM Rouen B17, 1–8, *Observations sur une méchanique ou moulin,* n.d. [1788], AD Seine-Maritime C2213 and *Rapport des Travaux de la Commission intermédiaire de Haute-Normandie depuis le 20 Décembre 1787 jusqu'au 27 Juillet 1790* (Rouen: 1790), AD Seine-Maritime C2114*. See also the extensive correspondence on the subject in AN F12 658a.

146. *Pétition de l'assemblée provinciale de Normandie,* [April?] 1788, AN F12 658a and *Procès-Verbal des séances de l'Assemblée provinciale de la généralité de Rouen tenue aux Cordeliers de cette Ville, aux mois de Novembre et Décembre 1787,* AD Seine-Maritime C2112*.

147. This is according to the following: Charles Engrand on pp. 156–164 and 194–198 of Hubscher 1986; *Lettre de Derveloy à d'Agay,* February 22, 1789, AD Somme C1675; d'Agay, *Lettre à Tolozan,* March 5, 1789, AD Somme C403; *Mémoire sur l'état actuel de la filature,* April 1790, AN F12 1341. For a contrary interpretation, see Harris 1998, 382–383.

148. *Arrêt du Conseil du Commerce,* October 24, 1786, AN F12 * 30, Jean-Baptiste de Cretot fils, Preton de Premale, J.B. Péton et neveux, Veuve de Fontenay et fils, *Mémoire,* October 24, 1787, AN F12 1342, Goy, *Mémoire général,* December 15, 1787, AN F12 1365, Jean-François Tolozan, *Mémoire,* 1792, AN F12 1340, Young 1969, 422.

149. Dardel 1966, 128–130.

150. *Pétition des Négociants et Fabricans de Troyes au Comités d'Agriculture et Commerce et Finances,* August 12, 1791, AN F12 1342, Jean-Marie Roland de la Platière, *Mémoire,* December 19, 1792, AN F12 1340, *Etat-Général des différentes branches de commerce et d'industrie renfermés dans la généralité*

de Soissons, December 29, 1789, AN F12 678, Thermeau 2002, 242; Hollander and Pageot 1989, 19; Forster 1980, 147–148; Kaplow 1964, 100; Ramet and Cau 1994, 591. This list is far from exhaustive, it provides only a first approximation.

151. Guillaume Hall, *Lettre*, October 27, 1786, AN F12 1340, plus other information on Milne from the same carton. Duc de Liancourt, *Lettre au Controleur Général*, December 20, 1786, AN, F12 1339, Pierre Laurens Daly, *Mémoire*, September 1, 1790, AN F12 652, Mokyr 1990, 107.

152. Lewis 1993, 35; Gillispie 1980, I, 407.

153. Lérue 1875, 10; Hirsch 1991, 169, Péronnet 1988, 71–80; Lewis 1993, 35–36; Garçon 1998, 21, Maillet 1996, 28–29, 102; *Rapport des Travaux de la Commission intermédiaire de Haute-Normandie* and Goy, *Mémoire général*, December 15, 1787 plus the reports on coal mines in AD Seine-Maritime C912.

154. Roland de la Platière, *Mémoire*, December 19, 1792, AN F12 1340; Doyon and Liaigre 1963; Ballot 1978; Gillispie 1980.

155. McCloy 1952, 101. My objection is that this enumeration does not count the large number of machines destroyed in 1789. See chapter 4 of the present volume.

156. For a list of groups that demanded the suppression of the Treaty in the bailliage of Rouen, see Bouloiseau 1957–58, I: 252, n. 106.

157. See, for example, the memoir written for the Academy of Science: *Encouragements des Manufactures et du Commerce de chaque généralité*, n.d. [1789], CNAM 12-120.

Chapter 4

1. Berg 1980, 1. Berg amended her view radically in a subsequent piece (1988).

2. Macleod (2004, 111–126) has recently shifted the terms of any discussion of a British machinery question away from Berg by illustrating the Continental origins of much British technology. See also Greasley and Oxley 1997.

3. There are, however, some important exceptions. For the Anglo-American technological world in the nineteenth century, see the sources cited by MacKenzie (1998, 41–42).

4. Thompson 1963; Randall 1991.

5. MacKenzie's discussion of this problem (1998, 34–36) is particularly thoughtful.

6. The emphasis on interacting "dual revolutions" was made most forcefully by Eric Hobsbawm (1962). See also C. Jones 1998.

7. A condensed version of this material was published as "Machine-Breaking in England and France during the Age of Revolution" (Horn 2005).

8. Thompson 1971.

9. Hobsbawm 1964, 7–26, esp. 9–13.

10. Thompson 1963, 452–602. For a recent view, see Archer 2000, 48–49.

11. Rudé 1964, 1970.

12. Landes 1969, 123. My formulation of the situation is derived from Berg 1988, 52.

13. The figure of 400 labor disputes for the period 1717–1800, of which 138 were after 1780, has become a minimum standard (Rodney Dobson, cited on p. 44 of Archer 2000).

14. This 1563 law (known as the statute of artificers, or as 5 Elizabeth) stipulated the length of the workday and gave the justices, country sheriffs and mayors the power to fix wages annually at the Easter quarter sessions. Even the occasional enforcement of such a measure was far beyond the capacities of the Bourbon states much less the Valois. See Moher 1988, 77.

15. Randall 1988; Berg 1994; Daunton 1995; Nuvolari 2002; Rule 1993.

16. Randall 1991, 151.

17. Rudé 1964, 71.

18. Hobsbawm 1964, 14.

19. Rudé 1970, 249–250.

20. Archer 2000, 45.

21. Ibid.; Berg 1994, 254; Rule 1986, 278.

22. Christie 1982, 173.

23. Thomis 1970, 16.

24. Berg 1988, 62–63, 67; Hobsbawm 1964, 16; Rudé 1964, 71; Archer 2000, 42.

25. Randall 1991, 249, 289.

26. Rudé 1964, 79.

27. "Luddites," 1971, 39.

28. This figure comes from Darvall 1934 (cited by Rudé 1964, 92). Curiously, no recent commentator has provided even a tentative estimate of the value of the destruction from these two waves of Luddism.

29. Rudé 1964, 80, 89.

30. "Luddites," 1971, 47.

31. Rudé 1964, 150.

32. See Archer 2000, 15–21, 54–55.

33. For a review of Mokyr's widely separated statements on early-nineteenth-century machine-breaking, see Nuvolari 2002, 393–394, 402–407.

34. MacKenzie (1998, 41–43) cited several studies attesting to such influence, including Lazonick 1981 and Sandberg 1969.

35. Daunton 1995, 499–501.

36. Rudé 1964, 85–90, 255.

37. Rudé 1970, 28; 1964, 83–84.

38. Archer 2000, 9, 87.

39. Orth 1991; Moher 1988. All English studies of the use of criminal law to discipline the working classes follow Hay 1975.

40. Thompson 1963, 451, 474, 529, 544–545, 605. The quotation is from the final page. See also Randall 1991, 248.

41. Thompson 1963, 493.

42. Wells 1991, 1983. See also Royle 2000.

43. For a few examples of this consensus, see Christie 1982; Dickinson 1977; Clark 1985; Crouzet 1990, 262–294; Mori 2000, 92–103, 133–152; Stevenson 1992, 326–330; Archer 2000, 89–93.

44. The only exception I have found is Darvall's outdated monograph *Popular Disturbances and Public Order in Regency England* (1934).

45. Thompson 1963, 807–808; Rudé 1964, 252, 264.

46. Rule 1993, 112–138, esp. 118–122; Rosenband 2004.

47. Randall 1991, 248; 1986, 15.

48. For some clues on how to conduct this reassessment, see Pollard 1979 (reprinted in Pollard 1999).

49. Mokyr 1990, 255; Landes 1969, 123.

50. Hobsbawm 1964, 21; Randall 1991, 289.

51. Christie 1982, 173.

52. Hobsbawm 1964, 21; Rudé 1964, 90. Nuvolari (2002, 397, 417) claimed that the 1792 destruction of the Grimshaw factory in Manchester was the "main determinant of the delayed adoption of this technique in the weaving industry." He was perhaps overly optimistic about the effects of industrial action.

53. Archer (2000, 75–78) skillfully synthesized a vast literature.

54. Rudé 1970, 27–28.

55. This discussion is heavily dependent on the insights of Sidney Pollard (1965, 1979). See also Chapman 1967, 174–209.

56. Rule 1981, 131–144; Rule 1986, 269–278; Randall 1991.

57. Hobsbawm 1964, 17–18.

58. See Crafts 1983.

59. Rudé 1964, 218; Dobson 1980, appendix.

60. Moher 1988, 87–90

61. Archer 2000, 86. According to Moher (1988, 74), a judicial approach to combinations as criminal conspiracies was inherent in official policy until 1824.

62. Hobsbawm 1964, 18–19.

63. Berg 1988, 62.

64. Moher 1988, 83. On the development of economic ideologies related to the machine and industrialization, see Berg 1980. Richard Biernacki (1995, 255–258) emphasized the difference between the ideology of liberalism and the lived reality for the workers.

65. Hobsbawm 1964, 11; Rudé 1964, 260; Pollard 1965, 234.

66. Cited by Rudé (1964, 268) from *Tableau de Paris* 6: 22–25. For a recent reading of contemporary views on both sides of the Channel, see Monnier 2005.

67. From the article "Émeutes," cited on pp. 111 and 112 of Popkin 1999.

68. Reddy (1984, 56–60) clearly understood machine-breaking in 1789 as an attempt by laborers to hold on to a world of production doomed to be displaced by the machine, in large measure through competition from England. He also had a much more negative view of the "modernity" of French entrepreneurs and officials who he portrayed as incapable or at best unwilling to follow the trail blazed by the English.

69. Manuel 1938, 180–183.

70. Hobsbawm 1964, 14.

71. Daryl M. Hafter, "Women Who Wove," in Hafter 1995, 50–55; Garden 1970, 586–568; Levasseur 1901, II: 767–768.

72. Rudé 1970, 69; Potofsky 1993; Kaplan 1979, 35, 69–70; Sonenscher 1987, 77, 81.

73. Dobson 1980, appendix; Hobsbawm 1964, 11; Rudé 1964, 125.

74. Thermeau 2002, 267–268; *Procès-verbal des officiers municipaux de la ville de Saint-Étienne*, January 12, 1790, AM Saint-Étienne 2F16; Gras 1906, 150, Galley 1903, 58, 71; Alder 1997, 215. My account of technological change contradicts Alder's.

75. Cynthia A. Bouton is completing an investigation of riotous behavior titled *The Politics of Provisions: France 1580s–1850s*.

76. Rudé 1970, 75; 1964, 125; Lefebvre 1973, 48; Pierre Bernadau, *Oeuvres Complètes de Pierre Bernadau de Bordeaux*, vol. 5, *Premier Recueil des Tablettes Manuscrites (mai 1787 à novembre 1789)*, BM Bordeaux ms. 713/5: 251, 262, 327; Ballot 1978, 20.

77. Vernier 1909, I, 192–193; Lemarchand and Mazauric 1989, 142–145. See also Picard 1910; Reddy 1984.

78. This version depends heavily on an important revisionist article: Rosenband 1997. See also Rudé 1964, 97–98, 126, 256, 265; Kaplan 1994, 355–383.

79. The literature on 1789 is vast. The best place to dive into it is Doyle 1999.

80. See Allinne 1981. Except where otherwise noted, I rely on his account for Normandy.

81. Mollat 1979, 286.

82. Jean-Baptiste Horcholle, *Evenements de la Révolution française à Rouen de 1789 à 1801*, n.d. [1801], BM Rouen Y128*; Poullain 1905, 5.

83. Reddy 1984, 60.

84. *Journal de Normandie* 87 (October 31, 1789), 397; Delécluse 1985, 85; *Jugement souverain prévôtal et en dernier ressort*, October 20, 1789, AD Seine-Maritime 220 BP14.

85. Chassenge 1991, 190; Gullickson 1986, 89–90.

86. Ruhlmann 1948, chapter 7.

87. Jean-Baptiste Horcholle, *Evenements de la Révolution française à Rouen de 1789 à 1801*, n.d. [1801], BM Rouen Y128*.

88. *Journal de Normandie* 58 (July 22, 1789), 255–256, Mollat 1979, 286–287; Chassagne 1991, 188–190.

89. Ballot 1978, 20.

90. *Rapport des Travaux de la Commission intermédiaire de Haute-Normandie*, 200.

91. Ibid.

92. Reddy 1984, 59–60.

93. Faure 1956, 54; Galley 1903, 58.

94. Fournial 1976, 164–165; Lefebvre 1973, 73, 182; Galley 1903, 61–63.

95. This paragraph is based on Alder's account of this event and uses his translation (1997, 214–215). Additional details are from these sources: Galley 1903, 58, 74–75; Tézenas du Montcel 1903), 464; Descreux 1868, 317–318; Schnetzler 197), 53; Jacques Sauvade, *Mémoire*, July 2, 1789, AC Saint-Étienne Ms 328 2 (2) [1 Mi 11]. These references also apply to the destruction of Sauvade's workshop.

96. For November's events and the quotation, see Alder 1997, 215–216; Thermeau 2002, 19.

97. Jacques Sauvade, *Mémoire*, July 2, 1789, AC Saint-Étienne Ms 328 2 (2) [1 Mi 11].

98. Descreux 1868, 318.

99. Lefebvre 1973; Chaudron 1923.

100. The standard account is Hunt 1978. See also Horn 2004, chapter 3.

101. Babeau 1873–74, I, 221, 232.

102. Hunt 1978, 95.

103. Maudhuit 1957, II: chapter 6; *Jugement Prévôtal et en dernier ressort à Troyes*, November 27, 1789, BM Troyes, cabinet local 50; Abbé Trémet, "Notes historiques de ce qui s'est passé à Troyes, 1770–1790," BM Troyes ms. 2-2322.

104. Babeau 1873–74, I: 243; Ricommard 1934, 38.

105. Maudhuit 1957, II: 196–197; Darbot 1979, 16.

106. *Pétition des Négociants et Fabricans de Troyes au Comités d'Agriculture et Commerce et Finances*, August 12, 1791, AN F12 1342.

107. Jacques-Edme Beugnot, "Discours," November 3, 1790, *Procès-verbal des séances du Conseil-général du département de l'Aube (1790)*, AN F1CIII Aube 2.

108. [Louis-Ézechias] Pouchet, *Essai sur les avantages locaux du département de l'Aube, et sur la prospérité nationale, ou Adresse à Mes Concitoyens du Département de l'Aube*, 1791, AN F12 652.

109. Ballot 1978, 21.

110. Jean-Nicolas Feugé, *Compte de la situation politique du département de l'Aube pendant le mois de nivôse an 8*, 15 Pluviôse, Year VIII (February 4, 1800), AN F1CIII Aube 3.

111. Darbot 1979, 31–32; Colomès 1943, 85–87; Ricommard 1934, 38–39.

112. Alder 1997, 246–247; Ballot 1978, 21–22; Manuel 1938, 180–183; Belmont 1998; Garçon 1998; Reynard 2001; Bergeron 1997, 122–141. On the absence of machine-breaking, see Marseille and Margairaz 198).

113. Jacques-François Martin, *Circulaire*, May 19, 1792, AD Somme L496.

114. *Mémoire sur les Encouragements à accorder au Commerce par le directoire du département de la Somme*, 22 Floréal, Year IV [May 11, 1796], AD Somme L496.

115. *La Décade du département de la Somme* 24: 2 (30 Fructidor, An VIII [September 17, 1800]; Nicolas Quinette, *Lettre au Ministre de l'Intérieur*, April 24, 1806, AD Somme M80003.

116. Poullain 1905, 19–20.

117. *Rapport des Travaux de la Commission intermédiaire de Haute-Normandie*, 166–167, 178.

118. *Note pour servir de supplement au Mémoire de M. De Maurey sur les moyens de perfectionner les arts mécaniques*, slsd [1790], AD Seine-Maritime C2120.

119. Dubuc 1975, 134–138.

120. Burstin 1989, 369–379.

121. *Pétition à l'Assemblée National pour les Ouvriers-Rubaniers de la Ville de Paris*, November 6, 1791, AN F12 1430.

122. Kaplan 1979, 2001; Sonenscher 1987, 77. Local studies could multiply the list nearly ad infinitum.

123. Lewis 1993, 44–48, 65; Liu 1994, 37–39; Minard 1998, 297; Rosenband 1998, 1999, 2000b.

124. See Crafts 1983, 61, 71, 82; Samuel 1992, 26–44; MacKenzie 1998, 34–41; Inkster 1991, 67.

125. The abolition of the *corporations* is usually considered separately as the consequence of the same attitudes that led to renunciations of rights on August

4–5, 1789, yet the laws that abolished the *corporations* and eliminated the workers' right to organize were passed only months apart.

126. Sonenscher 1989a, 387.

127. Rosenband (2004) demonstrates how laboring-class customs played an essential part in the process of legal restriction on the ability of workers to organize on both sides of the Channel.

128. Those not familiar with these events should see Doyle 1989.

129. See Krantz 1988.

130. Rudé found a goodly number of such public statements after Prairial. The quotation is from pp. 156 and 157 of Rudé 1970.

131. Sibalis n.d., 11–32, esp. 11.

132. Sewell 1980, 171–186; Truant 1994, 194–231.

133. Stone cutters, carpenters, wheelwrights, tailors, hat makers, shoe makers, joiners, blacksmiths, curriers, cabinet makers, farriers, bakers, and locksmiths. See *Lettre du Commissaire-Général de Police au Préfet de la Gironde*, 17 Floréal, Year XIII (May 7, 1805), AD Gironde, 1M 334. I treat this subject more thoroughly in "Coalitions, Compagnonnage and Competition: Bordeaux's Labor Market (1775–1825)," delivered at the French Historical Studies meeting in March 2001.

134. See Burstin 2005; Monnier 1981; Rudé 1964. See also Magraw 1992a, 19.

135. Ballot 1978, 40.

136. Manuel 1938 180–211, esp. 185–186, 202–206. The following paragraphs are based on this article. Manuel seems to have thoroughly utilized the primary sources reprinted in Bourgin and Bourgin 1912, 1921, and 1941.

137. Such machines were protested at Clermont-L'Hérault, at Lodève, at Alençon, at Eupen, and at Brünn between 1817 and 1824. A final protest took place at Verviers in Belgium in 1830. See Gayot 2001, 209–237. esp. 236–237; Gayot 1998, 429.

138. Gayot 1998, 429.

139. On the fate of the woollens trade in the south and the deindustrialization of the region, see Johnson 1995 and Thomson 1982.

140. Jean-Pierre Chaline makes this point explicitly for Normandy (Mollat 1979, 330).

141. The revolt led to a more repressive law on associations, enacted April 10, 1834 (Bezucha 1974).

142. For a provocative discussion of the emergence of English nationalism and how the state successfully deployed it to redirect or refocus economic, social, and political problems, see Colley 1992.

143. Berg 1980.

144. This is a theme in Mokyr 2002.

Chapter 5

1. On the methodological evolution of French history, see Hunt 1986; Stoianovich 1976; Furet 1983.

2. Aftalion 1990, 9.

3. For a few examples, see Becchia 2000; Chassagne 1980; Kaplow 1964; Monnier 1981; Woronoff 1984.

4. Margairaz 1988; Miller 1999. See also Hoffman and Rosenthal 2000.

5. Hirsch 1991; Hirsch and Minard 1998; Reddy 1984; Gayot and Hirsch 1989.

6. Biard 2003; Kennedy 1982, 1988, 2000, especially volumes 2 and 3; Lucas 1973.

7. Alder 1997; Bret 2002; Gillispie 1980, 2004; Richard 1922.

8. Palmer 1941.

9. On the social and occupational backgrounds of this heterogeneous group, see Soboul 1980.

10. On the early effects of the grain Maximums, see Miller 1999, 150–162.

11. See *Prix des Marchandises & Denrées, Journées et Main-d'oeuvre dans la département de Rhône-et-Loire*, n.d. [1794], AN F12 1547C.

12. Soboul 1966, 161–182.

13. Gross 1997.

14. On trade, see Bouloiseau 1983, 107–108; Lefebvre 1954, 239–276; Palmer 1941, 225–253. See also the legal rhetoric in the *Décret de la Convention nationale*, 18 Vendémiaire, Year II [October 9, 1793], AD Gironde 8J 429.

15. Cited on p. 11 of Richard 1922.

16. Alder devoted a chapter to events in Paris during this period, emphasizing "that the technological life and political struggle are mutually constitutive" (1997, 289). I focus on the broader outline of French industrial policy obscured by Alder's misleading overemphasis on "the machine in the Revolution."

17. Richard 1922, 7–14.

18. *ACSP*, October 1, 1793, *ACSP*, October 2, 1793, both in AN AF*II 121, and Lazare Carnot's report on arms making on 13 Brumaire, Year II [November 3, 1793] reprinted in the *Réimpression de l'ancien moniteur* 18 (Paris: Plon, 1847), 336–338.

19. *ACSP*, August 31, 1793, AN AF*II 121; *ACSP*, 10 Brumaire, Year II [October 31, 1793], AN AFII 214 (1834: 20); *ACSP*, 19 Frimaire, Year II [December 9, 1793], AN AFII 214 (1835: 26); *ACSP*, 29 Frimaire, Year II [December 19, 1793], AN AFII 214 (1836: 34); Alder 1997, 266; Richard 1922, 17, 22.

20. This cooperation extended beyond setting wages. Arbitration boards were established twice to determine the price to be paid for raw materials requisi-

tioned by Paris. *ACSP*, September 11, 1793, *ACSP*, 24 Frimaire, Year II [December 14, 1793], both in AN AF*II 121.

21. The preceding three paragraphs are based on the following sources: *ACSP*, September 20, 1793, AN AF*II 121; *Assemblée des Commissaires des sections de Paris et de ceux nommés par le ministre de la guerre pour la fixation du prix du travail de chaque partie de la fabrication des fusils*, Year II [1793], AN F12 1309; Alder 1997, 270; Richard 1922, 33–39. Alder's condensed version of events differs from mine and from Richard's on some points of fact.

22. *ACSP*, 21 Prairial, Year II [June 9, 1794], AN AFII 215 (1846: 49).

23. The Ministry of War began as the supervisory body. It was superseded by the Commission of Arms, Powders and Mines on 11 Germinal Year II [March 31, 1794]. Oversight in Paris was exercised by the Administration des armes portatives (Small Arms Administration), and then the Council for the Administration of Arms until 28 Brumaire, Year III [November 18, 1794] when the Commission of Arms assumed the task.

24. *ACSP*, 21 Frimaire, Year II [December 13, 1793], AN AFII 214 (1835: 34).

25. *Extrait des registres du Comité de Salut public de la Convention nationale*, 22 Frimaire, Year II [December 12, 1793], AN AFII 214 (1835: 42) and *ACSP*, 29 Frimaire, Year II.

26. See Burstin 1994, 271–293.

27. On the relationship of transport to supply problems, see Le Roux 1996.

28. See Alder 1997, 264; Lazare Carnot's report on arms making on 13 Brumaire, Year II [November 3, 1793] reprinted in the *Réimpression de l'ancien moniteur* 18 (Plon, 1847).

29. *Circulaire du Comité de Salut public*, 19 Germinal, Year II [April 8, 1794], AN AFII 215 (1844: 17) and *ACSP*, 8 Floréal, Year II [April 27, 1794], AN AFII 215 (1845: 18). There were already two representatives fulfilling this function. There is some disparity concerning Legendre's occupation. He may have been an iron master, in which case, his selection for this task sent a clear message (Patrick 1972, 253; Richard 1922, 28–29, 31, 60–61, 67, 178.

30. Gillispie 2004, 404–405.

31. Palmer 1941, 241–242.

32. *Reglement pour les ateliers de Chaillot et Arrêté confirmatif du Comité de Salut public*, 16 Floréal, Year II [May 5, 1794], AN AFII 215 (1845: 36); Richard 1922, 229.

33. Gillispie 2004, 423–424.

34. *ACSP*, 17 Frimaire, Year II [December 7, 1793]; *ACSP*, 7 Nivôse, Year II [December 27, 1793]; *ACSP*, 9 Nivôse, Year II [December 29, 1793], in AN AF*II 121, *ACSP*, 20 Messidor, Year II [July 8, 1794], AN AFII 215 (1848: 20), and Alder 1997, 271–272.

35. Alder 1997, 285–288.

36. Louis-Bernard Guyton, *Rapport fait au nom du Comitéé de salut public par L.-B. Guyton, à la séance du 14 pluviôse an III, sur l'état de situation des arsenaux et de l'armement des armées de terre et de mer de la République* (Paris: Imprimerie Nationale, 1795). Alder (1997, 288) makes several small errors in Guyton's statistics.

37. Richard 1922, 40; Alder 1997, 272, 285, 288. Alder's figures on the cost of gun locks do not agree with Richard's or with contemporary figures.

38. Lefebvre pointed this out (1964, 104). See also Palmer 1941, 226.

39. *Arrêt du Comité de Salut public*, 26 Brumaire, Year III [November 16, 1794], AN AF*II 127.

40. See Alder 1987, 266–268, 275, 283, 286. Alder illustrated how important the Maubeuge workers were to Paris' total output. He, however, vastly overemphasized the novelty of post-Thermidorean unrest in the workshops. Such activity was *normal* as his own work illustrated. What had changed was the willingness of the government to permit popular agitation. He provided useful occupational analysis of the armsworkers. See also Alder's depiction of productivity for the locksmiths of the section of the Quinze-vingts. The passage quoted here is cited on p. 329 of Richard 1922.

41. A decree dated August 19, 1792 made Maubeuge, Saint-Étienne, Tulle, Moulins, and Klingenthal national arms manufactures to be run by entrepreneurs assisted by administrative councils, under the supervision of the central government's agents. Other arms manufactures, including Autun, were sponsored and run by local authorities (Richard 1922, 105, 135).

42. Alder (1997, 238–248, 277, 280–281) argued that the *Atelier de perfectionnement* which functioned from 1794 to 1796 adopted the goal of interchangeability and drew many parallels to Blanc. Yet Alder's own account of the Atelier's goals emphasized machine production and the division of labor rather than true interchangeability. See also Ballot 1978, 501; Gillispie 2004, 424–426; Richard 1922, 149–151. My findings also cast doubt on a number of Joel Mokyr's arguments about technological development—arguments that were based on Alder's work. See Mokyr 1994a, esp. 246–247.

43. *ACSP*, May 2, 1793, AN AFII 214 (1831: 2); *Extrait du Procès-verbal de la Convention nationale*, June 7, 1793, AN AFII 214 (1831: 8).

44. Even after the Maximum was discontinued, the workers at this manufacture could not afford enough food. In the summer of the Year III [1795], they had to quadruple their take-home pay so they could eat. *Extrait des registres du Comité de Salut public de la Convention nationale*, 5 Germinal, Year II [March 25, 1794], AN AFII 215 (1843: 19), *Circulaire du Comité de Salut public*, 19 Germinal, Year II [April 8, 1794], AN AFII 215 (1844: 17), *ACSP*, 30 Germinal, Year II [April 19, 1794], AN AFII 215 (1844: 51), *ACSP*, 14 Floréal, Year II [May 3, 1794], AN AFII 215 (1845: 31), *ACSP*, 20 Floréal, Year II [May 9, 1794], AN AFII 215 (1845: 47), *ACSP*, 6 Floréal, Year III May 5, 1795], AN AFII 216 (1865: 12); Crook 1991, 161; Richard 1922, 139.

45. *ACSP*, 30 Prairial, Year II [June 18, 1794], AN AF II 78 (580); *Extrait du Registre des Arrêtés du Comité de Salut Public de la Convention Nationale, du vingt-deux Messidor, l'an deuxième de la République Française, une et indivisible*, AD Somme L496.

46. *ACSP*, 15 Prairial, Year II [June 3, 1794], AN AF II 78 (574); Richard 1922, 336.

47. *ACSP*, 23 Prairial, Year II [June 11, 1794], AN AFII 215 (1846: 51); *ACSP*, 10 Messidor, Year II [June 28, 1794], AN, AF II 78 (574).

48. For a parallel story regarding gunpowder production, see Bret 1996, 261–274; Richard 1922, 423–468, 487–541, 554–579.

49. On the mobilization of scientists, see Dhombres and Dhombres 1989, 48–62; Gillispie 2004, 339–444.

50. These three had collaborated in 1788 on a related subject. (See chapter 2 of the present volume.) Vandermonde also published *Procédés de la fabrication des armes blanches*, and Monge wrote a *Description de l'art de fabriquer les canons* in 1793 (*ACSP*, September 7, 1793; *ACSP*, 26 Frimaire, Year II [December 16, 1793], both in AN AF*II 121; Lazare Carnot, report on arms making on 13 Brumaire, Year II [November 3, 1793] reprinted in the *Réimpression de l'ancien moniteur* 18 (Plon, 1847). Jean-Henri Hassenfratz, Monge, and Jacques-Constantin Périer offered a public course on how to make musket and cannon barrels (Gillispie 2004, 395, 396).

51. The documents are in AN AF*II 123. See also Galley 1903, 212; Richard 1922, 78–88; Alder 1997, 272–288.

52. *ACSP*, 19 Ventôse, Year II [March 9, 1794], AN AF*II 123; Richard 1922, 218.

53. *Lettre de la Commission des armes, poudres et exploitation des mines de la république aux représentants du peuple composant le Comité de Salut public*, 26 Nivôse, Year III [January 16, 1795], AN AFII 216 (1859: 36).

54. Ballot 1978, 509.

55. Harris 1998, 173–359.

56. For example, the iron master Nicolas Rambourg from the Allier was sent to Rives in the Isère along with his foreman Belloy to learn how to make steel (*ACSP*, September 18, 1793, AN AF*II). Citizens Courby and Brasset set up a foundry, but first they were sent to Klingenthal in the Bas-Rhin to learn the process. Local manufacturers were ordered to let them in and assist them (*ACSP*, 20 Messidor, Year II [July 8, 1794], AN AFII 215 (1848: 18); Ballot 1978, 509; Richard 1922, 218).

57. *ACSP*, 12 Brumaire, Year III [November 2, 1794], AN AF*II 127.

58. Cementation steel was covered in some districts and not in others, even within the same département. See *Arreté du conseil général du district de Gournay portant le tableau du Maximum*, October 12, 1793 and *Arrêté du*

conseil général du district de Dieppe, 30 Vendémiaire, Year II [October 21, 1793], both in AD Seine-Maritime L440.

59. *ACSP*, 20 Floréal, Year II [May 9, 1794], AN AFII 215 (1845: 47); *ACSP*, 12 Vendémiaire, Year III [October 3, 1794], AN AF*II 127. *Adresse*, 19 Brumaire, Year III [November 9, 1794], AN F12 1547E, *Rapport au Comité d'Agriculture et des Arts par la Commission d'Agriculture et des Arts*, 11 Ventôse, Year III [March 2, 1795], AN F12 1556; *Observations de la Commission sur le Rapport présenté le 16 Prairial* [Year III, June 3, 1795] *sur les aciers et les limes de Romenil*, AN F14 4485; *Projet d'Arrête du Comité de Salut public*, 13 Thermidor, Year III [August 1, 1795], AN AFII 78 (580: 49); *ACSP*, 2 Fructidor, Year III [August 20, 1795], AN AFII 78 (580: 50); Ballot 1978, 509, 512, 514; Richard 1922, 222–227.

60. The two preceding paragraphs are based on these sources: *ACSP*, 27 Nivôse, Year II [January 16, 1794], AN AF*II 121; *Rapport au Comité d'agriculture et des arts de la Convention nationale*, 18 Pluviôse, Year III [February 7, 1795], AN F12 1556; Ballot 1978, 498–499; Richard 1922, 593–607.

61. Brunet, *Rapport*, 15 Prairial, Year II [June 3, 1794], Tissot, *Rapport*, 26 Fructidor, Year II [September 12, 1794] and *Copie de la lettre de l'Agent national du district d'Annecy à la Commission de l'Agriculture et des Arts à Annecy*, 22 Frimaire, Year III [December 12, 1794], all in AN F12 1556.

62. Lheritier, *Rapport au Comité de Salut public et d'Agriculture et des Arts de la Convention nationale par la Commission d'Agriculture et des Arts*, n.d. [Year III, 1795], AN F12 1556 and *Arrête du Comité de finances de la Convention nationale*, 1 Floréal, Year III [April 21, 1795], AN F12 1559, *Registre des Procès-verbaux des Séances des Membres du Conservatoire des Arts et Métiers*, 12 Thermidor, An IV–5 Brumaire, Year IX, CNAM 10-483, 78, *Notices relatives à tous ceux qui en 1806, ou dans les années précédantes, ont obtenu une médaille d'or*, n.d. [1806], AN F12 985.

63. *Rapport du Citoyen Regnier, Inspecteur comptable des armes portatives au Ministre de la Guerre*, 26 Prairial, Year IV [June 13, 1796], CNAM D207.

64. Joseph Gaudin fils, *Observations sur la fabrication de l'acier par cémentation sur les frais de cette manipulation et sur le dechet qu'elle entraine*, 7 Pluviôse, Year II [January 26, 1794], Bibliothèque de CNAM 286. *Extrait du Registre des procès-verbaux d'Inspection du magasin des armes blanches*, 5 Pluviôse, Year III [January 25, 1795], CNAM D46, *Rapport de Citoyen Pradier l'aîné, commissaire pour la fabrication des armes blanches sur la fabrication des aciers*, 27 Pluviôse, Year III [February 16, 1795], CNAM D46, and Jumelin et Leblanc, *Rapport concernant le C. Pradier*, 14 Floréal, Year III [May 4, 1795], CNAM D46.

65. Alder 1997, 272–288.

66. On Brienne, see the report by chemist and former conventionnel Joseph Bosc, *Compte politique de la situation du département de l'Aube*, 3 Frimaire, Year VII [November 23, 1798], AN F1CIII Aube 3. On Saint-Étienne, see

Mémoire historique sur la manufacture d'armes de Saint-Étienne, Year VII [1798], AD Loire L835.

67. Other examples include soap. See Darcet, Lelievre et Pelletier, *Rapport sur la fabrication des Savons,* 12 Nivôse, Year III [January 2, 1795], CNAM X19.

68. See Lucas 1973 and Palmer 1941. Biard 2003 is the first collective biography of this influential group.

69. Patrick 1972, 184, 283–284, 329, 357; Tézenas du Montcel 1952, 121–194; Galley 1903, 217–257, 410–414, 576–577.

70. This document is reproduced in Noël Pointe-d'Armeville, *Compte rendu à la Convention nationale et au peuple souverain,* in *Convention nationale: Comptes rendus des Représentants (Nivôse-Fructidor an II),* BN Le39-63, 10, 93.

71. Noël Pointe, *Compte rendu à la Convention nationale et au peuple souverain,* 13 Fructidor, Year II [August 30, 1794], in *Convention Nationale: Recueil des pièces (Fructidor an II–Frimaire an III),* BN Le39-81, 75, 243, 298–299, 319.

72. Pointe, *Compte rendu,* 13 Fructidor, Year II, 58, 172–173.

73. Pointe, *Compte rendu,* 13 Fructidor, Year II, 67.

74. At this same time, Claude-Joseph Ferry of the Ardennes, another of the troubleshooting representatives, currently in the same region, imprisoned an entrepreneur for not filling his production quota. He tried to get Pointe to decide his fate, but Pointe refused until more reports came in, then he freed this entrepreneur as well. Noël Pointe, *Compte rendu à la Convention nationale,* July 19, 1793, in *Convention nationale: Comptes rendus des Représentants (Juillet-Octobre 1793),* BN Le39-28, 57–60 and Pointe, *Compte rendu,* 13 Fructidor, Year II, 231–232, 250, 270, 283, 286.

75. Pointe, *Compte rendu,* 13 Fructidor, Year II, 20–21, 38, 110, 296, and 304–305. In Saint-Étienne, the local authorities demanded on 10 Germinal, Year II (April 21, 1794) that an inspector-general be appointed (Galley 1903, 576).

76. Pointe, *Compte rendu,* 13 Fructidor, Year II, 49, 58, and 296.

77. Gillispie 2004, 428.

78. Pointe, *Compte rendu,* July 19, 1793, 2; Pointe, *Compte rendu,* 13 Fructidor, Year II, iv–v, 218; Richard 1922, 285.

79. Noël Pointe, *Compte Rendu à la Convention nationale par Noël Pointe, Représentant du peuple près les fonderies du Creusot, de Pont-de-Vaux, et autres établissements dans les départements de Saone-et-Loire, l'Aine, Jura, Haute-Saone, Doubs, et Côte-d'Or,* Messidor, Year III [1795], BHVP 8-11149, 54.

80. Noël Pointe-d'Armeville, *Compte rendu à la Convention nationale et au peuple souverain,* in *Convention nationale: Comptes rendus des Représentants (Nivôse-Fructidor an II),* BN Le39-63, 19; Pointe, *Compte rendu,* 13 Fructidor, Year II, 86.

81. Pointe, *Compte rendu,* 13 Fructidor, Year II, 5–6, 12, 21, 31–32, 296.

82. Pointe, *Compte rendu*, 13 Fructidor, Year II, 6, 9, 177

83. Pointe-d'Armeville, *Compte rendu*, 15–16, 28, 79–81, 87, 89, 136–150, 158–160, 162–169; Pointe, *Compte rendu*, 13 Fructidor, Year II, 217, 308, 314–317; Tézenas du Montcel 1952, 191.

84. Pointe, *Compte Rendu*, Messidor, Year III, 2–10, 44–45.

85. The previous two paragraphs are based on Pointe, *Compte Rendu*, Messidor, Year III, 57–58.

86. Pointe, *Compte Rendu*, Messidor, Year III, 12–18, 29–30, 44, 51, and 55.

87. Pointe, *Compte Rendu*, Messidor, Year III, 59–60.

88. Pointe, *Compte Rendu*, Messidor, Year III, 87.

89. Tézenas du Montcel 1952, 187, 194.

Chapter 6

1. Hirsch and Minard 1998, 146–150. For the germ of this idea and its application, see Hirsch 1993, 159–166; Minard 2002, 203–212.

2. Kaplan 1979; Kaplan 2001, chapters 12–15.

3. Sonenscher 1989b, 328–376, esp. 352–353 and 365.

4. Sewell 1980, 92–113.

5. Horn 2004, chapter 3.

6. Reddy (1984) and Miller (1999) made valiant attempts to resolve this dilemma, but they did not put the issue in industrial terms.

7. McCloy 1952, 171–172.

8. Habakkuk 1962.

9. MacLeod 1988; Hilaire-Pérez 2000.

10. Mokyr (1990, 247–259) lists many groundbreaking British inventors who lost their patent rights.

11. See Gillispie 1980, 461.

12. Gillispie 2004, 198–199.

13. Claude Anthelme Costaz, *Mémoire sur les moyens qui ont amené le grand développement que l'industrie française à pris depuis vingt ans* (Paris: Firmin Didot, 1816), 31–32.

14. Shinn (1992, 533–566) argues that centralization characterized revolutionary France in scientific matters.

15. *Procès-verbaux de l'Assemblée nationale contenant les Séances depuis le premier Janvier 1792, l'an quatrième de la liberté et compris le 31 du même mois et suivi d'une Table des Matières* (Paris: de l'Imprimerie nationale, 1792), IV: 251 (January 22, 1792). On the Bureau and its membership, see de Place 1988 and Gillispie 2004.

16. McCloy 1952, 171–173; *Extrait des registres des délibérations du Conseil exécutif provisoire*, 9 Pluviôse, Year II [January 28, 1794], AD Gironde 3L 206.

17. I do not know whether these sums are what the Bureau suggested or what was actually disbursed. In 1790, the Constituent Assembly budgeted 2 million *livres* to be spent annually to encourage industry and innovation. The initial sum placed at the Bureau's disposal was 300,000 *livres* (de Place 1988, 140–141).

18. Unless otherwise cited, the material is derived from reports and correspondence contained in three registers: *Registre des Rapports faits au Bureau de Consultation concernant les Artistes (25 Janvier 1792 à 12 Septembre 1792*, CNAM 10-481, *Procès-verbaux des séances du Bureau consultatif des Arts et Manufactures, 2e registre, an six*, CNAM 10-487; *Registre des Procès-verbaux des Séances des Membres du Conservatoire des Arts et Métiers*, 12 Thermidor, An IV-5 Brumaire, Year IX, CNAM 10-483.

19. Contrast this emphasis with that of the restored Academy of Science (Chabot 2000; Gillispie 2004).

20. *Procès-verbaux des séances du Bureau consultatif des Arts et Manufactures, 2e registre, an six*, 95–97.

21. The best example of this politicization is Lodève (Johnson 1995, 13–14).

22. For this attitude, see the rapport written by Jean-Henri Hassenfratz regarding the files of Raoul, on May 15, 1793 in *Rapports pour le Ministre de l'Intérieur du Bureau de Consultation*, n.d. [1793], CNAM 10-389, 126.

23. Chassagne 1991, 220–229; Ballot 1978, 99–102.

24. See de Place 1988, 151.

25. *Procès-verbaux des séances du Bureau consultatif des Arts et Manufactures, 2e registre, an six*, CNAM 10-487, 201.

26. *Procès-verbaux des séances du Bureau consultatif des Arts et Manufactures, 2e registre, an six*, CNAM 10-487, 219.

27. Bret 2002, 98–113; Gillispie 1980, chapter 6.

28. This literature is voluminous. Major recent contributions include Alder 1997, Bret 2002, Dhombres and Dhombres 1989, and Gillispie 1997.

29. On this controversial question, those with a different perspective tend to look backward at the Revolution while focusing on engineering, statistical evaluations and the highest levels of technical education rather than production. See, for example, Brian 1994; Fox and Guagnini 1993; Fox 1996; Rusnock 2002. For a similar view of the nature of the French Revolutionary achievement, see Belhoste 2003; Barber 1994; Darnton 1968; Gillispie 1980, 2004; Rémond 1946.

30. *Rapport au Comité de Salut public extrait du registre des délibérations de la Commission d'Agriculture et de Commerce de la Convention nationale*, 24 Prairial, Year II [June 12, 1794], AN F12 1330; *Rapports pour le Ministre de l'Intérieur du Bureau de Consultation*, n.d. [1792–93], CNAM 10-390; *Rapport pour le Ministre de l'Intérieur du Bureau de Consultation*, n.d. [1793], CNAM 10-389, 327–330.

31. *Rapport pour le Ministre de l'Intérieur du Bureau de Consultation*, n.d. [1793], CNAM 10-389. See the reports on Perrin (January 23, 1793), Adrien Delarche (July 31, 1793), and Rivey (19 Brumaire, Year II [November 9, 1793]) (Gillispie 2004, 207).

32. Alexandre-Théophile Vandermonde, *Rapport fait par ordre du Comite de Salut Public sur les Fabriques et le Commerce de Lyon* published in the *Journal des Arts et Manufactures publié sous la direction de la Commission exécutive d'Agriculture det des Arts* 1:1 (15 Brumaire, Year III [November 5, 1794]).

33. *Registre des Rapports*, 389, May 30, 1792; *Registre des Procès-verbaux*, 86, 4 Brumaire, Year VII; 99, 27 Pluviôse, Year VII; 179, 28 Germinal, Year VII; 180, 4 Floréal, Year VIII; *Procès-verbaux des séances* 2, 6 Floréal, Year VI; 36, 6 Messidor, Year VI; 87, 16 Vendémiaire, Year VII; 97, 2 Brumaire, Year VII; 99, 4 Brumaire, Year VII; 162–166, 14 Ventôse, Year VII; 200–202, 16 Floréal, Year VII. 219, 12 Prairial, Year VII; 296, 24 Pluviôse, Year VIII; 344, 28 Messidor, Year VIII. The quotations are from 16 Floréal and 12 Prairial, Year VII.

34. Scion of a family that had overseen the manufacture at Aubusson since 1733, Pierre served as inspector from 1773 to 1791. He had close family ties to the local economic and political elite. Philippe Minard graciously provided this information.

35. See Deyon and Guignet 1980; Gillispie 1980, 390–412.

36. The three preceding paragraphs are based on *Procès-verbaux des séances* 26-9, 22 Prairial, Year VI.

37. Sabel and Zeitlin (1985) point out the lack of a contradiction here.

38. Bret 2002, 131. Even at the start, the Bureau disbursed much less money to inventors than was foreseen or budgeted by the legislature (Hilaire-Pérez 2000, 287).

39. McCloy 1952, 182; de Place 1988, 159.

40. Dhombres and Dhombres 1989, 93–150.

41. On the attitudes of French scientists in this period, see Dhombres and Dhombres 1989.

42. McCloy 1952, 193.

43. Margairaz 2005.

44. On the concept of political transparency, see Furet 1981, 19, 52.

45. Ozouf 1988.

46. Woloch 1988, 371–387.

47. In industrial terms, this problem paralleled the French state's difficulties in overseeing the food supply. The government firmly preferred regulation to unleashing revolutionary violence. See Miller 1999.

48. Thaurin 1859, 2.

49. Depping 1893, 9–10. Depping believed that the idea for an industrial exposition came from a proposal by the Marquis Mazade d'Avèze made the previous year.

50. Ozouf 1988, 122.

51. *Réimpression de l'Ancien Moniteur* 29:1 (Imprimerie Nationale, 1847), September 26, 1798, 402–403.

52. Nicolas-Louis François de Neufchâteau, *Circulaire aux Administrations centrales de Département aux Commissaires du Directoire exécutif près de ces Administrations et aux Bureaux consultatifs de Commerce*, 24 Vendémiaire, Year VII (October 15, 1798), AN F12 985.

53. All quotations in this paragraph are from the *Réimpression de l'Ancien Moniteur* 29:1, 402–403.

54. Nicolas-Louis François de Neufchâteau, *Circulaire aux Administration centrales de Départemens et Commissaires du Directoire exécutif près de ces Administrations*, 9 Fructidor, Year VI (August 26, 1798), AN F12 985.

55. Bouin and Chanut 1980, 24.

56. Nicolas-Louis François de Neufchâteau, *Marche et Cérémonies à observer le 3e Jour complémentaire, à l'ouverature de l'Exposition publique des Produits de l'industrie française*, September 1798, AN F12 985.

57. François de Neufchâteau, *Circulaire aux Administrations centrales de Département aux Commissaires du Directoire exécutif près de ces Administrations et aux Bureaux consultatifs de Commerce*, 24 Vendémiaire, Year VII (October 15, 1798), AN F12 985.

58. Depping 1893, 14–15.

59. Depping 1893, 36.

60. Bouin and Chanut 1980, 24–25.

61. *Première exposition des produits de l'industrie française*, Year VI (1798), BHVP 12-963815. Bouin and Chanut (1980, 24) incorrectly mention 110 exhibitors.

62. There are nine attestations for Raoul's files from mechanics, instrument makers, watch makers, and locksmiths (J. H. Hassenfratz, *Rapport sur le Citoyen Raoul*, May 21, 1793, CNAM L364). Raoul remained the premier file maker in France well into the Consulate (*Procès-verbaux des séances de la Section des Arts du Conseil général d'Agriculture, Arts et Commerce du Ministre de l'Intérieur*, 4 Messidor, Year IX-26 Ventôse, Year XIII, CNAM 10-486, 170).

63. Depping 1893, 25–26.

64. McCloy 1952, 78–80.

65. Such goods did not represent the bulk of French manufactured exports even of textiles (Butel 1993, 86–87, 220–239).

66. See Roche 1991 and the many works cited in Fairchild 1993a.

67. Depping 1893, 25–28.

68. Simian 1889, 41.

69. Bouin and Chanut 1980, 25.

70. Depping 1893, 25–26.

71. *Première exposition des produits de l'industrie française*, Year VI (1798), BHVP 12-963815.

72. Crosland 1969, 138–139.

73. François de Neufchâteau, *Circulaire*, 24 Vendémiaire, Year VII (October 15, 1798).

74. *Pétition au Ministre de l'Intérieur*, 2 Vendémiaire, Year VII (September 23, 1798), AN F12 985; Jean-Antoine Chaptal, *Circulaire aux Préfets des départemens*, 1 Floréal, Year IX (April 21, 1801), AD Loire 75M 1.

75. Hirsch 1991, 83; Johnson 1995, 11–16; Johnson 1993, 37–62.

76. On the parallels to the grain trade, see Miller 1999, 163–195.

77. Nicolas-Marie Quinette, *Lettre à l'administration centrale du département de la Somme*, 9 Fructidor, Year VII (August 27, 1799), AD Somme L496.

78. *Procès-verbal de la Fête anniversaire de la fondation de la République célébrée à Paris le 1er Vendémiaire an 8*, AD Somme L406.

79. This section was made possible by grant SBR 9310699 to Margaret Jacob from the National Science Foundation for which I was a named researcher. A number of these points on the importance of Chaptal and the cultural components of his industrial vision have been published in Horn and Jacob 1998.

80. In this manner, Chaptal embodied what Mokyr (2002, 28–77, esp. 31, 36, 52, 64, and 74) has termed the "Industrial Enlightenment."

81. On Chaptal's life and career, see Péronnet 1988 and the sources cited by Horn and Jacob (1979, 20–24, 47–49).

82. Chaptal asserted that it had left "our industry enslaved" (Jean-Antoine Chaptal, "Quelques réflexions sur l'industrie en général, à l'occasion de l'exposition des produits de l'industrie française en 1819," cited in Péronnet 1988, 243–246).

83. The French scientific elite was far from united in believing that theory should be wedded so throughly to practice. This is a major theme of the Dhombres in *Naissance*.

84. Bergeron 1993, 364, 370–372.

85. Nor was this perspective solely to be found in Chaptal's later writing. They are also to be found in his *Catéchisme à l'usage des bons patriotes* (Montpellier: de Tournel, 1790), 70–71. See Horn and Jacob 1998. See also Jouvenel 1942, 146–148.

86. Gillispie 2004, 616–622.

87. Pigeire 1932, 275.

88. Jean-Antoine Chaptal, *Rapport et project de loi sur l'instruction publique* (Paris: Imprimerie nationale, Year IX [1801]), 4.

89. Nicole Dhombres emphasizes this point (Péronnet 1988, 138–140).

90. Bergeron 1993, 416.

91. Chaptal, *Rapport et project de loi sur l'instruction publique* 92–93 and *De l'industrie française*, 415–420. See also Fontanon 1998, 279.

92. Kargon (1977) depicts how a similar institution worked in the context of a local industrial society in Britain. More generally, see Harrison 1961.

93. Pigeire 1932, 274.

94. Jean-Antoine Chaptal, *Circulaire aux Préfets des départemens*, 29 Thermidor, Year XIII [August 17, 1805], AD Somme, M80028, and *Arrêté portant organisation d'une École d'arts et métiers à Compiègne*, in the *Bulletin des Lois* 7: 220–262, 3rd series (Year XI [1802]), 484–494; Le Brun, *Notice sur les Écoles impériales d'Arts et Métiers*, 1863, AN F17 14317; Pigeire 1932, 239, 270–275, 350–354.

95. Chaptal, *Rapport et projet*, 93.

96. Chaptal 1893, 93; Péronnet 1988, 148–149.

97. Five additional branches opened later in the century. All later became engineering schools, graduating significant proportions of the entire profession. See Ballot 1978, 32–33; Day 1987, 70–76. The quotation is from p. 71 of the latter.

98. See Ballot 1978, 30–32; Péronnet 1988, 196–204.

99. Chaptal (1893, 100) claimed credit. Others think it was Joseph-Marie Dégerando (Gillispie 2004, 629).

100. The Society has extensive records. In two meeting with the oversight committee of its archives in June 1995, I was refused access unless I agreed to switch my research topic to France's economic relations with the United States. It is to be hoped that the new president will change such practices. The statutes are cited by Pigeire (1932, 399–400). For the activities of the Society, see E. J. Guillard-Senaineville, *Notice sur la Société d'Encouragement pour l'industrie nationale*, 1818 [1814], AN F12 2333; Péronnet 1988, 191–195.

101. Chaptal 1893, 274–277.

102. See Péronnet 1988, 244. On the initial willingness of the Bonapartist regime to conciliate its opponents, particularly recalcitrant workers, see Horn 2002.

103. Rosenband (2004) situated such movements in their proper international and sectoral context.

104. Woronoff 1994, 199; Bergeron 1993, 438–439. The quote is from Claude Anthelme Costaz, *Mémoire sur les moyens qui ont amené le grand développement que l'industrie française à pris depuis vingt ans* (Didot, 1816), 16–17.

105. The two preceding paragraphs are based on the following: Bergeron 1993, 438–444; Costaz, *Mémoire sur les moyens*, 17–19; Pigeire 1932, 170.

106. Pigeire 1932, 404–407, Bergeron 1993, 463–500; Horn 2006a.

107. Chaptal, *Circulaire*, 1 Floréal, Year IX.

108. Ibid.

109. *Exposition publique des produits de l'Industrie française et catalogue des productions industrielles*, Vendémiaire, Year X (September 1801), AN F12 2200. Bouin and Chanut (1980, 27) incorrectly cited 299 exhibitors.

110. Colmont 1855, 11.

111. *Procès-verbal des opérations du Jury nommé par le Ministre de l'intérieur pour examiner les Produits de l'Industrie française mis à l'Exposition des jours complémentaires de la neuvième année de la République*, Vendémiaire, Year X (September 1801), BHVP 8-950288.

112. Colmont 1855, 22; Bouin and Chanut 1980, 28.

113. *Noms et demeures des fabricans et artistes admis à l'Exposition*, 1802, AD Somme M80004

114. Lorion 1968, 127.

115. Ibid., 127.

116. Pigeire 1932, 407.

117. Bergeron 1993, 256–257; Horn and Jacob 1998.

118. The two preceding quotations are from Scipion Mourgue, *Rapport au Ministre de l'Intérieur*, 1 Messidor, Year XI (June 21, 1803), AN F12 654. Gayot 1998.

119. Ballot 1978, 34.

120. Pigeire 1932, 424–425.

121. Laissus 1989, 285. For confirmation, see Dhombres and Dhombres 1989, 345–447, 773–804.

Chapter 7

1. Crouzet (1990, 295–317, esp. 304–306) emphasizes that the rate of France's growth in output fell further behind Britain's in the period 1790–1810 but does not mention the depth of the trough in French production during the 1790s. As a result, even though France's growth in output for the first decade of the century was greater than Britain's, that fact disappears because of the statistical scale.

2. For Napoleon's views of the rivalry with England, see Crouzet 1990, 285–290.

3. Jean-Baptiste-Nompère de Champagny, *Circulaire du Ministre de l'Intérieur aux Messieurs les Préfets*, February 22, 1806, AN F12 985.

4. Le Normand and Moléon 1824, I: 64; Jouvenel 1942, 119. See also Bourguet 1988; Perrot 1977.

5. Nicolas Quinette, *Lettre à Son Excellence Le Ministre de l'Intérieur*, February 26, 1806, AN F12 985.

6. Lefebvre 1969a, 236–237; Jouvenel 1942, 196–201.

7. Industrial expositions were warmly received in the provinces. When no national exposition was planned for 1803, local authorities (usually the Chamber of Commerce and/or the Consultative Chamber plus the Prefect) organized departmental or municipal expositions beginning with Rouen and the Seine-Inférieure in November 1802. The Calvados (Caen), in lower Normandy, held two expositions and the Rhône-et-Loire (Lyon), Maine-et-Loire (Angers) and Aube (Troyes) each held one between 1803 and 1805.

8. Quinette, *Lettre à Son Excellence Le Ministre de l'Intérieur.*

9. Nicolas Quinette, *Circulaire*, February 19, 1806, AN F12 985.

10. Gras 1906, xxviii.

11. Jean-Baptiste-Nompère de Champagny, *Liste des Membres qui doivent composer le Jury National*, 1806, AN F12 985.

12. Lorion 1968, 130. I cannot find the document he cited.

13. See Hafter 1984, esp. 317–318, and the bundle of reports in AN F12 985.

14. Liu 1994, 84–86, 89–95.

15. Lomuller 1978, 110.

16. For the views of several influential entrepreneurs, see Chassagne 1980, 103.

17. Colmont 1855, 30; Hafter 1984, 323.

18. Kemp 1971, 100–102.

19. Crouzet (1958, I: 211) emphasized that none of these threats materialized.

20. For a contrary interpretation, see Lefebvre 1969a, 198.

21. Marzagalli 1999, 94–103.

22. Between 1807 and 1812, Great Britain issued 44,346 licenses at £13–14 each (Lefebvre 1969b, 109).

23. The Milan Decree dated November 23, 1807 repeated the terms of a Decree issued at Fontainbleau on October 13, 1807. I refer to the Milan version.

24. Lefebvre 1969b, 11.

25. Broers 1996, 94–98, 144–163; Boudon 2000, 301.

26. Bergeron 1970, 469.

27. Crouzet 1990, 315. This tendency lasted well into the nineteenth century.

28. Jouvenel 1942, 252–253.

29. Ellis 1981, 102–103; Colin de Sussy, *Circulaire aux départemens*, n.d. [1806], AN F12 533.

30. Broers' (1996, 96) chief examples of this policy were the land "donations" that underlay the recreation of the nobility and the Continental System.

31. Ellis 1981, 116–117.

32. Cited by Ellis (1981, 126). Gaudin made this pronouncement in September 1807. The correspondence is in AN AFIV 1080.

33. Bergeron 1970; 478–480; Lefebvre 1969a, 195; General Jean-Girard Lacuée de Cessac, *Mémoire, sur les moyens de détruire quelques causes principales de la préférence que les peuples des États-Unis donnent aux manufactures anglaises sur les manufactures françaises et de diminuer l'effet de quelques autres,* January 26, 1807, AN F12 502.

34. See AD Seine-Maritime 1M 150 for 1807.

35. Ellis 1981, 118.

36. *Notice concernant les prêts accordés aux fabricants et manufacturiers en exécution des décrets des 27 mars et 11 mai 1807,* October 8, 1809, AN F12 1559; Mollien 1898, II: 461–465.

37. On the steady increase in the number of prohibited articles, see Lefebvre 1969a, 196–197. For the protectionist debates of the Consulate and early Empire, see Chassagne 1991, 261–268.

38. Lefebvre 1969b, 10.

39. Lefebvre 1969b, 5–6.

40. Ellis 1981, 1–25.

41. Jouvenel 1942, 215–311.

42. For contemporary views of the "réunions," see *Exposé de la situation de l'Empire présenté par S. Ex. M. le Comte de Montalivet,* Ministre de l'Intérieur, au Corps Législatif, le 25 Février 1813, BN Le50-530.

43. *Procès-verbal de la distribution des prix fondés par S.M. l'Empereur, pour l'encouragement des fabriques et manufactures du épartement de la Roër,* August 1, 1813, AN F12 1569. The essential work on the Rhineland is Kisch 1989.

44. François Bardel, *Feuille de Renseignemens sur les cotons et les tissus,* March 1807, AN F12 533.

45. See *Statistique Industrielle,* 1807, AN F12 1568, with departmental responses. See also Chassagne 1991, 268–270.

46. Bergeron (1981, 175–176; 1970. 495–496) surveyed the figures and the literature on this subject. Notably, Ernest Labrousse found a 25% increase in industrial production in 1803–1812 versus 1781–1790, and Jean Marczewski posited a 3% annual increase in industrial production 1796–1812. Maurice Lévy-Leboyer (1968, 788–807) argued for an industrial deceleration in 1790–1810.

47. Chassagne 1991, 269; Clause 1974, II: 994–1014; Daly 2001, 174–177; Ellis 1981, 165–197; Gras 1906, 578; Hirsch 1991, 272–273; Ricommard 1934, 40–42.

48. Lefebvre 1969a, 170; Guillard-Senaineville, *Notice sur la Société d'Encouragement pour l'industrie nationale,* 1818 [1814], AN F12 2333.

49. See "Tableau des Importations et Exportations depuis 1788 jusqu'en 1809," reprinted in Guillard-Senaineville, *Notice.*

50. *Observations de Lièvin Bauwens à Gand sur une Letter de S. Excell. Mr. le Sénateur François de Neufchâteau*, March 15, 1808, AN F12 533.

51. *Mémoire sur les Établissemens formés par M. Ternaux ainé, tant en France que chez l'étranger, et qui depuis l'an 7, contribuent à l'amélioration de l'Industrie et de l'Agriculture*, 1807, AN F12 618.

52. Cited by Chassagne (1991, 268).

53. Herbinot de Mauchamps 1837, 264.

54. Barker 1969, 188.

55. France fell further behind since British cotton consumption increased 400% from 1790 to 1810, said Crouzet. This hind-sighted view eliminated the trough experienced by France during the revolutionary decade and minimized the extent and rapidity of growth under the Consulate and Empire. See Crouzet 1990, 304–306; Ellis 1981, 151; Lefebvre 1969a; Lyons 1994, 270–272, 277.

56. As the naturalization of merino sheep and theincreased cultivation of substitute crops for colonial goods (potatoes, indigo, and sugar beets) illustrates, agriculture was not neglected. On investment, see Lévy-Leboyer 1964, 448–452.

57. Other improvements in the chemical sector could be found in the enormous development of Marseille's manufacture of artificial soda from marine salt to substitute for the natural form traditionally imported from Spain (Bergeron 1981, 181–182).

58. Mokyr 1990, 252–259. On innovation under Napoleon, see Crosland 1967, 30–40. For contemporary thoughts on these matters, see *Rapports à l'Empereur sur le progrès des sciences des lettres et des arts depuis 1789* published in two volumes by Belin in 1989. One volume was edited by Jean Dhombres, one by Yves Laissus.

59. Ellis 1981, 187–188.

60. Bergeron 1981, 177–180; Chassagne 1991, 261–370; Gayot 1998, 321–333; Hirsch 1991, 285–287; Johnson 1995, 15–22; Kaplow 1964, 112–126.

61. On Rawle, see Chassagne 1991, 307–310.

62. Crouzet 1990, 344.

63. Bergeron 1981, 175–179; Chassagne 1991, 339; Daly 2001, 169–171.

64. This point has been made frequently in comparisons of early industrialization between the United States and Great Britain, but does not often appear in the literature on France even when the growth of waterpower is noted. For an example, see Woronoff 1994, 207–211. An exception is Chassagne 1991, 337–340. Greenberg (1982) emphasized the continuing importance and indeed the preference for water over steam power by manufacturers in textiles, iron smelting and flour milling on both side of the Atlantic until the late 1830s. See also Hills 1986, 1970. Hunter (1975) criticizes the Anglo-centeredness of discussions of power and the sainted position of Watt while pointing out the supplementary aspect of steam to water power. I owe these references to conversations with Pat Malone.

65. *État des manufactures et autres établissements qui versent leurs produits dans le commerce et qui supposent un certain dégré d'industrie situés dans la ville de Paris*, February 1807, AN F12 1569.

66. Bergeron 1981, 175–181; Johnson 1995, 15–22; Lyons 1994, 273; Woronoff 1994, 191–194. For a contemporary view, see Costaz 1816, 11.

67. On how successful French mercantilist policy was under Napoleon, see *Exposé de la situation de l'Empire présenté par S. Ex. M. le Comte de Montalivet*, Ministre de l'Intérieur, au Corps Législatif, le 25 Février 1813, BNF Le50-530. For a review of French agricultural output and productivity, see Hoffman 1996 and Postel-Vinay 1989. Though generally optimistic about French agriculture, Hoffman and Postel-Vinay emphasized the stagnation of technical improvement in the period.

68. Butel 1992, 293–310; Crouzet 1990, 300–301. See also the sections on Bordeaux in Marzagalli 1999.

69. See Szotstak 1991.

70. See Ellis 1981, chapters 3–5.

71. From *Les effets du blocus continental* (1809), cited by Lefebvre (1969b, 114). Lefebvre's dusty volume remains the best English-language account of the twists and turns of the Blockade which reveals how out of fashion the subject is.

72. The statistics in the following paragraphs are based on Crouzet 1958, II: 882–895. Unless otherwise noted, the following chronological survey is based on Crouzet 1958, I: 274–282, 319–321; II: 707–729. For a more theoretical and slightly more recent survey, see Hueckel 1985, v–lvi.

73. Crouzet 1990, 48–93.

74. Lyons 1994, 217.

75. Cited on p. 84 of Evans 1984.

76. Christie 1982, 311–315; Lyons 1994, 216–217.

77. Mollien 1898, III: 8.

78. Revealingly for those interested in the transnational aspect of the relationship of masters and men, parallel problems of labor regulation were much discussed by the General Council of Manufactures in 1810–1812 just as the Luddite movement got under way in Britain. See Gille 1961, 2–19.

79. Thompson 1963, 553.

80. Lyons 1994, 217, 277.

81. Deane and Cole 1967, 152, 185, 187, 204, 210, 216, 225, 282.

82. Mori 2000, 153–154; Crouzet 1990, 283.

83. For a brief review of structural change, see Heywood 1992, 8–13.

84. Harley and Crafts 2000, 820. Verley (1991) situated these structural changes in the transformation of international exchange so different from the "world economies" of the eighteenth century.

85. For France, see Verley 1997, 312. For Britain, see Deane and Cole 1967, 78, 103, 166; Harley 1999, 172; Hoffman 1955, appendix.

86. *Lettre du Préfet du département de la Seine-Inférieure au Ministre de l'Intérieur*, May 28, 1809, AN F12 1561.

87. Kaplow 1964, 123–124; Jouvenel 1942, 343–344.

88. Broers 1996, chapters 3–5; Lefebvre 1969b, 118–119.

89. Lefebvre 1969b, 121–130.

90. Mollien 1898, II: 462; Jouvenel 1942, 316.

91. Jouvenel 1942, 294–298.

92. Dufraisse 1966, 532.

93. The licenses began many prominent nineteenth-century fortunes (Mollien 1898, II: 443).

94. Barker 1969, 185–213.

95. Jouvenel 1942, 399–401. Jouvenel thought that the self-blockade was fundamentally changed in July 1810, which he called the "decisive month" for the French economy under Napoleon.

96. The Councils and local communities of entrepreneurs objected regularly to duties on imported raw materials. See Gille 1961, 23–29.

97. Jouvenel 1942, 395.

98. Dufraisse 1966, 523–529.

99. Bergeron 1981, 175–176.

100. There was contemporary support for this concept among entrepreneurs and scientists who believed that recent improvements in wool and linen spinning technology made France more competitive. See for example the report of the chemist Jean-Baptiste Vitalis in *Séance publique de la Société libre d'émulation de Rouen tenue le 22 Juin 1811* (Rouen: P. Periaux, 1811), AD Seine-Maritime RH 12/3, 69–73.

101. Napoleon's view—reported in AN AF*IV 170 (October 22, 1810) and cited in Barker 1969—was in a liasse that was "introuvable." See also May 1939.

102. Mollien 1898, II: 461–462.

103. Jouvenel 1942, 400–401.

104. Viennet 1947.

105. Mollien 1898, III: 31–32.

106. Mollien 1898, III: 10–14.

107. *Exposé de la situation de l'Empire présenté par S. Ex. M. Le Comte de Montalivet*, 33–35.

108. Péronnet 1988, 461–462.

109. Miller 1999, 198–235.

110. Costaz 1843, II: 272–273.

111. On the tariffs of the first restoration, see Ponteil 1966, 71–73.

112. Local studies document the beat of this economic rhythm. A valuable resource for an important industrial region is Inizan 1996–97, which gives number of workers, machines, output by type of fabric and value in several key industries for the period 1810–1815. See also Daly 2001, 178; Kaplow 1964, 125.

113. Vidalenc 1958b, 461.

114. From a bureaucratic standpoint, perhaps the most telling differences were the vast budget shortfalls and enormous debt (Ponteil 1966, 48).

115. The French version was enacted in July 1819 (Bertier de Sauvigny 1966, 219).

116. Dunham 1955, 390.

117. For the protectionist arguments, see Chamber of Commerce of Rouen, *Observations sur la nécessité de maintenir la prohibition des Produits de Fabrique étrangère et d'assurer l'exécution de l'article 59 de la Loi du 28 Avril 1816*, January 1817, AD Seine-Maritime 8M3.

118. Ponteil 1966, 71–73.

119. Bertier de Sauvigny 1966, 219–229.

120. Asselain 1984, 136–138.

121. Lewis 1993, 286–287. Political considerations came even more to the fore under Charles X (1824–1830).

122. On the French cotton industry during the Restoration, see Chassagne 1991, 373–655. On the development of other niches, see Reddy 1984, 94–100.

123. Becchia 2000, 488–500; Gayot 1998, 422–436; Johnson 1995, 23–45.

124. Bezucha 1974, 13–25; Ballot 1978, 369–382; Daumas 1968, 680–682.

125. Lewis 1993, 270–273.

126. Fournial 1976, 189, 205, Gille 1961, 148–149, Gras 1906, 143; Jacques Schnetzler 1973, 78; Thermeau 2002, 241–251.

127. Ballot 1978, 375; Fournial 1976, 177; Thermeau 2002, 246–251.

128. This law was passed April 21, 1810. The correspondence regarding the adjudication of concessions and the oversight of entrepreneurs in the Loire during the first part of the nineteenth century is in AN F14 4381.

129. Accampo 1989.

130. Despite such treatment, between 1,300 and 1,400 British workers came to France in the period 1814–1824 (Bertier de Sauvigny 1966, 223).

131. The preceding five paragraphs are based on the following accounts: Descreux 1868, 36–51, 168–179; Fournial 1976, 178–187; Gras 1900, 75; Gras 1980, xxxvi–xxxvii, 12–13, 386–389; Schnetzler 1973, 59–78; Thermeau 2002, 287–289; Verney-Carron 1999, 87–95.

132. Strapped municipalities or départements sometimes limited expenditures on such institutions. The central government urged that their budgets be restored.

See *Circulaire de Laine, Ministre de l'Intérieur aux Préfets* of December 15, 1816 reprinted in *Circulaires, instructions et autres actes émanés du Ministre de l'Intérieur, ou relatifs à ce département, de 1797 à 1821 inclusivement,* second edition, volume 3, 1816–1819 (Paris: Imprimerie royale, 1823), AD Seine-Maritime RR 19/11, 132–133.

133. Colmont 1855, 38–69. National expositions were held regularly from this point. The Restoration convened expositions in 1823 and 1827, and the July Monarchy followed in 1834, 1839, and 1844. The Second Republic also hosted one in 1849, all before the Crystal Palace Exhibition in London in 1851 belatedly brought the institution to the full attention of the English-speaking world.

Chapter 8

1. Rosenband 2004.

2. Levasseur 1903 remains the essential reference on these measures and subsequent legal extrapolations.

3. *Lettre du Ministre de l'Intérieur au Préfet du département du Rhône,* June 30, 1821, AN F12 4648. Other correspondence is in this box. For Paris, see Préfet de Police, *Lettre à M. Anisson Duperon, auditeur au Conseil d'État, Inspecteur Général de l'Imprimerie Impériale,* November 4, 1812, AN AJ 17/2. Michael Sibalis graciously shared the latter document with me.

4. Magraw 1992b, 19.

5. See Kaplan 1979, 33–51.

6. Truant 1994, 194, 204–205.

7. Sibalis 1988, 718–730; Truant 1994, 25–30, 204–210.

8. Charles Delacroix, *Lettre,* 28 Floréal, Year XIII (May 18, 1805), AD Gironde 1M334. The following account is based on a dozen letters from Floréal and Prairial Year XIII by Delacroix, Gaudet, Bourguignon, and police agents.

9. *Lettre du Commissaire-Général de Police au Préfet de la Gironde,* 17 Floréal, Year XIII, AD Gironde 1M 334. The rest of this paragraph is also based on this document. In a private communication regarding an earlier draft of some of this material, Michael Sibalis suggested that not all these organizations were formal compagnonnages.

10. I suspect that Bordeaux's labor environment was a major cause of the city's lack of technological innovation during the revolutionary era.

11. AD Gironde 1M 334 and 4M 210; Bourgin and Bourgin 1912, 1921, 1941. See also Desgraves and Dupeux 1969, 49.

12. d'Hauterive 1913.

13. Truant 1994, 241–243.

14. Tudesq 1969, 49.

15. Truant (1994, 2, 205–220, 234) found compagnonnages in the départements of the Aude, Côte-d'Or, Gironde, Rhône, Maine-et-Loire, Loiret, Eure-et-Loire, Gard, Seine, Haute-Garonne, Loir-et-Cher, Ille-et-Villaine, Nièvre, Drôme, Saône-et-Loire, Bouches-du-Rhône, Indre-et-Loire, Loire-Atlantique, Charente-Maritime, Hérault, Var, Vaucluse, and Vienne.

16. Rosenband 1998, 11–19.

17. Levasseur 1903, 506–507.

18. Sewell 1980, 163–173.

19. Magraw 1992b, 38–39.

20. Agulhon 1970, 116–119; Truant 1994, 207, 234.

21. Bourgin and Bourgin 1912, 295; 1921. 1–5, 90, 167, 212. For some police reports on worker coalitions during the Restoration, see Baron de Maissemy, *Lettre*, June 18, 1812, AD Somme 99M 80054. AN F7 9787. See also *Le Moniteur Universel* 173: 230, August 18, 1825.

22. Bourgin and Bourgin 1921, 208–210; Mourgue, *Rapport au Ministre de l'Intérieur*; Becchia 2000, 455–476; Gayot 1998, 429; Magraw 1992b, 41.

23. Evrard 1947, 343–348.

24. Vidalenc 1981, 117–118; Chassagne 1991, 503–510; Alexandre 1983, 45–47, 53. The primary sources are in AD Seine-Maritime 10M 330, 2U 1623, 2U 563, 2U 565, and J841 and in AN F7 9787. See the extensive coverage in the *Journal de Rouen et du Département de la Seine-Inférieure (Politique et Littéraire)*, vol. 60 (July–December 1825), AD Seine-Maritime JPL 3/71; also see documents in Bourgin and Bourgin 1941.

25. For a contemporary worker's views of employers, see Noiret 1836.

26. Pierre Bernadau, *Oeuvres Complètes de Pierre Bernadau de Bordeaux*, tome 10, *Sixième Recueil des Tablettes Manuscrites de l'Ecouteur Bordelais de 1821 à 1831 inclusivement* (1832), Bibliothèque de Bordeaux, ms. 713/10, 294–295.

27. Truant 1994, 19–20.

28. Agulhon 1970, 118.

29. Calonne 1906, 139–140; Deyon 2000, 92.

30. Hirsch 1991, 357.

31. Dunham 1955, 270–271.

32. For a parallel and earlier example, see Rosenband 2004.

33. Sibalis 1988, 718–730.

34. Auvray, *Lettre du préfet du département de la Sarthe au Ministre de l'Intérieur*, 13 Messidor, Year XI, AN F12 1560. Other relevant documents are here. See also Hirsch 1991, 252 and Levasseur 1903, 540–545.

35. Hirsch 1991, 572–578 and Levasseur 1903, 549.

36. *Lettre de Girandeau, Durand le jeune et Compagnie et Pierre Boutin, Laporte et Compagie, marchands de fer à son Excellence, le Ministre de l'Intérieur*, 23 Prairial, Year XII, AN F12 1560.

37. *Rapport que le Préfet du département de la Seine a présenté au Conseil du Commerce, Arts et Agriculture près ce même département, un protjet de loi rélatif aux Manufactures et aux gens de travail de toutes professions*, n.d. [Year X], AN F12 2366.

38. Sibalis 1988, 720.

39. Claire H. Crowston, Steven Kaplan, and Gilles Postel-Vinay are nearly finished with an important study titled *Apprenticeship in France, 1650–1914*.

40. Bergeron 1993, 415–420, 424–438.

41. This material is in AN F12 1560. On Bordeaux, see *L'Indicateur, ou Journal du Commerce, de Nouvelles, de Littérature et d'Annonces*, #181 to 350 (March 22, 1805–September 23, 1805), #248 and #249 of May 28–29, 1805 reprint an *Ordonnance concernant l'état des portefaix et autres gens de peine*, promulgated by the commissaire général de police and approved by General Charles Delacroix, Prefect of the Gironde. For the regulations, see *Circulaire de Laine, Ministre de l'Intérieur aux Préfets*, July 3, 1818 in *Circulaires, instructions et autres actes émanés du Ministre de l'Intérieur, ou relatifs à ce département, de 1797 à 1821 inclusivement*, second edition, 3rd vol. (Paris: Imprimerie royale, 1823), 338.

42. *Lettre du Ministre des Manufactures et du Commerce, Comte de l'Empire, à Son Excellence le Ministre de l'Intérieur, Comte de l'Empire*, September 19, 1812, AN F12 1560. There is additional material on other places with corporations des arts et métiers such as Florence.

43. *Lettre des membres composant la Chambre de Commerce d'Amiens au Préfet du département de la Somme*, August 14, 1821, AD Somme 99M 80067/5 and *Observations du Chambre de Commerce de la ville d'Amiens*, October 6, 1821, BM Amiens 2F33 (1). Clause 1974, II: 994–995 and Hirsch 1991, 358–359.

44. Bergeron (1981, 101) calls these organizations "companies" and emphasized the need to ensure Paris' food supply by exploiting the resources of the annexed territories, which required large-scale financial resources.

45. Sibalis 1988, 724–729.

46. Many are cited in *Adresse des délégués des signataires de la requête au Roi à Messieurs les membres de la Chambre du Commerce à Amiens*, September 29, 1817, AD Somme 99M 107676. See also the *Projet de règlement concernant les fabriques et ouvriers du département de la Seine-Inférieure avec les observations y relatives*, September 5, 1817, AD Seine-Maritime 10M2.

47. *Rapport rélatif au rétablissement demandé des Corps de Marchands et des Communautés d'arts et métiers, et contenant la Discussion des Mesures pro-*

posées pour la police du Commerce, des Manufactures et des Ateliers, 1819, AD Seine-Maritime 10M1.

48. See the complaints voiced by small-scale manufacturers about "les grands" cited by Magraw (1992b, 35–36).

49. d'Hauterive 1963, 173–175.

50. Gotteri 1999, 39.

51. *Lettre du Préfet de Police de Paris au Ministre de l'Intérieur*, June 26, 1815, AN F12 1560.

52. Coornaert 1941, 99.

53. Gossez 1984.

54. *Avis de la Chambre de Commerce d'Amiens*, March 22, 1822, AD Somme 99M 107687.

55. Gross 1997.

56. *Traité portant établissement d'une bourse commune et d'un fonds de secours entre M.M. les teinturiés maîtres en coton existant dans la ville ou banlieue d'Amiens*, July 6, 1821, AD Somme 99M 107687 and a notarial act before Antoine-Joseph-Marie Lyon, October 10, 1821, AD Somme 99M 80067/5.

57. These tactics were reviewed in the *Extrait du registre des Délibérations de la Chambre de Commerce d'Amiens*, April 18, 1832, AD Somme 99M107687.

58. AD Somme 99M 80067/5.

59. *Réponse des teinturiers d'Amiens au Rapport fait contr'eux par la Chambre de Commerce de la ville au sujet de leur association pour la formation d'une caisse de secours*, n.d. [October 1821], BM Amiens 2F 33 (1).

60. *Lettre de la Maire d'Amiens au préfet du département de la Somme*, October 29, 1821 and *Lettre des membres composant la Chambre de Commerce d'Amiens au Préfet du département de la Somme*, December 11, 1821, both in AD Somme 99M 80067/5. See also *Observations du Chambre de Commerce de la ville d'Amiens*, October 6, 1821, BM Amiens 2F33 (1) and the entries for December 6 and December 20, 1821 in the *Registre aux délibérations de la Chambre de Commerce d'Amiens 27 mai 1819 à 2 novembre 1831*, AD Somme 99M 107695. Entry for February 20, 1822, in *Registre aux délibérations de la Chambre de Commerce d'Amiens 27 mai 1819 à 2 novembre 1831*.

61. The quotations are from *Lettre du Ministre de l'Intérieur au Préfet du département de la Somme*, October 11, 1821 and *Lettre du Ministre de l'Intérieur au Préfet du département de la Somme*, December 28, 1821. See also *Lettre des membres composant la Chambre de Commerce d'Amiens au Préfet du département de la Somme*, April 10, 1822, all in AD Somme 99M 80067/5 and *Lettre du Ministre de l'Intérieur au Préfet du département de la Somme*, October 18, 1821, BM Amiens 2F 33 (1).

62. Hubscher 1986, 211–214; Calonne 1900, 141–150.

63. *Lettre du Préfet de la Somme à la Président de la Chambre de Commerce*, September 29, 1832, AD Somme 99M 107687.

64. *Extrait du registre des Délibérations de la Chambre de Commerce d'Amiens,* April 18, 1832, AD Somme 99M 107687; Hubscher 1986, 211; Calonne 1900, II: 141.

65. The law was not very effective. *Circulaire du Comte de Sussy, Ministre des manufactures et du commerce aux Préfets,* November 9, 1812 in *Circulaires, instructions et autres actes émanés du Ministre de l'Intérieur, ou relatifs à ce département, de 1797 à 1821 inclusivement,* second edition, second vol. (Paris: Imprimerie royale, 1822), 410.

66. On theft of raw materials by workers, see Clause 1974, II: 996–997, 1010; Dewerpe and Gaulupeau 1990, 44–47; Johnson 1995, 27–28.

67. Petitions from Lyon and Carcassonne are in AN F12 2337 others are in the AN F12 2366. For Sedan, see Mourgue, *Rapport au Ministre de l'Intérieur.*

68. Jean-Antoine Chaptal, *Lettre au Préfet du département de la Loire,* 12 Fructidor, Year IX, AD Loire 81M1.

69. Mourgue, *Rapport au Ministre de l'Intérieur.* See also Clause 1974, II: 997.

70. Claude-Anthelme Costaz, *Rapport présenté au Ministre de l'Intérieur,* April 7, 1809, and Jean-Baptiste Nompère de Champagny, *Circulaire du Ministre de l'Intérieur aux Messieurs les Préfets,* April 20, 1807, both in AN F12 2337.

71. See the *Avis aux fabricans et maîtres d'ateliers de la ville d'Amiens,* November 12 1814, BM Amiens 2F 33 (1) issued by the mayor.

72. Claude-Anthelme Costaz, *Rapport présenté au Ministre de l'Intérieur,* June 19, 1807, AN F12 2337, Baron de Maissemy, *Lettre du Préfet de la Somme aux membres composant la Chambre de Commerce d'Amiens,* March 8, 1813, AD Somme 99M 107668 and Reddy 1984, 71.

73. Bourgin and Bourgin 1912, 1; and *Lettre du Ministre de l'Intérieur au Monseigneur le Garde des Sceaux et Ministre de la Justice,* March 7, 1821. The manufacturers of Louviers had difficulty maintaining quality and preventing fraud: *Mémoire pour MM. les Manufacturiers de draps à Louviers . . . ,* 1820, AN F12 2366.

74. Bergeron 1993, 405–408. The originals and Costaz's report are in AN F12 2337.

75. Bourgin and Bourgin 1912, 89–92.

76. For these practices from the perspective of workers, see Noiret, *Mémoires d'un Ouvrier Rouennais* and the *Exposé des fraudes qui se commettent journellement dans la fabrication, la refabrication et les mélanges des Indigos, avec les moyens de les découvrir,* by Pugh, a member of Rouen's Society of Emulation, 9 Thermidor, Year IX [July 28, 1801], in the Reports collected in AD Seine-Maritime RH 12/1.

77. Mauchamps 1837, 212–214.

78. *Circulaire du Comte de Sussy,* November 9, 1812. The same was true in other industries such as commercial firearms, cutlery, and hardware. Regulations were issued on September 5 and December 14 of 1810. Counterfeiting through

adulteration or substituting less costly and less effective raw materials was also common in soap making, the manufacture of beer, wine, alcohol, and such mainstays of the French economy as vinegar and olive oil. Claude Anthelme Costaz, *Mémoire sur les moyens qui ont amené le grand développement que l'industrie française à pris depuis vingt ans* (Paris: Firmin Didot, 1816), 22 and *Circulaire du Comte de Montalivet, Ministre de l'Intérieur aux Préfets*, March 27, 1810 in *Circulaires, instructions et autres actes émanés du Ministre de l'Intérieur*, vol. 2, 177–178.

79. Nicolas Quinette, *Situation es filatures*, May 14, 1809, AN F12 1561. For a few examples from the old regime, see François-Marie-Bruno, comte d'Agay, *Arrêté*, March 19, 1774, AD Somme C325, *Copie de la lettre écrite par MM. les Président, et Syndic de la Chambr du commerce de Picardie aux officiers municipaux d'Amiens le 10 Decembre 1773*, AD Somme C322, and Louis Villard, *Supplément au Mémoire sur le Bureau de marque et autres objets, dressé le mois d'aout dernier*, n.d. [1780], AD Somme C356.

80. See the primary sources cited by Dunham 1955, 259–262.

81. Bergeron 1993, 450–451; Daly 2001, 188; Levasseur 1903, 390–393.

82. Several memoirs in AN F12 652 detail the old-regime origins of this issue. See Bergeron 1993, 452–453.

83. *Circulaires, instructions et autres actes émanés du Ministre de l'Intérieur, ou relatifs à ce département, de 1797 à 1821 inclusivement*, second edition, 3rd vol. (Paris: Imprimerie royale, 1823), 110–113 and the *Avis aux fabricants de tissus et tricots de coton et de laine et de touts autres tissus*, July 3, 1818, AN F12 4792.

84. *Circulaires, instructions et autres actes émanés du Ministre de l'Intérieur*, vol. 3, 197–198, 314–320, 391–392, 448–462 and *Circulaires, instructions et autres actes émanés du Ministre de l'Intérieur, ou relatifs à ce département, de 1797 à 1825 inclusivement*, second edition, volume 5 (Paris: Imprimerie royale, 1829), 259–260.

85. *Affiche d'une Arrêté de M. le Préfet de la Somme*, February 20, 1817, BM Amiens 2F 49 (2).

86. Under Napoleon, French law distinguished between *contrebande* and *fraude*. The former applied to commerce in illegal goods and the latter referred to avoiding the duties on legally imported goods. For contraband, the penalty was usually harsher, but the vagaries of French law meant that the dividing line between the two was usually blurred (Daly 2001, 189–190; Ellis 1981, 201–202).

87. See Dufraisse 1961.

88. Collet, *Lettre au Ministre de la Police par le Directeur général des Douanes*, 9 Frimaire, Year X, AN F7 3023.

89. d'Hauterive 1908, 448–497. On mobile columns, see Brown 1997.

90. Cited by Marzagalli (1999, 209).

91. Herbouville, *Note sur l'affaire de la maison de commerce Veuve Lombaert et d'Heyden fils, d'Anvers*, 6 Brumaire, Year X, AN F7 3023. There are other, similar observations in this box.

92. For details of this network, see d'Hauterive 1908, 428–429; Vidalenc 1958b, 453; Dufraisse 1961, 213–217.

93. Dufraisse 1961, 218–232.

94. *Rapport au Premier Consul*, 8 Fructidor, Year X and Charles Louis Huguet, *Lettre au Talleyrand, Ministre des Rélations extérieures*, 23 Vendémiaire, Year X both in AN F7 3023; d'Hauterive 1908, 244.

95. d'Hauterive 1963, 153.

96. Ibid., 267–268.

97. Ibid., 338–339, 448.

98. Jouvenel 1942, 349–352.

99. d'Hauterive 1908, 146, 155. Such examples could be multiplied almost ad infinitum.

100. Gotteri 1997, 32.

101. Broers 1996, 177 and Jouvenel 1942, 353–356.

102. Gotteri 2003, 4–5.

103. Gotteri 2004, 47.

104. On Hamburg, see Marzagalli 1999, 192–207.

105. On Heligoland, see Marzagalli 1999, 176–180.

106. d'Hauterive 1908, 89, 309–310; 1911, 256, 391; Gotteri 2004, 333–334.

107. Daly 2001, 192; Dufraisse 1961, 212–213.

108. On this subject and on the role of Emden in 1804–05, see Bergeron 1970.

109. d'Hauterive 1908, 275–276, 313, 349, 419.

110. d'Hauterive 1911, 45; d'Hauterive 1913, 143; Gotteri 1997, 516.

111. Pierre Bernadau, *Oeuvres Complètes de Pierre Bernadau de Bordeaux*, tome 8, *Quatrième Recueil des Tablettes Manuscrites de l'Ecouteur Bordelais de Septembre 1802 à 1813 inclus* (1814), Bibliothèque de Bordeaux ms. 713/8, 486; Gotteri 2001, 29; Marzagalli 1999, 131–155, 183.

112. On Livorno. see Margazalli 1999.

113. Crouzet 1958 I: 139; Ellis 1981, 203–208.

114. d'Hauterive 1911, 264–265, 271; Gotteri 2003, 523.

115. Ibid., 469; Gotteri 2001, 530.

116. d'Hauterive 1908, 392; 1913, 278, 400.

117. d'Hauterive 1908, 39–40; 1911, 249; Ellis 1981, 207–208; Gotteri 1998, 621; Jouvenel 1942, 348–349

118. d'Hauterive 1908, 418–419; 1911, 40; Gotteri 2000, 215–216.

Chapter 9

1. This paragraph is based on the following works: Bairoch 1963, 325–326; Bertier de Sauvigny 1966, 222; Day 1987, 13–14, 19, 76–78; Dhombres and Dhombres 1989, 795–804; Dunham 1955, 399–419; Ponteil 1966, 62–83.

2. Horn 2002, 235.

3. *Procès-verbal des opérations du Jury nommé par le Ministre de l'intérieur pour examiner les Produits de l'Industrie française mis à l'Exposition des jours complémentaires de la neuvième année de la République*, Vendémiaire, Year X, BHVP, 8-95028; *Procès-verbal des opérations du Jury nommé par le Ministre de l'intérieure pour examiner les Produits de l'Industrie française mis à l'Exposition des jours complémentaires de la dixième année de la République* (Paris: Imprimerie de la République, Vendémiaire, Year XI), AN AD XIX D1; "Observations sur l'exposition des produits de l'industrie française et sur la distribution des prix," in *Registre de rapports et correspondance 1806–1808*, March 3, 1808, CNAM 10-482.

4. Reddy 1984, 91–93; Woronoff 1994, 217.

5. The next four paragraphs are based on the following accounts: Ballot 1978, 278–279; Beury 1983, 135–144; 1984, 23–24, 38–45; Colomès 1943, 91–103; Daumas 1968, 691–703; Heywood 1974, 54, 66–68, 133; Ricommard 1934, 38–49, 60–69. On the development of this industry in the twentieth century, see Chenut 2005.

6. Beury 1984, 21–22, 48–84. The quotations are cited by Colomès (1943, 104–105).

7. Beury 1984, 97.

8. Mathias 1979, 14.

9. Thompson 1963, 605.

Bibliography

Accampo, Elinor. 1989. *Industrialization, Family Life and Class Relations: St. Chamond, 1815–1914*. University of California Press.

Aftalion, Florin. 1990. *The French Revolution: An Economic Interpretation*. Cambridge University Press.

Agulhon, Maurice. 1970. *Une Ville Ouvrière au temps du socialisme utopique: Toulon de 1815 à 1851*. La Haye.

Alder, Ken. 1997. *Engineering the Revolution: Arms and Enlightenment in France, 1763–1815*. Princeton University Press.

Alexandre, Alain. 1983. *Le Houlme d'hier ou Regards sur le passé d'une commune de la banlieu rouennaise*. Self-published.

Allinne, Jean-Pierre. 1981. "À propos des bris de machines textiles à Rouen pendant l'été 1789: Émeutes anciennes ou émeutes nouvelles?" *Annales de Normandie* 31, no. 1: 37–58.

Archer, John E. 2000. *Social Unrest and Popular Protest in England 1780–1840*. Cambridge University Press.

Ashworth, William J. 2003. *Consuming the People: Trade, Production, and the English Customs and Excise 1643–1842*. Oxford University Press.

Asselain, Jean-Charles. 1984. *Histoire économique de la France du XVIIIe siècle à nos jours*, volume 1. Seuil.

Babeau, Albert. 1873–74. *Histoire de Troyes pendant la révolution*. Dumoulin.

Baczko, Bronislaw. 1992. "Instruction publique." In *Dictionnaire critique de la Révolution française*, ed. F. Furet and M. Ozouf. Flammarion.

Bairoch, Paul. 1963. *Révolution industrielle et sous-développement*. Société d'édition d'enseignement supérieur.

Ballot, Charles. 1978 [1923]. *L'Introduction du machinisme dans l'industrie française*, ed. C. Gével. Slatkine.

Barber, Giles. 1994. *Studies in the Booktrade of the European Enlightenment*. Pindar.

Barker, Richard J. 1969. "The Conseil général des Manufactures under Napoleon (1810–1814)." *French Historical Studies* 6, no. 2: 185–213.

Baugh, David A. 2004. "Naval power: What gave the British Navy superiority?" In *Exceptionalism and Industrialisation: Britain and Its European Rivals, 1688–1815*, ed. L. Prados de la Escosura. Cambridge University Press.

Becchia, Alain. 2000. *La draperie d'Elbeuf (des origines à 1870)*. Université de Rouen.

Belhoste, Bruno. 2003. *La formation d'une technocratie: l'École polytechnique et ses élèves de la Révolution au Second Empire*. Belin.

Belmont, Alain. 1998. *Des ateliers au village: les artisans ruraux en Dauphiné sous l'Ancien Régime*. Presses Universitaires de Grenoble.

Berg, Maxine. 1980. *The Machinery Question and the Making of Political Economy 1815–1848*. Cambridge University Press.

Berg, Maxine. 1988. "Workers and machinery in eighteenth-century England." In *British Trade Unionism 1750–1850: The Formative Years*, ed. J. Rule. Longman.

Berg, Maxine. 1994. *The Age of Manufactures, 1700–1820: Industry, Innovation and Work in Britain*, second edition. Routledge.

Berg, Maxine. 2005. *Luxury and Pleasure in Eighteenth-Century Britain*. Oxford University Press.

Berg, Maxine, and Helen Clifford, eds. 1999. *Consumers and Luxury: Consumer Culture in Europe. 1650–1850*. Manchester University Press.

Bergeron, Louis. 1970. "Problèmes économiques de la France napoléonnienne." *Revue d'histoire moderne et contemporaine* 17: 469–505.

Bergeron, Louis. 1972. "Remarques sur les conditions du développement industriel en Europe occidentale à l'époque napoléonienne." *Francia* 1: 546–647.

Bergeron, Louis. 1981. *France under Napoleon*. Princeton University Press.

Bergeron, Louis. 1997. "The businessman." In *Enlightenment Portraits*, ed. M. Vovelle. University of Chicago Press.

Bergeron, Louis, ed. 1993 [1819]. *Chaptal: De l'industrie française*. Imprimerie Nationale.

Bertier de Sauvigny, Guillaume de. 1966. *The Bourbon Restoration*. University of Pennsylvania Press.

Beury, André. 1983, 1984, 1986. *Troyes de 1789 à nos jours*, three volumes. Librairie Bleue.

Bezucha, Robert J. 1974. *The Lyon Uprising of 1834: Social and Political Conflict in the Early July Monarchy*. Harvard University Press.

Biard, Michel. 2003. *Missionaires de la République: Les représentants du peuple en mission (1793–1795)*. Éditions du CTHS.

Biernacki, Richard. 1995. *The Fabrication of Labor: Germany and Britain, 1640–1914*. University of California Press.

Billion, Pierre. 1941. *La Communauté unie des marchands de Troyes (1696–1791)*. J.-L. Paton.

Black, Jeremy. 1994. *British Foreign Policy in an Age of Revolutions 1783–1793.* Cambridge University Press.

Bosher, John. 1964. *The Single Duty Project: A Study of the Movement for a French Customs Union in the Eighteenth Century.* Athlone.

Bosher, John. 1988. *The French Revolution.* Norton.

Bossenga, Gail. 1988. "Protecting merchants: Guilds and commercial capitalism in eighteenth-century France." *French Historical Studies* 15, no. 4: 693–703.

Bossenga, Gail. 1991. *The Politics of Privilege: Old Regime and Revolution in Lille.* Cambridge University Press.

Boudon, Jacques-Olivier. 2000. *Histoire du Consulat et de l'Empire.* Perrin.

Bouin, Philippe, and Christian-Philippe Chanut. 1980. *Histoire française des foires et des expositions universelles.* Nesle.

Boulle, Pierre H. 1975. "Marchandises de traite et développement industriel dans la France et l'Angleterre du XVIIIe siècle." *Revue française d'histoire d'Outre-Mer* 62, no. 226–7: 309–330.

Bouloiseau, Marc, ed. 1957–58. *Cahiers de doléances du tiers état du bailliage de Rouen pour les États généraux de 1789.* Presses Universitaires de France.

Bouloiseau, Marc. 1983. *The Jacobin Republic 1792–1794.* Cambridge University Press.

Bourde, André J. 1953. *The Influence of England on the French Agronomes, 1750–1789.* Cambridge University Press.

Bourgin, Georges, and Hubert Bourgin. 1912, 1921, 1941. *Le Régime de l'Industrie en France de 1814 à 1830: Les Patrons, les Ouvriers et l'État.* Alphonse Picard et fils.

Bourguet, Marie-Noëlle. 1988. *Déchiffrer la France, La statistique départementale à l'époque napoléonnienne.* Éditions des archives contemporaines, Ordres Sociaux.

Bowden, Witt. 1919. "The English manufacturers and the Commercial Treaty of 1786 with France." *American Historical Review* 25, no. 1: 18–35.

Bret, Patrice. 1996. "The organization of gunpowder production in France, 1775–1830." In *Gunpowder: The History of an International Technology,* ed. B. Buchanan. Bath University Press.

Bret, Patrice. 2000. "Des essais de la Monnaie à la recherche et à la certificaiton des métaux: un laboratoire modèle au service de la guerre et de l'industrie (1775–1830)." *Annales historiques de la Révolution française* 320: 137–148.

Bret, Patrice. 2002. *L'État, l'armée, la science: L'invention de la recherche publique en France (1763–1830).* Presses Universitaires de Rennes.

Brewer, John. 1989. *The Sinews of Power: War, Money and the English State, 1688–1783.* Knopf.

Brian, Eric. 1994. *La mesure de l'État: administrateurs et géomètres au XVIIIe siècle.* A. Michel.

Broers, Michael. 1996. *Europe under Napoleon, 1799–1815.* Arnold.

Brose, Eric Dorn. 1998. *Technology and Science in the Industrializing Nations 1500–1914.* Humanities Press.

Brown, Howard G. 1997. "From organic society to security state: The war on brigandage in France, 1797–1802." *Journal of Modern History* 69: 661–695.

Burstin, Haim. 1989. "Travail, entreprise et politique à la Manufacture des Gobelins pendant la période révolutionnaire." In *La Révolution française et le développement du capitalisme*, ed. G. Gayot and J.-P. Hirsch. *Revue du Nord 5*, hors-serie.

Burstin, Haim. 1993. "Unskilled labor in Paris at the end of the eighteenth century." In *The Workplace before the Factory: Artisans and Proletarians, 1500–1800*, ed. T. Safley and L. Rosenband. Cornell University Press.

Burstin, Haim. 1994. "Problems of work during the Terror." In *The Terror*, ed. K. Baker (volume 4 of *The French Revolution and the Creation of Modern Political Culture*). Pergamon.

Burstin, Haim. 2005. *Une Révolution à l'oeuvre: Le faubourg Saint-Marcel (1789–1794).* Champ Vallon.

Butel, Paul. 1992. "Traditions et mutations du commerce français du règne de Louis XVI à la fin du Premier Empire." In *Fleurieu et la Marine de son temps*, ed. U. Bonnel. Economica.

Butel, Paul. 1993. *L'Économie française au XVIIIe siècle.* SEDES.

Cahen, Leon. 1939. "Une nouvelle interprétation du traité Franco-Anglais de 1786–1787." *Revue Historique* 185, no. 2: 257–285.

Calonne, Albéric de. 1899–1906. *Histoire de la ville d'Amiens*, three volumes. Piteux Frères.

Chabot, Hugues. 2000. "Le Tribunal de la science. Les Rapports négatifs à l'Académie des Sciences comme illustrations d'un scientifiquement (in)correct (1795–1835)." *Annales historiques de la Révolution française* 320: 173–182.

Chandler, Alfred D., Jr. 1991. "Creating competitive capability: Innovation and investment in the United States, Great Britain, and Germany from the 1870s to World War I." In *Favorites of Fortune: Technology, Growth, and Economic Development since the Industrial Revolution*, ed. P. Higonnet et al. Harvard University Press.

Chapman, Stanley D. 1967. *The Early Factory Masters: The Transition to the Factory System in the Midlands Textile Industry.* David & Charles.

Chaptal, Jean-Antoine. 1893. *Mes Souvenirs sur Napoléon par le Comte Chaptal publiés par son arrière-petit-fils le vicomte Antoine Chaptal.* Plon.

Chassagne, Serge. 1980. *Oberkampf: Un entrepreneur capitaliste au Siècle des Lumières.* Aubier.

Chassagne, Serge. 1991. *Le coton et ses patrons: France, 1760–1840.* Éditions de l'école des hautes études en sciences sociales.

Chaudron, Émile. 1923a. *L'Assistance publique à Troyes à la fin de l'ancien régime et pendant la Révolution 1770–1800.* Éditions de la "vie universitaire."

Chaudron, Émile. 1923b. *La Grande peur en Champagne méridionale.* Éditions de la "vie universitaire."

Chenut, Helen Harden. 2005. *The Fabric of Gender: Working-Class Culture in Third Republic France, Troyes 1880–1939.* Pennsylvania State University Press.

Christie, Ian R. 1982. *Wars and Revolutions: Britain, 1760–1815.* Harvard University Press.

Clark, J. C. D. 1985. *English Society 1688–1832: Ideology, Social Structure and Political Practice during the Ancien Regime.* Cambridge University Press.

Clark, William, Jan Golinski, and Simon Schaffer, eds. 1999. *The Sciences in Enlightened Europe.* University of Chicago Press.

Clause, Georges. 1974. Le Département de la Marne sous le Consulat et l'Empire (1800–1815). Doctoral thesis, Université de Paris IV.

Colley, Linda. 1992. *Britons: Forging the Nation, 1707–1837.* Yale University Press.

Colmont, Achille de. 1855. *Histoire des Expositions des produits de l'industrie française.* Guillaumin.

Colomès, André. 1943. *Les Ouvriers du Textile dans la Champagne Troyenne 1730–1852.* Domat-Montchrestien.

Coornaert, Émile. 1941. *Les corporations en France avant 1789.* Gallimard.

Costaz, Claude-Anthelme. 1843. *Histoire de l'administration en France de l'agriculture, des arts utiles, du commerce, des manufactures, des subsistances, des mines et des usines, accompagnée d'observations et de vues, et terminée par l'exposé des moyens qui ont amené le grande essor pris par l'industrie française, depuis la Révolution,* third edition. Bouchad-Huzard.

Crafts, Nicholas F. R. 1983. "British economic growth, 1700–1831: A review of the evidence." *Economic History Review* 36: 177–199.

Crafts, Nicholas F. R. 1985. *British Economic Growth during the Industrial Revolution.* Clarendon.

Crafts, Nicholas F. R. 1998. "Forging ahead and falling behind: The rise and relative decline of the first industrial nation." *Journal of Economic Perspectives* 12, no. 2: 193–210.

Crook, Malcolm. 1991. *Toulon in War and Revolution: From the Ancien Régime to the Restoration, 1750–1820.* St. Martin's Press.

Crosland, Maurice, ed. 1967. *The Society of Arcueil: A View of French Science at the Time of Napoleon I.* Harvard University Press.

Crosland, Maurice, ed. 1969. *Science in France in the Revolutionary Era.* Harvard University Press.

Crouzet, François. 1958. *L'économie britannique et le blocus continental (1806–1813).* Presses Universitaires de France.

Crouzet, François. 1967. "England and France in the eighteenth century: A comparative analysis of two economic growths." In *The Causes of the Industrial Revolution in England*, ed. R. Hartwell. Methuen.

Crouzet, François. 1990. *Britain Ascendant: Comparative Studies in Franco-British Economic History*. Cambridge University Press.

Crouzet, François. 1996. "France." In *The Industrial Revolution in National Context: Europe and the USA*, ed. M. Teich and R. Porter. Cambridge University Press.

Crowston, Clare H. 2000. "Engendering the guilds: Seamstresses, tailors and the clash of corporate identities in Old Regime France." *French Historical Studies* 23, no. 2: 335–392.

Crowston, Clare H. 2001. *Fabricating Women: The Seamstresses of Old Regime France, 1675–1791*. Duke University Press.

Cullen, Louis. 2000. "La crise économique de la fin de l'Ancien Régime." In *L'économie française du XVIIIe au XXxe siècle*, ed. J.-P. Poussou. Presses Universitaires de France.

Dakin, Douglas. 1939. *Turgot and the Ancien Régime in France*. Methuen.

Daly, Gavin. 2001. *Inside Napoleonic France: State and Society in Rouen, 1800–1815*. Ashgate.

Darbot, Jean. 1979. *La Trinité, première manufacture de bas au métier de Troyes*. CDDP.

Dardel, Pierre. 1966. *Commerce et navigation à Rouen et au Havre au XVIIIème siècle: Rivalité croissante entre ces deux ports—La Conjoncture*. Société libre d'émulation de la Seine-Maritime.

Darnton, Robert. 1968. *Mesmerism and the End of Enlightenment in France*. Harvard University Press.

Darvall, Francis O. 1934. *Popular Disturbances and Public Order in Regency England*. Oxford University Press.

Daumas, Maurice. 1968. *Histoire générale des techniques*, volume 3: *L'Expansion du machinisme 1725–1860*. Quadrige/Presses Universitaires de France.

Daunton, Martin J. 1995. *Progress and Poverty: An Economic and Social History of Britain 1700–1850*. Oxford University Press.

Daviet, Jean-Pierre. 1993. *L'économie préindustrielle 1750–1840*. La Découverte.

Davis, Natalie Zemon. 1975. *Society and Culture in Early Modern France*. Stanford University Press.

Day, Charles. 1987. *Education for the Industrial World: The Ecoles d'Arts et Métiers and the Rise of French Industrial Engineering*. MIT Press.

Deane, Phyllis, and W. A. Cole. 1967. *British Economic Growth 1688–1959*, second edition. Cambridge University Press.

Delécluse, Jacques. 1985. *Les Consuls de Rouen, Marchands d'hier, entrepreneurs d'aujourd'hui: Histoire de la Chambre de Commerce et d'Industrie de Rouen des origines à nos jours*. Éditions du P'tit Normand.

Delécluse, Jacques. 1989. *Une polemique industrielle à la veille de la Révolution de 1789: Les marchands rouennais et l'activité textile face au traité de commerce avec l'Angleterre.* Cahiers des Études Normandes 3: 1–93.

de Place, Dominique. 1988. "Le Bureau de consultation pour les Arts, Paris 1791–1796." *History and Technology* 5: 139–178.

Depping, Guillaume. 1893. *La Première exposition des produits de l'industrie française en l'an VI (1798).* Alphonse Picard.

Desan, Suzanne. 1989. "Crowds, community, and ritual in the work of E. P. Thompson and Natalie Davis." In *The New Cultural History,* ed. L. Hunt. University of California Press.

Descreux, Denis. 1868. *Notices biographiques stéphanois.* Constantin.

Desgraves, Louis, and Georges Dupeux, eds. 1969. *Bordeaux au XIXe siècle.* Delmas.

de Vries, Jan. 1994. "The Industrial Revolution and the Industrious Revolution." *Journal of Economic History* 54, no. 2: 249–270.

Dewerpe, Alain, and Yves Gaulupeau. 1990. *La Fabrique des prolétaires: Les ouvriers de la manufacture d'Oberkampf à Jouy-en-Josas (1760–1815).* Presses de l'École normale supérieure.

Deyon, Pierre. 2000. "L'industrie amiénoise au XIXe siècle et les séductions du protectionnisme." *Revue du Nord* 82, no. 334: 91–102.

Deyon, Pierre, and Philippe Guignet. 1980. "The royal manufactures and economic and technological progress in France before the industrial revolution." *Journal of European Economic History* 9: 611–632.

d'Hauterive, Ernest. 1908–1963. *La Police secrète du premier empire.* Perrin.

Dhombres, Jean, ed. 1989. *Rapports à l'Empereur sur le progrès des sciences, des lettres et des arts depuis 1789,* volume 1. Belin.

Dhombres, Jean, ed. 1991. *La Bretagne des savants et des ingénieurs 1750–1825.* Éditions Ouest-France.

Dhombres, Nicole, and Jean Dhombres. 1989. *Naissance d'un nouveau pouvoir: sciences et savants en France 1793–1824.* Payot.

Dickinson, H. T. 1977. *Liberty and Property: Political Ideology in Eighteenth Century Britain.* Holmes and Meier.

Dobson, Rodney. 1980. *Masters and Journeymen: A Pre-history of Industrial Relations 1717–1800.* Croom Helm.

Donaghay, Marie. 1978. "Calonne and the Anglo-French Commercial Treaty of 1786." *Journal of Modern History* 50, no. 3: 1157–1184.

Donaghay, Marie. 1982. "Textiles and the Anglo-French Commercial Treaty of 1786." *Textile History* 13, no. 3: 205–224.

Donaghay, Marie. 1984. "The best laid plans: French execution of the Anglo-French commercial treaty of 1786." *European History Quarterly* 14: 401–422.

Donaghay, Marie. 1990. "The exchange of products of the soil and industrial goods in the Anglo-French commercial treaty of 1786." *Journal of European Economic History* 19, no. 2: 377–401.

Doyle, William. 1988. *Origins of the French Revolution*, second edition. Oxford University Press.

Doyle, William. 1989. *The Oxford History of the French Revolution*. Clarendon.

Doyle, William. 1999. *Origins of the French Revolution*, third edition. Oxford University Press.

Doyon, André, and Lucien Liaigre. 1963. "L'hôtel de Mortagne après la mort de Vaucanson (1782–1837)." *Histoire des enterprises* 11: 5–23.

Dubuc, André. 1975. "L'Industrie textile en haute normandie au cours de la Révolution et de l'Empire." In *Le Textile en Normandie: Études diverses*, ed. A. Dubuc. Société libre d'émulation de la Seine-Maritime.

Dufraisse, Roger. 1961. "Contrebandiers normands sur les bors du rhin à l'époque napoléonienne." *Annales de Normandie* 9, no. 3: 209–232.

Dufraisse, Roger. 1966. "Régime douanier, blocus, système continental: Essai de mise au point." *Revue d'Histoire économique et sociale* 44: 518–543.

Dunham, Arthur Louis. 1955. *The Industrial Revolution in France 1815–1848*. Exposition Press.

Egret, Jean. 1975. *Necker: Ministre de Louis XVI 1776–1790*. Honoré Champion.

Ellis, Geoffrey. 1981. *Napoleon's Continental Blockade: The Case of Alsace*. Clarendon.

Engrand, Charles. 1986. "Recherche, réalisation et rupture des équilibres politiques, religieux et culturels (XVIIe–XVIIIe siècles)." In *Histoire d'Amiens*, ed. R. Hubscher. Privat.

Erdman, David V. 1986. *Commerce des lumières: John Oswald and the British in Paris, 1790–1793*. University of Missouri Press.

Evans, Eric J. 1984. *The Forging of the Modern State: Early Industrial Britain 1783–1870*. Longman.

Evrard, Fernand. 1947. "Les ouvriers du textile dans la région rouennaise (1789–1802)." *Annales historiques de la Révolution française* 106, October–December: 333–352.

Fairchild, Cissie. 1993a. "Consumption in early modern Europe: A review article." *Comparative Studies in Society and History* 35, no. 4: 850–858.

Fairchild, Cissie. 1993b. "The production and marketing of *populuxe* goods in eighteenth-century Paris." In *Consumption and the World of Goods*, ed. J. Brewer and R. Porter. Routledge.

Faure, Pétrus. 1956. *Histoire du mouvement ouvrier dans le département de la Loire*. Dumas.

Ferguson, Dean T. 2000. "The body, the corporate idiom, and the police of the unincorporated worker in early modern Lyons." *French Historical Studies* 23, no. 4: 546–575.

Flammeront, Jules, ed. 1898. *Remonstrances du Parlement de Paris au XVIIIe siècle*, volume 3. *1768–1788*. Imprimerie National.

Floud, Roderick, and Donald McCloskey, eds. 1994. *The Economic History of Britain since 1700*, second edition, volume 1: *1700–1860* Cambridge University Press.

Fontanon, Claudine. 1998. "Le Conservatoire national des arts et métiers (1794–1820)." *La France n'est-elle pas douée pour l'industrie?* ed. L. Bergeron and P. Bourdelais. Belin.

Forster, Robert. 1980. *Merchants, Landlords, Magistrates: The Depont Family in Eighteenth-Century France*. Johns Hopkins University Press.

Fournial, Étienne, ed. 1976. *Saint-Étienne: Histoire de la Ville et de ses Habitants*. Horvath.

Fox, Robert, ed. 1996. *Technological Change*. Harwood.

Fox, Robert, and Anna Guagnini, eds. 1993. *Education, Technology and Industrial Performance in Europe, 1850–1939*. Cambridge University Press.

Fox-Genovese, Elisabeth. 1976. *The Origins of Physiocracy: Economic Revolution and Social Order in Eighteenth-Century France*. Cornell University Press.

Furet, François. 1981. *Interpreting the French Revolution*. Cambridge University Press.

Furet, François. 1983. "Beyond the *Annales*." *Journal of Modern History* 55, no. 3: 389–410.

Galley, Jean-Baptiste. 1903. *L'Élection de Saint-Étienne à la fin de l'ancien régime*. Ménard.

Gallinato, Bernard. 1992. *Les Corporations à Bordeaux à la fin de l'ancien régime: Vie et mort d'un mode d'organisation du travail*. Presses Universitaires de Bordeaux.

Garçon, Anne-Françoise. 1998. *Mine et métal 1780–1880: Les non-ferreux et l'industrialisation*. Presses Universitaires de Rennes.

Garden, Maurice. 1970. *Lyon et les lyonnais au XVIIIe siècle*. Société d'édition "Les Belles-Lettres."

Garrioch, David. 1986. *Neighbourhood and Community in Paris, 1740–1790*. Cambridge University Press.

Gayot, Gérard. 1998. *Les draps de Sedan 1646–1870*. Éditions de l'École des Hautes Études en Sciences Sociales.

Gayot, Gérard. 2001. "Les 'Ouvriers les plus nécessaires' sur le marché du travail des manufactures des draps aux XVIIe–XVIIIe siècles." In *Les ouvriers qualifiés de l'industrie*, ed. G. Gayot and P. Minard. Imprimerle de l'Université Charles-de-Gaulle-Lille.

Gayot, Gérard, and Jean-Pierre Hirsch, eds. 1989. *La Révolution française et le développement du capitalisme*. *Revue du Nord*. Collection histoire.

Gerbaux, Fernand, and Charles Schmidt, eds. 1906. *Procès-Verbaux des comités d'agriculture et de commerce de la constituante de la législative et de la convention*. Imprimerie Nationale.

Gerschenkron, Alexander. 1962. *Economic backwardness in Historical Perspective: a Book of Essays*. Belknap.

Gille, Bertrand. 1961. *Le Conseil général des manufactures (Inventaire analytique des procès-verbaux) 1810–1829*. SEVPEN.

Gillispie, Charles Coulston. 1980. *Science and Polity in France at the End of the Old Regime*. Princeton University Press.

Gillispie, Charles Coulston. 1997. *Pierre-Simon Laplace (1749–1827): A Life in Exact Science*. Princeton University Press.

Gillispie, Charles Coulston. 2004. *Science and Polity in France: The Revolutionary and Napoleonic Years*. Princeton University Press.

Gossez, Rémi. 1984. *Un ouvrier en 1820: Manuscrit inédit de Jacques Étienne Bédé*. Presses Universitaires de France.

Gotteri, Nicole. 1997–2004. *La Police secrète du premier empire*. Honoré Champion.

Gras, Louis-Joseph. 1900. *Histoire de la Chambre consultative des arts et manufactures de Saint-Étienne (1804–1833)*. Théolier.

Gras, Louis-Joseph. 1906. *Histoire de la Rubanerie et des Industries de la Soie à Saint-Étienne et dans la région stéphanoise*. Théolier.

Gras, Louis-Joseph. 1908. *Histoire Économique de la Métallurgie de la Loire*. Théolier.

Greasley, David, and Les Oxley. 1997. "Endogenous growth or 'big bang': Two views of the First Industrial Revolution." *Journal of Economic History* 57, no. 4: 935–949.

Greenberg, Dolores. 1982. "Reassessing the power patterns of the Industrial Revolution: An Anglo-American comparison." *American Historical Review* 87, no. 5: 1237–1261.

Grenier, Jean-Yves. 1996. *L'économie d'Ancien Régime: Un monde de l'échange et de l'incertitude*. Albin Michel.

Gross, Jean-Pierre. 1997. *Fair Shares for All: Jacobin Egalitarianism in Practice*. Cambridge University Press.

Guignet, Philippe. 1977. *Mines, Manufactures et Ouvriers du Valenciennois au XVIIIe siècle*. Arno.

Gullickson, Gay L. 1986. *The Spinners and Weavers of Auffay: Rural Industry and the Sexual Division of Labor in a French Village, 1750–1850*. Cambridge University Press.

Habakkuk, H. J. 1962. *American and British Technology in the Nineteenth Century: The Search for Labour-Saving Inventions.* Cambridge University Press.

Hafter, Daryl M. 1984. "The business of invention in the Paris Industrial Exposition of 1806." *Business History Review* 58: 317–335.

Hafter, Daryl M., ed. 1995. *European Women and Preindustrial Craft.* Indiana University Press.

Hafter, Daryl M., ed. 1997. "Female masters in the ribbonmaking guild of eighteenth-century Rouen." *French Historical Studies* 20, no. 1: 1–14.

Hardman, John. 1993. *Louis XVI.* Yale University Press.

Hardman, John. 1995. *French Politics 1774–1789: From the Accession of Louis XVI to the Fall of the Bastille.* Longman.

Hardman, John, and Munro Price, eds. 1998. *Louis XVI and the comte de Vergennes: correspondance 1774–1787.* Voltaire Foundation.

Harley, C. Knick. 1999. "Reassessing the Industrial Revolution: A macro view." In *The British Industrial Revolution: An Economic Perspective,* second edition, ed. J. Mokyr. Westview.

Harley, C. Knick, and N. F. R. Crafts. 2000. "Simulating the two views of the British Industrial Revolution." *Journal of Economic History* 603, no. 3: 819–841.

Harris, J. R. 1992. *Essays in Industry and Technology.* Variorum.

Harris, J. R. 1998. *Industrial Espionage and Technology Transfer: Britain and France in the Eighteenth Century.* Ashgate.

Harris, Robert D. 1979. *Necker: Reform Statesman of the Ancien Régime.* University of California Press.

Harrison, John Fletcher Clews. 1961. *Learning and Living, 1790–1860: A History of the Adult Education Movement.* Routledge and Kegan Paul.

Hay, Douglas. 1975. "Property, authority and the criminal law." In *Albion's Fatal Tree: Crime and Society in Eighteenth-Century England,* ed. D. Hay. A. Lane.

Heywood, Colin. 1974. An Economic and Social History of Troyes and the Department of the Aube under the July Monarchy, 1830–1848. Doctoral dissertation, University of Reading.

Heywood, Colin. 1992. *The Development of the French Economy, 1750–1914.* Cambridge University Press.

Hilaire-Pérez, Liliane. 2000. *L'invention technique au siècle des Lumières.* Albin Michel.

Hills, Richard L. 1970. *Power in the Industrial Revolution.* Manchester University Press.

Hills, Richard L. 1986. "Steam and waterpower: Differences in transatlantic approach." In *The World of the Industrial Revolution: Comparative and International Aspects of Industrialization,* ed. R. Weible. Museum of American Textile History.

Hirsch, Jean-Pierre. 1991. *Les deux rêves du commerce: Entreprise et institution dans la région lilloise (1760–1860)*. Éditions de l'École des hautes études en sciences sociales.

Hirsch, Jean-Pierre. 1993. "L'effet Le Chapelier dans la pratique et les discours des entrepreneurs français." In *Naissance des libertés économiques: Le décret d'Allard et la loi Le Chapelier*, ed. A. Plessis. Institut d'histoire de l'industrie.

Hirsch, Jean-Pierre, and Philippe Minard. 1998a. "'Laissez-faire nous et protégez-nous beaucoup': Pour une histoire des pratiques institutionelles dans l'industrie française, XVIIIe–XIXe siècles." In *La France n'est-elle pas douée pour l'industrie?* ed. L. Bergeron and P. Bourdelais. Belin.

Hirschman, Albert O. 1995. *A Propensity to Self-Subversion*. Harvard University Press.

Hobsbawm, Eric. 1962. *The Age of Revolution, 1789–1848*. New American Library.

Hobsbawm, Eric. 1964. "The machine breakers." In *Laboring Men: Studies in the History of Labour*. Anchor Books.

Hoffman, Philip T. 1996. *Growth in a Traditional Society: The French Countryside 1450–1815*. Princeton University Press.

Hoffman, Philip T., and Jean-Laurent Rosenthal. 2000. "New work in French economic history." *French Historical Studies* 23, no. 3: 423–454.

Hoffman, Walther G. 1955. *British Industry, 1700–1950*. Blackwell.

Hogge, George, ed. 1861. *The Journal and Correspondence of William, Lord Auckland*. Richard Bentley.

Hollander, Paul, and Pierre Pageot. 1989. *La Révolution française dans le Limousin et la Marche*. Privat.

Horn, Jeff. 1995. "The limits of centralization: Municipal politics in Troyes during the L'Averdy reforms." *French History* 9, no. 2: 153–179.

Horn, Jeff. 2002. "Building the New Regime: Founding the Bonapartist state in the Department of the Aube." *French Historical Studies* 25, no. 2: 225–263.

Horn, Jeff. 2004. *«Qui parle pour la nation?» Les élections et les élus de la Champagne méridionale, 1765–1830*. Société des études robespierristes.

Horn, Jeff. 2005. "Machine-breaking in England and France during the Age of Revolution." *Labour/Le Travail* 55, no. 2: 143–166.

Horn, Jeff. 2006a. "La reception normand des négociations commerciales avec l'Angleterre 1801–2." In *Les traités et les relations entre les peuples. Constructions, enjeux et réceptions: le cas franco-anglais (1713–1802)*, ed. P. Dupuy and R. Mornieux. Presses Universitaires de Rouen.

Horn, Jeff. 2006b. "Privileged enclaves: Entrepreneurial opportunities in eighteenth-century France." *Proceedings of the Western Society for French History* (in press).

Horn, Jeff, and Margaret Jacob. 1998. "Jean-Antoine Chaptal and the cultural roots of French industrialization." *Technology and Culture* 39, no. 4: 671–698.

Hubscher, Ronald, ed. 1986. *Histoire d'Amiens*. Privat.

Hueckel, Glenn Russell. 1985. *The Napoleonic Wars and their Impact on Factor Returns and Output Growth in England, 1793–1815*. Garland.

Hufton, Olwen H. 1974. *The Poor of Eighteenth-Century France 1750–1789*. Clarendon.

Hunt, Lynn Avery. 1978. *Revolution and Urban Politics in Provincial France: Troyes and Reims, 1786–1790*. Stanford University Press.

Hunt, Lynn Avery. 1986. "French history in the last twenty years: The rise and fall of the Annales paradigm." *Journal of Contemporary History* 21: 209–224.

Hunter, Louis C. 1975. "Waterpower in the century of the steam engine." In *America's Wooden Age*, ed. B. Hindle. Sleepy Hollow Restorations.

Inizan, Guillaume. 1996–97. L'Industrie en Seine Inférieure pendant la Révolution et l'Empire. Mémoire de Maîtrise, Université de Rouen.

Inkster, Ian. 1991. *Science and Technology in History: An Approach to Industrial Development*. Rutgers University Press.

Isambert, François-Antoine, et al. 1833. *Recueil général des anciennes lois françaises depuis l'année 420 jusqu'à la Révolution de 1789*. Imprimerie nationale.

Jacob, Margaret C. 1997. *Scientific Culture and the Making of the Industrial West*. Oxford University Press.

Jacob, Margaret C. 2000. "Commerce, industry, and the laws of Newtonian science: Weber revisited and revised." *Canadian Journal of History* 35: 272–292.

Jacob, Margaret C., and Larry Stewart. 2004. *Practical Matter: Newton's Science in the Service of Industry and Empire, 1687–1851*. Harvard University Press.

Jeremy, David J. 1977. "Damming the flood: British government efforts to check the outflow of technicians and machinery, 1780–1843." *Business History Review* 51: 1–34.

Johnson, Christopher H. 1993. "Capitalism and the state: Capital accumulation and proletarianization in the Languedocian woolens industry, 1700–1789." In *The Workplace before the Factory*, ed. T. Safley and L. Rosenband. Cornell University Press.

Johnson, Christopher H. 1995. *The Life and Death of Industrial Languedoc 1700–1920: The Politics of Deindustrialization*. Oxford Univeristy Press.

Jones, Colin. 1998. "Bourgeois revolution revivified: 1789 and social change." In *The French Revolution*, ed. G. Kates. Routledge.

Jones, Colin. 2002. *The Great Nation: France from Louis XV to Napoleon 1715–99*. Columbia University Press.

Jones, Peter M. 1995. *Reform and Revolution in France: The Politics of Transition, 1774–1791.* Cambridge University Press.

Jones, Peter M. 2003. *Politics and Rural Society: The Southern Massif Central c 1750–1880.* Cambridge University Press.

Jouvenel, Bertrand de. 1942. *Napoléon et l'économie dirigée: le blocus continental.* Les Éditions de la toison d'or.

Julia, Dominique. 1981. *Les Trois Couleurs du tableau noir: La Révolution.* Belin.

Kaplan, Steven L. 1979. "Réflexions sur la police du monde du travail, 1700–1815." *Revue historique* 261, no. 1: 17–77.

Kaplan, Steven L. 1994. "Guilds, 'false workers,' and the Faubourg Saint-Antoine." In *Edo and Paris,* ed. J. McClain et al. Cornell University Press.

Kaplan, Steven L. 2001. *La fin des corporations.* Fayard.

Kaplow, Jeffry. 1964. *Elbeuf during the Revolutionary Period: History and Social Structure.* Johns Hopkins University Press.

Kaplow, Jeffry. 1972. *The Names of Kings: The Parisian Laboring Poor in the Eighteenth Century.* Basic Books.

Kargon, Robert H. 1977. *Science in Victorian Manchester: Enterprise and Expertise.* Johns Hopkins University Press.

Kemp, Tom. 1971. *Economic Forces in French History: An Essay on the Development of the French Economy 1760–1914.* Dennis Dobson.

Kennedy, Michael L. 1982, 1988, 2000. *The Jacobin Clubs in the French Revolution.* Princeton University Press.

Kisch, Herbert. 1989. *From Domestic Manufacture to Industrial Revolution: The Case of the Rhineland Textile Districts.* Oxford University Press.

Krantz, Frederick, ed. 1988. *History from Below: Studies in Popular Protest and Popular Ideology.* Blackwell.

Labourdette, Jean-François. 1990. *Vergennes: Ministre principal de Louis XVI.* Desjonquères.

Lacour-Gayet, Robert. 1963. *Calonne: Financier, Réformateur, Contre-révolutionnaire 1734–1802.* Hachette.

Laissus, Yves, ed. 1989. *Rapports à l'Empereur sur le progrès des sciences, des lettres et des arts depuis 1789,* volume 2. Belin.

Landes, David S. 1969. *The Unbound Prometheus: Technological change and industrial development in Western Europe from 1750 to the present.* Cambridge University Press.

Landes, David S. 1999a. "The fable of the dead horse; or, the Industrial Revolution revisited." In *The British Industrial Revolution: An Economic Perspective,* second edition, ed. J. Mokyr. Westview.

Landes, David S. 1999b. *The Wealth and Poverty of Nations: Why Some Are So Rich And Some Are So Poor.* Norton.

Lanza, Janine. 1996. Family Making and Family Breaking: Widows in Early Modern France. PhD thesis, Cornell University.

Lazonick, William. 1981. "Production relations, labor productivity, and choice of technique: British and U.S. cotton spinning." *Journal of Economic History* 41, September: 491–516.

Lazonick, William. 1991. "What happened to the theory of economic development." In *Favorites of Fortune*, ed. P. Higonnet et al. Harvard University Press.

Lefebvre, Georges. 1954. *Études sur la Révolution française.* Presses Universitaires de France.

Lefebvre, Georges. 1964. *The French Revolution from 1793 to 1799.* Columbia University Press.

Lefebvre, Georges. 1969a. *Napoleon: From 18 Brumaire to Tilsit, 1799–1807.* Columbia University Press.

Lefebvre, Georges. 1969b. *Napoleon: From Tilsit to Waterloo 1807–1815.* Columbia University Press.

Lefebvre, Georges. 1973. *The Great Fear of 1789: Rural Panic in Revolutionary France.* Vintage.

Legrand, Robert. 1995. *Révolution et Empire en Picardie: Économie et finances.* F. Paillart.

Lemarchand, Guy, and Claude Mazauric. 1989. "Le concept de la liberté d'entreprise dans une région de haut développement économique: la Haute-Normandie 1787–1800." In *La Révolution française et le développement du capitalisme*, ed. G. Gayot and J.-P. Hirsch. Collection histoire.

Le Normand, Louis-Sébastien le, and J.-G.-V. de Moléon. 1824. *Description des expositions des produits de l'industrie française, faites à Paris depuis leur origine jusqu'à celle de 1819 inclusivement.* Bachelier.

Le Roux, Thomas. 1996. *Le Commerce intérieur de la France à la fin du XVIIIe siècle: Les contrastes économiques régionaux de l'espace française à travers les archives du Maximum.* Nathan.

Lérue, J.-A. De. 1875. *Notice sur Descroizilles (François-Antoine-Henri) Chimiste, né à Dieppe et sur les membres de sa famille.* Ch.-F. Lapierre.

Levasseur, Émile. 1901. *Histoire des classes ouvrières et de l'industrie en France avant 1789*, second edition. Arthur Rousseau.

Levasseur, Émile. 1903. *Histoire des classes ouvrières et de l'industrie en France de 1789 à 1870*, second edition. Arthur Rousseau.

Lévy-Leboyer, Maurice. 1964. *Les Banques européenes et l'industrialisation internationale dans la première moitié du XIXe siècle.* Presses Universitaires de France.

Lévy-Leboyer, Maurice. 1968. "La croissance économique en France au XIXe siècle: Résultats préliminaires." *Annales E. S. C.* 23: 788–807.

Lewis, Gwynne. 1993. *The Advent of Modern Capitalism in France 1770–1840: The Contribution of Pierre-François Tubeuf.* Clarendon.

Liu, Tessie P. 1994. *The Weaver's Knot: The Contradictions of Class Struggle and Family Solidarity in Western France, 1750–1914*. Cornell University Press.

Lomuller, L. M. 1978. *Guillaume Ternaux 1763–1833: créateur de la première intégration industrielle française*. Les Éditions de la Cabro d'Or.

Lorion, André. 1968. "Les Expositions de l'Industrie française à Paris (1798–1806)." *Revue de l'institut napoléon* 108: 125–131.

Louat, André, and Jean-Marc Servat. 1995. *Histoire de l'industrie française jusqu'en 1945: Une industrialisation sans révolution*. Bréal.

Lucas, Colin. 1973. *The Structure of the Terror: the example of Javogues and the Loire*. Oxford University Press.

"Luddites." 1971. "The Luddites in the period 1779–1830." In *The Luddites and Other Essays*, ed. L. Munby. Michael Katanka.

Lyons, Martyn. 1994. *Napoleon Bonaparte and the Legacy of the French Revolution*. Palgrave.

MacKenzie, Donald. 1998. *Knowing Machines: Essays on Technical Change*. MIT Press.

Macleod, Christine. 1988. *Inventing the Industrial Revolution: The English Patent System, 1660–1800*. Cambridge University Press.

Macleod, Christine. 2004. "The European origins of British technological predominance." In *Exceptionalism and Industrialisation*, ed. L. Prados de la Escosura. Cambridge University Press.

Magraw, Roger. 1992a. *The Age of Artisan Revolution, 1815–1871*. Blackwell.

Magraw, Roger. 1992b. *A History of the French Working Class*. Blackwell.

Maillet, Sébastien. 1996. Fonderie, métallurgie et manufactures d'armes à Rouen et dans sa région sous la Révolution française. Mémoire de maîtrise, Université de Rouen.

Manuel, Frank E. 1938. "The Luddite movement in France." *Journal of Modern History* 10, no. 2: 180–211.

Margairaz, Dominique. 1988. *Foires et marchés dans la France préindustrielle*. Éditions de l'EHESS.

Margairaz, Dominique. 2005. *François de Neufchâteau : biographie intellectuelle*. Publications de la Sorbonne.

Markovitch, Tihomir J. 1965–66. *L'industrie française de 1789 à 1964*. volume 4 of the *Histoire quantitative de l'économie française*. Institut de science économique appliquée.

Marseille, Jacques, and Dominique Margairaz, eds. 1988. *1789, au jour le jour: avec en supplément, l'almanach gourmand, l'almanach mondain, le regard de l'étranger*. A. Michel.

Marzagalli, Silvia. 1999. *Les Boulevards de la fraude: Le négoce maritime et le Blocus continental 1806–1813—Bordeaux, Hambourg, Livourne*. Presses universitaires de Septentrion.

Mathias, Peter. 1979. *The Transformation of England: Essays in the economic and social history of England in the eighteenth century*. Methuen.

Mathias, Peter. 1983. *The First Industrial Nation: An Economic History of Britain 1700–1914*, second edition. Routledge.

Mathias, Peter, and M. M. Postan, eds. 1978. *The Cambridge Economic History of Europe* VII: 1, *The Industrial Economies: Capital, Labour, and Enterprise: Britain, France, Germany, and Scandinavia*. Cambridge University Press.

Mauchamps, Herbinot de, ed. 1837. *Mémoires de M. Richard-Lenoir*. Delaunay.

Maudhuit, Bernard. 1957. Recherches sur le textile troyen au XVIIIe siècle d'après le fond Berthelin et Fromageot. Maîtrise d'Histoire Économique, Université de Reims.

May, Louis Philippe. 1939. "Une version inédite d'une allocution de Napoléon au sujet du Blocus continental." *Revue historique* 186: 264–272.

McCloy, Shelby T. 1952. *French Inventions of the Eighteenth Century*. Univesity of Kentucky Press.

Meyer, Jean. 1991. "Marine de guerre: Sciences et technologies (1750–1850)." In *La Bretagne des savants et des ingénieurs 1750–1825*, ed. J. Dhombres. Éditions Ouest-France.

Meyssonnier, Simone. 1989. "Aux origines de la science économique française: Le libéralisme égalitaire." In *La Révolution française et le développement du capitalisme*, ed. G. Gayot and J.-P. Hirsch. *Revue du Nord* 5 hors-série, Collection histoire.

Miller, Judith A. 1999. *Mastering the Market: The State and the Grain Trade in Northern France, 1700–1860*. Cambridge University Press.

Minard, Philippe. 1994. Les inspecteurs des manufactures en France, de Colbert à la Révolution. Doctoral thesis, Université de Paris I.

Minard, Philippe. 1998. *La fortune du colbertisme: État et industrie dans la France des Lumières*. Fayard.

Minard, Philippe. 2000. "Colbertism continued? The Inspectorate of Manufactures and strategies of exchange in eighteenth-century France." *French Historical Studies* 23, no. 3: 477–496.

Minard, Philippe. 2002. "État et économie en France après la Révolution: Quel libéralisme?" In *Terminée, la Révolution?* ed. M. Biard. Amis du Vieux Calais.

Moher, James. 1988. "From suppression to containment: Roots of trade union law to 1825." In *British Trade Unionism 1750–1850*, ed. J. Rule. Longman.

Mokyr, Joel. 1990. *The Lever of Riches: Technological Creativity and Economic Progress*. Oxford University Press.

Mokyr, Joel, ed. 1993. *The British Industrial Revolution: An Economic Perspective*. Westview.

Mokyr, Joel. 1994a. "Progress and inertia in technological change." In *Capitalism in Context*, ed. J. James and M. Thomas. University of Chicago Press.

Mokyr, Joel. 1994b. "Technological change, 1700–1830." In *The Economic History of Britain since 1700*, ed. R. Floud and D. McCloskey. Cambridge University Press.

Mokyr, Joel. 1999. "Editor's introduction: The New Economic History and the Industrial Revolution." In *The British Industrial Revolution*, second edition, ed. J. Mokyr. Westview.

Mokyr, Joel. 2002. *The Gifts of Athena: Historical Origins of the Knowledge Economy*. Princeton University Press.

Mokyr, Joel. 2005. "The intellectual origins of modern economic growth." *Journal of Economic History* 65, no. 2: 285–351.

Mollat, Michel, ed. 1979. *Histoire de Rouen*. Privat.

Mollien, François Nicholas. 1898. *Mémoires d'un ministre du trésor public 1780–1815*. Guillaumin.

Monnier, Raymonde. 1981. *Le faubourg Saint-Antoine, 1789–1815*. Société des études robespierristes.

Monnier, Raymonde. 2005. "Tableaux croisés chez Mercier et Rutlidge: Le *peuple* de Paris et le *plébéien* anglais." *Annales historiques de la Révolution française* 339, no. 1: 1–16.

Mori, Jennifer. 2000. *Britain in the Age of the French Revolution, 1785–1820*. Longman.

Morrisson, Christian, and Wayne Snyder. 2000. "The income inequality of France in historical perspective." *European Review of Economic History* 4: 59–93.

Mousnier, Roland E. 1979. *The Institutions of France under the Absolute Monarchy 1598–1789: Society and the State*. University of Chicago Press.

Murphy, Orville T. 1966. "DuPont de Nemours and the Anglo-French Commercial Treaty of 1786." *Economic History Review* 19, second series, no. 4: 541–564.

Murphy, Orville T. 1982. *Charles Gravier, Comte de Vergennes: French Diplomacy in the Age of Revolution: 1719–1787*. State University of New York Press.

Murphy, Orville T. 1998. *The Diplomatic Retreat of France and Public Opinion on the Eve of the French Revolution, 1783–1789*. Catholic University Press.

Musson, A. E., and Eric Robinson. 1969. *Science and Technology in the Industrial Revolution*. Manchester University Press.

Noiret, Charles. 1836. *Mémoires d'un Ouvrier Rouennais*. François.

Nuvolari, Alessandro. 2002. "The 'machine breakers' and the Industrial Revolution." *Journal of European Economic History* 31, no. 2: 393–426.

Nye, John V. C. 2002. "The importance of being late: French economic history, cliometrics, and the new institutional economics." *French Historical Studies* 23, no. 3: 423–439.

O'Brien, Patrick K. 1993. "Political preconditions for the Industrial Revolution." In *The Industrial Revolution and British Society*, ed. P. O'Brien and R. Quinault. Cambridge University Press.

O'Brien, Patrick K. 1994. "Central government and the economy, 1688–1815." In *The Economic History of Britain since 1700*, second edition, volume 1, ed. R. Floud and D. McCloskey. Cambridge University Press.

O'Brien, Patrick, and Caglar Keyder. 1978. *Economic Growth in Britain and France 1780–1914: Two Paths to the Twentieth Century*. Allen & Unwin.

O'Brien, Patrick K., and Roland Quinault, eds. 1993. *The Industrial Revolution and British Society*. Cambridge University Press.

O'Brien, Patrick K., Trevor Griffiths, and Philip Hunt. 1996. "Technological change during the First Industrial Revolution: The paradigm case of textiles, 1688–1851." In *Technological Change*, ed. R. Fox. Harwood.

Ogilvie, Sheilagh C., and Markus Cerman, eds. 1996. *European Proto-Industrialization*. Cambridge University Press.

Orth, John V. 1991. *Combination and Conspiracy: A Legal History of Trade Unionism, 1721–1906*. Clarendon.

Ozouf, Mona. 1988. *Festivals and the French Revolution*. Harvard University Press.

Palmer, Robert R. 1941. *Twelve Who Ruled: The Committee of Public Safety during the Terror*. Princeton University Press.

Palmer, Robert R. 1959. *The Age of the Democratic Revolution: A Political History of Europe and America, 1760–1800*. Princeton University Press.

Palmer, Robert R. 1985. *The Improvement of Humanity: Education and the French Revolution*. Princeton University Press.

Parker, Harold T. 1986. "Administrative activity." In *Proceedings of the 1984 Consortium on Revolutionary Europe, 1750–1850*, ed. H. Parker et al. University of Georgia Press.

Parker, Harold T. 1993. *An Administrative Bureau during the Old Regime: The Bureau of Commerce and Its Relations to French Industry from May 1781 to November 1783*. University of Delaware Press.

Patrick, Alison. 1972. *The Men of the First French Republic: Political Alignments in the National Convention of 1792*. Johns Hopkins University Press.

Payen, Jacques. 1969. *Capital et machine à vapeur au XVIIIe siècle: Les frères Périer et l'introduction en France de la machine à vapeur de Watt*. Mouton.

Péronnet, Michel, ed. 1988. *Chaptal*. Privat.

Perrot, Jean-Claude. 1977. *L'âge d'or de la statistique régionale française (an IV-1804)*. Société des études robespierristes.

Picard, Roger. 1910. *Les cahiers de 1789 et les classes ouvrières*. M. Rivière.

Pigeire, Jean. 1932. *La vie et l'oeuvre de Chaptal (1756–1832)*. Ses.

Pollard, Sidney. 1965. *The Genesis of Modern Management: A Study of the Industrial Revolution in Great Britain*. Penguin.

Pollard, Sidney. 1979. "Management and labor in Britain during the period of industrialization." In *Labor and Management: Proceedings of the Fourth Fiji Conference*, ed. K. Nakagawa. University of Tokyo Press.

Pollard, Sidney. 1999. *Labour History and the Labour Movement in Britain*. Ashgate.

Pomeranz, Kenneth. 2000. *The Great Divergence: China, Europe, and the Making of the Modern World Economy*. Princeton University Press.

Ponteil, Félix. 1966. *Les institutions de la France de 1814 à 1870*. Presses Universitaires de France.

Popkin, Jeremy, ed. 1999. *Panorama of Paris: Selections from* Le Tableau de Paris. Pennsylvania State University Press.

Postel-Vinay, Gilles. 1989. "A la recherche de la révolution économique dans les campagnes (1789–1815)." *Revue économique* 40: 1015–1046.

Poullain, A. C. 1905. *Analyses des délibérations de l'assemblée municipale et électorale du 16 Juillet au 4 Mars 1790 et du Conseil général de la Commune du 4 Mars 1790 au 25 Brumaire an IV (16 Novembre 1795)*. Lecerf.

Potofsky, Allan. 1993. The Builders of Modern Paris: The Organization of Labor from Turgot to Napoleon. Doctoral thesis, Columbia University.

Prados de la Escosura, Leandro, ed. 2004. *Exceptionalism and Industrialisation: Britain and Its European Rivals, 1688–1815*. Cambridge University Press.

Price, Munro. 1995. *Preserving the Monarchy: The comte de Vergennes, 1774–1787*. Cambridge University Press.

Price, Richard. 1999. *British Society 1680–1880: Dynamism, Containment and Change*. Cambridge University Press.

Price, Roger. 1987. *A Social History of Nineteenth-Century France*. Holmes & Meier.

Ramet, Henri, and Christian Cau. 1994. *Histoire de Toulouse*. Le Pérégrinateur.

Randall, Adrian. 1986. "The philosophy of Luddism: The case of the West of England woolen workers, ca. 1790–1809." *Technology and Culture* 27, no. 1: 1–17.

Randall, Adrian. 1988. "The industrial moral economy of the Gloucestershire weavers in the eighteenth century." In *British Trade Unionism 1750–1850*, ed. J. Rule. Longman.

Randall, Adrian. 1991. *Before the Luddites: Custom, Community and Machinery in the English Woollen Industry, 1776–1809*. Cambridge University Press.

Reddy, William H. 1984. *The Rise of Market Culture: The Textile Trade and French Society, 1750–1900*. Cambridge University Press.

Rémond, André. 1946. *John Holker. Manufacturier et grand fonctionnaire en France au XVIIIe siècle, 1719–1786*. Marcel Rivière.

Revel, Jacques. 1987. "Les corps et communautés." In *The Political Culture of the Old Regime*, ed. K. Baker. Pergamon.

Reynard, Pierre-Claude. 1999. "Early modern state and enterprise: Shaping the dialogue between the French monarchy and paper manufacturers." *French History* 13, no. 1: 1–25.

Reynard, Pierre-Claude. 2001. *Histoires de papier: la papeterie auvergnate et ses historiens*. Presses universitaires de Blaise-Pascal.

Richard, Camille. 1922. *Le Comité de Salut public et les fabrications de guerre sous la Terreur*. F. Rieder.

Ricommard, Jean. 1934. *La Bonneterie à Troyes et dans le département de l'Aube: Origines, évolution. caratèreres actuels*. Hachette.

Roche, Daniel. 1978. *Le siècle des lumières en province: académies et académiciens provinciaux, 1680–1789*. Mouton.

Roche, Daniel, ed. 1986. *Journal of My Life by Jacques-Louis Ménétra*. Columbia University Press.

Roche, Daniel. 1987. *The People of Paris: An Essay in Popular Culture in the Eighteenth Century*. University of California Press.

Roche, Daniel. 1991. *La culture des apparences: Une histoire du vêtement (XVIIe–XVIIIe siècle)*. Fayard.

Roche, Daniel. 2000. *History of Everyday Things: the Birth of Consumption in France, 1600–1800*. Cambridge University Press.

Rosanvallon, Pierre. 1992. "Physiocrats." In *Dictionnaire critique de la Révolution française: Idées*, ed. F. Furet and M. Ozouf. Flammarion.

Rose, J. Holland. 1908. "The Franco-British Commercial Treaty of 1786." *English Historical Review* 23, no. 3: 709–724.

Rosenband, Leonard N. 1997. "Jean-Baptiste Réveillon: A man on the make in Old Regime France." *French Historical Studies* 20, no. 3: 481–510.

Rosenband, Leonard N. 1998. "The perils of petty production: Pierre and Jean-Baptiste Serve of Chamalières." *Science in Context* 11, no. 1: 3–21.

Rosenband, Leonard N. 1999. "Social capital in the early Industrial Revolution." *Journal of Interdisciplinary History* 29, no. 3: 435–457.

Rosenband, Leonard N. 2000a. "The competitive cosmopolitanism of an Old Regime craft." *French Historical Studies* 23, no. 3: 455–476.

Rosenband, Leonard N. 2000b. *Papermaking in Eighteenth-Century France: Management, Labor and Revolution at the Montgolfier Mill, 1761–1805*. Johns Hopkins University Press.

Rosenband, Leonard N. 2004. "Comparing combination acts: French and English papermaking in the Age of Revolution." *Social History* 29, no. 2: 65–85.

Rosenband, Leonard N. 2005. "Never just business: David Landes, *The Unbound Prometheus*." *Technology and Culture* 46, no. 1: 168–176.

Rostow, W. W. 1960. *The States of Economic Growth*. Cambridge University Press.

Royle, Edward. 2000. *Revolutionary Britannia? Reflections on the threat of revolution in Britain 1789–1848*. Manchester University Press.

Rubinstein, W. D. 1993. *Capitalism, Culture and Decline in Britain, 1750–1990*. Routledge.

Rudé, George. 1964. *The Crowd in History: A Study of Popular Disturbances in France and England 1730–1848*. Wiley.

Rudé, George. 1970. *Paris and London in the Eighteenth Century: Studies in Popular Protest*. Viking.

Ruhlmann, Georges. 1948. *Les corporations, les manufactures et le travail libre à Abbeville au XVIIIe siècle*. Sirey.

Rule, John. 1981. *The Experience of Labour in Eighteenth Century English Industry*. St. Martin's Press.

Rule, John. 1986. *The Labouring Classes in Early Industrial England, 1750–1850*. Longman.

Rule, John. 1993. "Trade unions, the government and the French Revolution, 1789–1802." In *Protest and Survival: The Historical Experience—Essays for E. P. Thompson*, ed. J. Rule and R. Malcolmson. Merlin.

Rusnock, Andrea A. 2002. *Vital Accounts: Quantifying Health and Population in eighteenth Century England and France*. Cambridge University Press.

Sabel, Charles, and Jonathan Zeitlin. 1985. "Historical alternatives to mass production: Politics, markets and technology in nineteenth-century industrialization." *Past and Present* 108: 133–176.

Samuel, Ralph. 1992. "Mechanization and hand labour in industrializing Britain." In *The Industrial Revolution and Work in Nineteenth-Century Europe*, ed. L. Berlanstein. Routledge.

Sandberg, Lars C. 1969. "American rings and English mules: The role of economic rationality." *Quarterly Journal of Economics* 83: 25–43.

Saricks, Ambrose. 1965. *Pierre Samuel du Pont de Nemours*. University of Kansas Press.

Schmitt, Jean-Marie. 1980. *Aux Origines de la Révolution industrielle en Alsace: Investissements et relations sociales dans la vallée de Saint-Amarin au XVIIIe siècle*. Istra.

Schnetzler, Jacques. 1973. Les Industries et les hommes dans la région de St.-Étienne. Ph.D. thesis, Université de Lyon II.

Sée, Henri. 1930. "The Normandy Chamber of Commerce and the Commercial Treaty of 1786." *Economic History Review* 2: 308–313.

Sewell, William H., Jr. 1980. *Work and Revolution in France: The Language of Labor from the Old Regime to 1848.* Cambridge University Press.

Shapiro, Gilbert, and John Markoff. 1998. *Revolutionary Demands: A Content Analysis of the Cahiers de doléances of 1789.* Stanford University Press.

Shinn, Terry. 1992. "Science, Tocqueville, and the state: The organization of knowledge in modern France." *Social Research* 59, no. 3: 533–566.

Sibalis, Michael D. 1988. "Corporatism after the corporations: The debate on restoring the guilds under Napoleon I and the Restoration." *French Historical Studies* 15, no. 4: 718–730.

Sibalis, Michael D. n.d. "Parisian labour during the French Revolution" In Historical Papers/Communications Historiques: A Selection from the Papers presented at the Annual Meeting [of the Canadian Historical Association] Held at Winnipeg 1986.

Sickinger, Raymond L. 1999. "The coming of the French Revolution and the English textile trade, 1783–1792." In *Consortium on Revolutionary Europe,* ed. O. Connelly et al. Florida State University Press.

Simian, Charles. 1889. *François de Neufchâteau et les expositions.* A. Ghio.

Smith, Adam. 1775. *An Inquiry into the Nature and Causes of the Wealth of Nations.* W. Strahan and T. Cadell.

Smith, John Graham. 1979. *The Origins and Early Development of the Heavy Chemical Industry in France.* Clarendon.

Soboul, Albert. 1966. *Paysans, sans-culottes et Jacobins.* Clavreuil.

Soboul, Albert. 1980. *The Sans-Culottes, the Popular Movement, and Revolutionary Government, 1793–1794.* Princeton University Press.

Sonenscher, Michael. 1986. "Journeymen's migrations and workshop organization in eighteenth-century France." In *Work in France,* ed. S. Kaplan and C. Koepp. Cornell University Press.

Sonenscher, Michael. 1987. "Journeymen, the courts and the French trades 1781–1791." *Past and Present* 114: 77–109.

Sonenscher, Michael. 1989a. "Le droit du travail en France et en Angleterre à l'époque de la Révolution." In *La Révolution française et le développement du capitalisme,* ed. G. Gayot and J.-P. Hirsch. Collection histoire.

Sonenscher, Michael. 1989b. *Work and Wages: Natural law, politics and the eighteenth-century French trades.* Cambridge University Press,.

Stevenson, John. 1992. *Popular Disturbances in England 1700–1832,* second edition. Longman.

Stewart, Larry. 1992. *The Rise of Public Science: Rhetoric, Technology, and Natural Philosophy in Newtonian Britain, 1660–1750*. Cambridge University Press.

Stoianovich, Trian. 1976. *French Historical Method: The Annales Paradigm*. Cornell University Press.

Stone, Bailey. 1981. *The Parlement of Paris, 1774–1789*. University of North Carolina Press.

Szotstak, Rick. 1991. *The Role of Transportation in the Industrial Revolution: A Comparison of England and France*. McGill–Queen's University Press.

Tézenas du Montcel, Paul. 1903. *L'Assemblée du département de Saint-Étienne et sa Commission intermédiaire (8 Octobre 1787—21 Juillet 1790)*. Champion.

Tézenas du Montcel, Paul. 1952. *Le Forez sous la Terreur. Deux régicides: Claude Javogues et Noel Pointe, membres de la Convention*. Caveau Stéphanois.

Thaurin, Jacques-Michel. 1859. *Essai historique sur les expositions industrielles en général et en particulier sur celles qui ont eu lien à Rouen depuis 1803 avec des remarques critiques*. Librairie nouvelle.

Thermeau, Gérard. 2002. *À l'aube de la révolution industrielle: Saint-Étienne et son agglomération*. Université de Saint-Étienne.

Thomis, Malcolm I. 1970. *The Luddites: Machine-Breaking in Regency England*. Archon Books.

Thompson, E. P. 1963. *The Making of the English Working Class*. Vintage.

Thompson, E. P. 1971. "The moral economy of the English crowd in the eighteenth century." *Past and Present* 50: 71–136.

Thomson, J. K. J. 1982. *Clermont-de-Lodève: Fluctuations in the prosperity of a Languedocian cloth-making town*. Cambridge University Press.

Truant, Cynthia M. 1994. *The Rites of Labor: Brotherhoods of Compagnonnage in Old and New Regime France*. Cornell University Press.

Tudesq, André. "La Restauration. Renaissance et déceptions. 1969. In *Bordeaux au XIXe siècle*, ed. L. Desgraves and G. Dupeux. Delmas.

Tunzelmann, Nick von. 1994. "Technology in the early nineteenth century." In *The Economic History of Britain since 1700*, ed. R. Floud and D. McCloskey. Cambridge University Press.

Verley, Patrick. 1991. "La Révolution industrielle anglaise: Une révision (note critique)." *Annales ESC* 46, no. 3: 735–755.

Verley, Patrick. 1997. *La Révolution industrielle*. Gallimard.

Verney-Carron, Nicole. 1999. *Le ruban et l'acier: Les élites économiques de la région stéphanois au XIXe siècle (1815–1914)*. Université de Saint-Étienne.

Vernier, Jules-Joseph. 1909–1911. *Cahiers de doléances des bailliages de Troyes et de Bar-sur-Seine*. P. Nouel.

Vidalenc, Jean. 1958a. "Quelques remarques sur le rôle des Anglais dans la Révolution Industrielle en France particulièrement en Normandie, de 1750 à 1850." *Annales de Normandie* 8, no. 2: 273–290.

Vidalenc, Jean. 1958b. "Relations Economiques et Circulation en Normandie à la fin du Premier Empire (1810–1814)." *Annales de Normandie* 8, no. 4: 441–461.

Vidalenc, Jean. 1981. *Aspects de la Seine-Inférieure sous la Restauration 1814–1830*. CRDP.

Viennet, Odette. 1947. *Napoléon et l'industrie française: la crise de 1810–1811*. Plon.

Wells, Roger. 1983. *Insurrection: the British Experience, 1795–1803*. A. Sutton.

Wells, Roger. 1991. "English society and revolutionary politics in the 1790s: The case for insurrection." In *The French Revolution and British Popular Politics*, ed. M. Philp. Cambridge University Press.

Wiener, Martin. 1981. *English Culture and the Decline of the Industrial Spirit 1850–1890*. Cambridge University Press.

Williamson, Jeffrey E. 1984. "Why was British growth so slow during the Industrial Revolution?" *Journal of Economic History* 44: 687–712.

Woloch, Isser. 1988. "'Republican institutions,' 1797–1799." In *The Political Culture of the French Revolution*, ed. C. Lucas. Pergamon.

Wong, R. Bin. 1997. *China Transformed: Historical Change and the Limits of the European Experience*. Cornell University Press.

Wood, Ellen Meiksins. 1992. *The Pristine Culture of Capitalism: A Historical Essay on Old Regimes and Modern States*. Verso.

Woronoff, Denis. 1984. *L'industrie sidérurgique en France pendant la Révolution et l'Empire*. Éditions de l'EHESS.

Woronoff, Denis. 1994. *Histoire de l'industrie en France du XVIe siècle à nos jours*. Seuil.

Young, Arthur. 1969. *Travels in France during the years 1787, 1788 and 1789*, ed. J. Kaplow. Doubleday Anchor.

Maps

Old Regime France. Adapted by Richard A. Musal and Francisco Velez.

FRANCE IN 1790

········· Departmental Boundary
    . City or Town

0   40   80   120   160   200 Miles

France in 1790. Adapted by Richard A. Musal.

France in 1800. Adapted by Richard A. Musal.

Index